APPLIED SUPERCONDUCTIVITY

Edited by

VERNON L. NEWHOUSE

School of Electrical Engineering
Purdue University
Lafayette, Indiana

Volume II

ACADEMIC PRESS New York San Francisco London 1975

A Subsidiary of Harcourt Brace Jovanovich, Publishers

ACADEMIC PRESS, INC.
111 Fifth Avenue, New York, New York 10003

United Kingdom Edition published by
ACADEMIC PRESS, INC. (LONDON) LTD.
24/28 Oval Road, London NW1

Library of Congress Cataloging in Publication Data

Newhouse, Vernon L
Applied superconductivity.

Includes bibliographies.
1. Superconductors. I. Title.
TK7872.S8N42 621.39 74-1633
ISBN 0-12-517702-X (v. 2)

PRINTED IN THE UNITED STATES OF AMERICA

Contents

9. Future Prospects

Theodore Van Duzer

List of Contributors

Numbers in parentheses indicate the pages on which the authors' contributions begin.

THEODOR A. BUCHHOLD (489),* GE Research & Development Center, Schenectady, New York

WILLIAM H. HARTWIG (541), Department of Electrical Engineering, The University of Texas at Austin, Austin, Texas

Y. IWASA (387), Francis Bitter National Magnet Laboratory, Massachusetts Institute of Technology, Cambridge, Massachusetts

D. B. MONTGOMERY (387), Francis Bitter National Magnet Laboratory, Massachusetts Institute of Technology, Cambridge, Massachusetts

CORD PASSOW (541), Center for Nuclear Studies, Karlsruhe, Germany

THEODORE VAN DUZER (641), Department of Electrical Engineering and Computer Sciences, University of California, Berkeley, California

* Present address: 62 Wiesbaden, Uranusweg 6, West Germany.

Preface

Attempts to exploit the fascinating properties of superconductors started shortly after the original discovery of the effect by Onnes when he tried to produce dissipation-free high-field electromagnets. As is well known, these attempts failed owing to the relatively low critical field of the superconductors known at that time. Although later efforts to use superconductors were more successful, e.g., the use of magnetic-field-controlled switches by Casimir–Jonkers and de Haas in the early 1930s, applied superconductivity can only be said to have come of age in the 1960s when high-field superconducting magnets began to be used widely, even for experiments at room temperature. This very important application provides a good example of the impact of basic research on applied science, since the high critical field of Nb_3Sn, which made these magnets possible, was discovered in the course of fundamental researches aimed at elucidating the origins of superconductivity itself. An example of how applied research leads to improved fundamental understanding is exemplified by the fact that research on high-field superconductors led to the rediscovery of Abrikosov's work on class-II superconductors, which might otherwise have continued to be ignored for many years.

Since superconductors exhibit zero resistance at low frequencies, they are already important in the production of large magnetic fields and show promise in the production and transport of large quantities of electric power. Since they operate close to zero temperature where Johnson noise becomes small, they either already are or promise soon to become the most sensitive detectors of magnetic fields and of radiation at all frequencies. Furthermore since superconductive circuits can easily store single-flux quanta, they promise to become the most compact and, therefore, the fastest means of handling and storing information.

The current interest in pure superconductivity is proved by the award, in the last few years, of separate Nobel prizes for the theory of superconductivity as well as for superconductive and semiconductor tunneling. The speed of progress in the field of applied superconductivity is exemplified by the fact that high-field superconductors were unknown in 1962 when John Bremer published "Superconductive Devices," the initial work in this field, and that the Josephson effect, which is the basis of most of the radiation detection and magnetic-field measurement devices mentioned above, had not yet been discovered at the time of publication of this editor's book, "Applied Superconductivity," in 1964!

Since the subject of applied superconductivity has now grown to an extent where it can no longer be covered exhaustively by a single author, this treatise is divided into chapters on the various areas, each written by one or more authorities on the subject in question. The work is divided into two volumes, the first of which deals with electronic applications and radiation detection and contains a chapter on liquid helium refrigeration.

The second volume discusses magnets, electromechanical applications, accelerators, microwave and rf devices, and ends with a chapter on future prospects in applied superconductivity.

A corollary to being an authority on a subject is the many demands made upon one's time. Thus a deadline must often be secondary to other responsibilities. For reasons of this sort, the original versions of some chapters in this treatise were completed before others, and we were never able to obtain a chapter on high-power rotating machinery.

Each chapter in these two volumes can be read independently, and most assume very little or no background in the physics of superconductivity. The topics treated do not require the use of advanced quantum mechanics; thus the books should be accessible to students or research workers in any branch of engineering or physics. They are intended to serve both as a source of reference material to existing techniques and as a guide to future research. For those wishing to extend their background in the physics of superconductivity, some recent books on the subject, selected from a larger list kindly compiled by Arthur J. Bond, are given in the following Bibliography.

Bibliography

The books listed on the following page are those written recently from the experimentalists' point of view and can be recommended for those wishing to expand their background in the physics of superconductivity.

De Gennes, P. G., (1966). "Superconductivity of Metals and Alloys." Benjamin, New York.

Kulik, I. O. and Yanson, I. K. (1972). "Josephson Effect on Superconducting Tunneling Structures." Halsted, New York.

Kuper, C. G. (1968). "An Introduction to the Theory of Superconductivity." Oxford Univ. Press, London and New York.

Lynton, E. A. (1972). "Superconductivity." Halsted, New York.

Parks, R. D. (1972). "Superconductivity," Vols. 1 and 2. Dekker, New York.

Saint-James, D. (1969). "Type II Superconductivity." Pergamon, Oxford.

Savitsky, *et al.* (1973). "Superconducting Materials." Plenum, New York.

Solymar, L. (1972). "Superconductivity Tunneling and Applications." Wiley, New York.

Williams, J. E. C. (1970). "Superconductivity and Its Applications." Pion, New York.

Contents of Volume I

Chapter 6

High-Field Superconducting Magnets

Y. IWASA and D. B. MONTGOMERY

Francis Bitter National Magnet Laboratory
Massachusetts Institute of Technology
Cambridge, Massachusetts

I. Introduction

Magnetic force is one of the most fundamental forces in nature. It has played and will continue to play a vital role in the development of physics and technology. In the coming years its role in the fields of biology, medicine, transportation, power generation and distribution, and mining is expected to increase greatly.

A magnetic force is, of course, produced only when there is a magnetic field. Therefore, being able to determine how to produce, handle, and control a magnetic field efficiently and economically is important. This chapter deals with some aspects of the most efficient and economical method of producing large magnetic fields known to man today—the use of superconducting materials.

At the end of this chapter, we have listed a limited number of three types of general references: (1) books, which cover fundamental aspects of our subject matter, (2) review articles on superconducting magnets which should give the reader a historical and general background of the development of superconducting magnets, and (3) titles of conferences concerning superconducting magnets from which proceedings have been published. Although these proceedings are unfortunately rather inaccessible unless one has attended the conferences, they are essential to keeping up with the latest developments.

In selecting articles to be referred to in the text, we tried to give original papers in most cases. When good review articles existed, however, we gave them preference over the original papers. This practice was used particularly in the areas of magnet degradation and flux jumping.

II. Various Applications of Superconducting Magnets

A. Small Magnets

Many small superconducting magnets have been built in the past ten years (Bean and Schmitt, 1963) primarily as research tools in traditional areas of physics. Here by small magnets we mean solenoidal coils with stored magnetic energies of less than a few hundred kilojoules. Figure 1 shows one such magnet built commercially.† It is wound with Nb_3Sn tape and produces a central field of 108 kOe in a 3.8-cm-diam bore with a transport current of 146 A. There will no doubt be more of these built in the future for similar purposes, but with one difference: the maximum fields, which in the past rarely exceeded 125 kOe, will routinely reach 150 kOe or even higher.

Some of the most exciting areas of application for small superconducting magnets in the years to come, however, are to be found in biology and medicine: (1) nuclear magnetic resonance (NMR) spectroscopy of biomolecules (Ferguson and Phillip, 1967), (2) electronmicroscopy (Fernandez-Moran, 1965), and (3) intravascular navigation (Yodh *et al.*, 1968) and related areas of medicine (Frei, 1970, 1973). This is so chiefly because of the superconducting magnet's ability to produce fields which are not only high but also extremely homogeneous spatially and temporally stable.

† Courtesy of Intermagnetic General Corp., Guilderland, New York.

FIG. 1. A typical Nb_3Sn-tape-wound superconducting magnet. It produces a central field of 108 kOe with a current of 146 A. It has orthogonal "see-through ' midplane ports. Bore diameter is 3.8 cm; field homogeneity is 6 parts in 10^4 within a 1-cm-diam sphere. (Courtesy of Intermagnetics General Corp.)

By the NMR technique, it is possible to investigate biological molemules—their electron distributions, motions, and exchange phenomenon—in solution without interfering with their normal functions. Superconducting magnets of fields up to 80 kOe with a spatial homogeneity of better than one part in 10^8 over a 0.5-cm-diam sphere and a time stability of 4 ppm over a period of one month are already available commercially (Oxford Instruments, Oxford, England) for NMR spectroscopy. With the advancement of conductor technology, maximum fields will certainly be pushed to the vicinity of 150 kOe in the near future.

Time stability is very important for electronmicroscopy—it increases resolution capability and allows a longer exposure time at lower beam intensities. Superconducting magnet lenses (Fernandez-Moran, 1966) now have been utilized successfully in electron microscopy.

Channels in the body such as blood vessels are frequencly used as a means of access to parts of the body otherwise inaccessible except by major surgery. With tubes introduced into various vessels, a number of functions can be performed with relative ease and with a minimum of trauma to the patient. However, not all regions of the body which could benefit from such catheterization are presently reachable by the standard techniques. For example, controlled catheterization of the intracranial internal carotid

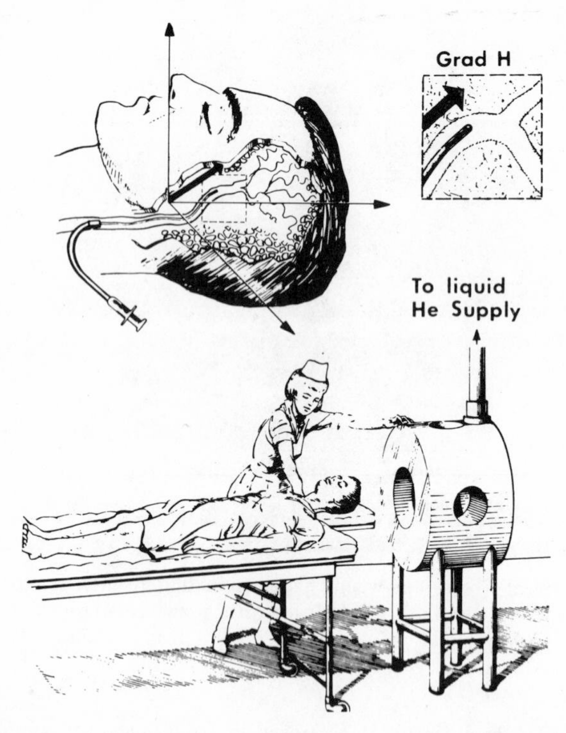

FIG. 2. Magnetic guidance system for permanent-magnet-tipped catheter proposed for neurosurgery applications by MIT Francis Bitter National Magnet Laboratory and Massachusetts General Hospital.

artery and its divisions has not thus far been achieved, because of the extreme tortuosity, narrowness, and irregularity of this vessel. An external magnet system capable of guiding a ferromagnetic-tipped catheter through such vessels would greatly facilitate many medical treatments, both in the head and elsewhere. Such a magnet could consist of three mutually orthogonal super-conducting coil pairs all contained in a single room-temperature-access cryostat, as shown in Fig. 2 (Montgomery *et al.*, 1969a).

B. LARGE MAGNETS

From an economic viewpoint, the greatest impact of superconducting magnets will probably be felt in the areas of technology where a high field is required on a large scale, e.g., (1) high-energy physics, (2) generators and motors, (3) transportation, (4) magnetic separation, and (5) fusion reactors. In the area of high-energy physics, conversion from conventional magnets to superconducting magnets started about eight years ago and has since been progressing steadily; in the areas of generators and motors, transportation, pollution control, and ore concentrators it has just started; full-scale fusion reactors, projected to be operational by or near the year 2000 (Tuck, 1971), are economically feasible only with the superconducting magnets providing the necessary magnetic field (except for proposed "laser-fusion reactors").

All of the applications to be described below are probably feasible technically and economically within the present superconducting materials technology. Any improvements in critical temperature, field, and current will make the systems even more attractive.

1. *High-energy physics*

In an accelerator installation, basically three types of magnets are required: (1) bubble-chamber magnets to identify and analyze particle interactions, (2) beam-transport magnets to select, guide, and focus beams, and (3) accelerator magnets to accelerate the particles. Since bubble-chamber magnets are usually large, and a very large number of beam-transport and accelerator magnets are required for each accelerator installation, we can appreciate why more and more superconducting magnets are being adopted for this purpose. The superconducting magnet is particularly suited for the first two applications, where normally it can be operated at constant field. Further progress is needed in the area of accelerator magnets before the superconducting magnet can be adopted fully to this time-varying application.

a. Bubble-chamber magnets. One of the largest bubble-chamber magnets ever built is the one designed and constructed at the Argonne National Laboratory and operating at the National Accelerator Laboratory (Purcell, 1972). The magnet has a winding bore of 14 ft with a central field of 30 kOe. The peak field on the conductor, multifilament Nb–Ti composite, is 51 kOe, and the operating current is 5000 A. The magnet consists of two coils, each coil containing 22 pancakes. Stored energy of the magnet is 396 MJ; the overall current density is 1885 A/cm^2. The maximum compressive force between coils is calculated to be 11,000 tons. The entire magnet with cryostat weights 140 tons, 88 tons of this being for the winding. Figure 3 shows a section of windings under construction.

Judging purely from the capital cost of such a magnet system alone, one can argue that it might be economically competitive to provide the magnetic field with resistive water-cooled copper conductors, powered by a 10-MW power supply. However, bubble chambers operate continuously, often for periods of years, and the power bill for such a magnet system

FIG. 3. Pancake winding for the National Accelerator Laboratory's 14-ft bubble-chamber magnet being constructed at the Argonne National Laboratory.

could conceivably reach $1,000,000 per year. The cost of operating a superconducting system is a small fraction of that.

b. Beam-transport magnets. Beam-transport magnets, dipole and quadrupole magnets, are used extensively for deflecting and focusing particles in an accelerator—as much as 200 such magnets are typically required for a machine. Because of the great number required, conversion of the magnets from standard to superconductive may be desirable in spite of their small individual size and power requirements. Aside from the saving in operation cost such conversion might offer, the superconducting magnet, because of its small size and high field, has an additional advantage over conventional iron-cored magnets in its short focal length. This is important when dealing with high-momentum, short-lived particles: only with such magnets can particles be focused before they decay. One example of such a superconducting dipole magnet is shown in Fig. 4.

c. Accelerator magnets. Future high-energy accelerators (1000 GeV and higher) require magnet rings, composed of many small magnets similar

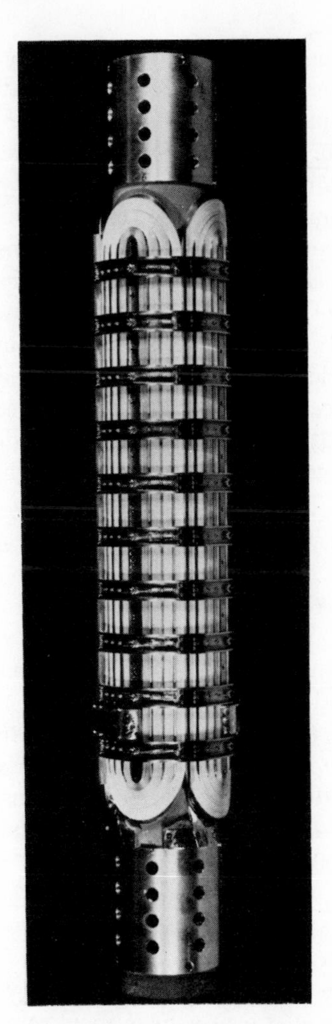

FIG. 4. A typical quadrupole magnet wound of Nb_3Sn tape used for focusing particles. Bore diameter is 10 cm, effective length 60 cm. The magnet produces a field gradient of 3.5 kOe/cm. (Courtesy of Brookhaven National Laboratory.)

to beam-transport magnets, of as much as 25 km in circumference when conventional iron-core magnets are used. This huge dimension can be reduced, by a factor of as much as five, if high-field superconducting magnets are used. Alternatively, the existing accelerators' energy levels can be improved by replacing the iron-core magnets with superconducting magnets.

There is, however, one important technical obstacle which has slowed the conversion of these magnets; they are pulsed rather than dc, and energy dissipation in these ac-lossy, high-field, high-current superconductors is nonnegligible. At present the best magnets still dissipate several watts of heat per meter of magnet length (Sampson *et al.*, 1970) (including the usual cryogenic losses) when operated at a frequency of 0.1 Hz; for a 5-km ring this amounts to a total power consumption of close to 25,000 W. It is not surprising, therefore, that one of the most concentrated areas of research in applied superconductivity currently is in the development of conductors with much better ac performance.

Recently, Blewett of Brookhaven National Laboratory has proposed a new machine called the Intersecting Storage Accelerator (ISABALLE) in which the advantage of superconducting magnets—their losslessness in dc—is fully utilized (Blewett and Hann, 1972). In this machine, the injected particles have an energy level of 30 GeV, are accelerated to 200 GeV in 100 sec (thus reducing the ac frequency to less than 0.01 Hz), and are kept at the top level for hours (dc) while low-collision cross-section particles collide to produce much higher energy particles.

2. *Generators and motors, transportation, magnet separation*

a. Generators and motors. Probably the first large-scale commercial application of superconductivity will be in the electric power industry with the replacement of the normal metal conductors and iron circuits in generators with superconducting field windings. In normal rotating machines copper wires are used to carry current, and iron is used to increase and control flux. Most prior increases in performance of these machines were due to improvements in insulation, cooling, and magnetic circuit designs. Present-day large machines rotating at 3600 rpm are limited to the 1000 to 1500 MVA range.

The use of superconducting wire, such as Nb–Ti, in the rotor field windings can increase the flux density from 20 kOe, the saturation field of the iron in conventional machines, to 50 kOe or more.

The use of superconducting field windings leads to the following improvements over conventional rotating machine technology. The high flux density in a relatively small volume considerably reduces the size of the rotor and, hence, of the surrounding armature. This reduction in size results in considerable improvement in mechanical stability and also reduces the cost. Since no magnetic circuit iron is needed to help produce and direct the magnetic field, more of the stator volume can be filled with normal conductor.

A machine with a 45-kVA rating using a rotating superconducting field winding was designed and tested at the Cryogenic Engineering Laboratory of the Massachusetts Institute of Technology (MIT) in 1969 (Thullen and Smith, 1970). The rotor produced a dipole field of approximately 32 kOe and rotated at 3600 rpm. The superconducting rotor and its liquid

FIG. 5. Overall view of the first successful alternator with a superconducting field winding rotor, designed and constructed at the MIT Cryogenic Engineering Laboratory. It has a rated output of 45 kVA.

helium Dewar rotated as a unit and were supplied with helium through a vacuum insulation tube with a rotatable joint. The machine could be run continuously, provided it was directly connected to a source of liquid helium. An overall view of the machine is shown in Fig. 5.

The MIT group has just completed a second experimental machine with a design rating in excess of 2 MVA (Smith *et al.*, 1972). The superconducting rotor is 20 cm in diameter and has an active length of 60 cm. A schematic diagram of the machine is shown in Fig. 6. At present, there appear to be no major technical obstacles to prevent the realization, within a decade or so, of steam-turbine generators of 5000 MVA or more.

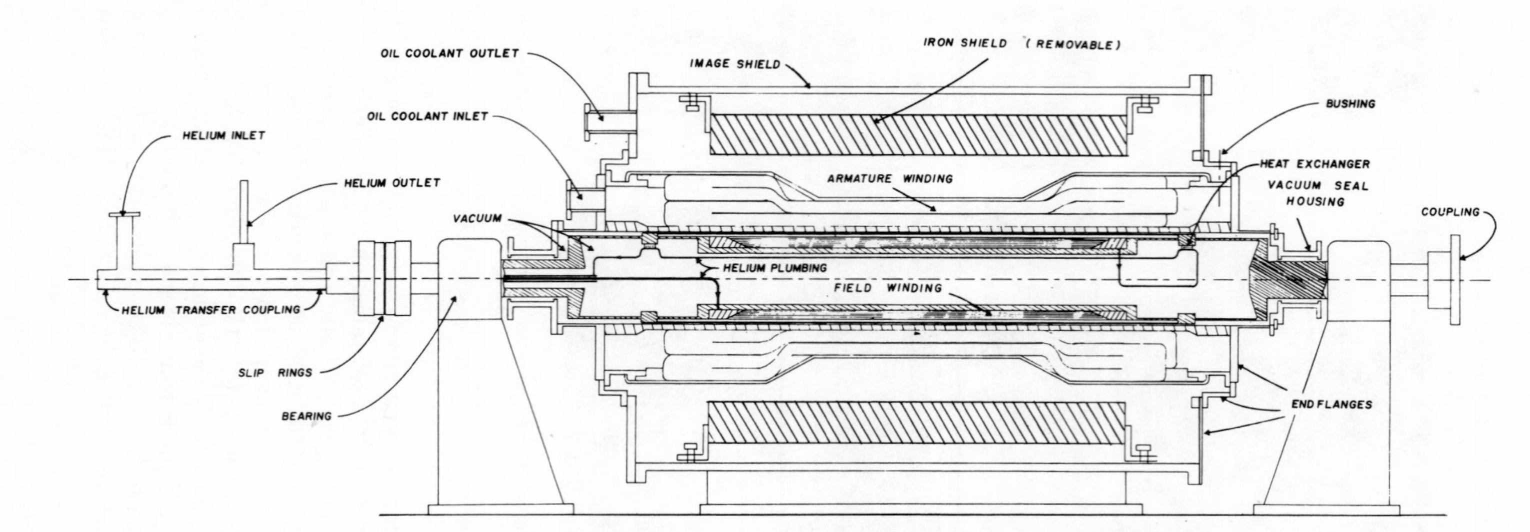

FIG. 6. Basic configuration of a 2-MVA alternator designed and presently under construction at the MIT Cryogenic Engineering Laboratory. Bore diameter is 20 cm, effective length 60 cm.

Another important application in rotating machinery is the use of superconductors in dc machines. Recently, International Research and Technology, Ltd. (United Kingdom) completed the design and construction of a homopolar, or Faraday, disk motor (Appleton, 1969). It consists of an outer superconducting solenoid producing a large magnetic flux and an inner normal-metal armature consisting of a Faraday disk which carried current radially between the inner and outer axes. A torque is produced on the Faraday disk due to the interaction of the field of the superconductor with the radial armature current. A preliminary 50-hp machine has been constructed, and, more recently, a 3000-hp motor has been run successfully and put into operation as a water pump in Fawley, England. It has been conjectured that a superconducting motor rotating at a speed of 100 to 1000 rpm can be designed with an output power between 10 and 100 MVA, which is an order of magnitude greater than the output of conventional machines of the same weight.

b. Transportation. Another large-scale application is that of magnetic levitation for high-speed ground transportation. The levitation is produced by repulsion between the magnetic field produced by magnets on board the vehicle and the induced eddy currents in the metallic guideway. The idea of using magnetic suspension for transportation is not new—as early as 1908 Bachelet proposed such a system (Polgreen, 1970). The idea, however, became more easily realizable after high-field, high-current superconductors became available in abundance and was revived in 1966 by Powell and Danby of the Brookhaven National Laboratory. Since then, many different schemes have been proposed; one of them, proposed by the MIT–Raytheon group, is briefly described here (Kolm and Thornton, 1972).

The system proposed has features which resemble an airplane rather than a train and hence is called magneplane. It is electromagnetic flight at ground level. A cylindrical vehicle containing superconducting coils is suspended about 30 cm above a trough-shaped guideway and is free to roll about its longitudinal axis and thus to assume the correct bank angle at curves. This self-banking feature is important at very high speeds, in the range of 200 to 500 km/hr. The propulsion is generated for the vehicle by the interaction between the superconducting dc magnetic loops on the vehicle and activated synchronous propulsion coils imbedded in the guideway. The propulsion is silent, requires no on-board vehicle power supply, and combines levitation with propulsion. A more detailed description of magnetic levitation by Buchhold can be found in Chapter 7.

c. Magnet separation. Magnetic separation is a very old technique that has been used for years in the mining industry—all of the iron ore mined in the Mesabi Range, for example, is enriched at the site by means of low-field conventional magnetic separators. The kaolin clay industry also uses magnetic separation to remove the dark stains in the clay deposits. Recently, though on a very limited scale, magnetic separation has been applied in secondary water treatment. As demonstrated recently at the MIT's Francis Bitter National Magnet Laboratory, a sample of water drawn from the Charles River can be processed to high clarity standards after simple magnetic seeding and a single pass through a high-field separator (de Latour, 1973). Most surprising, perhaps, is the fact that the coliform count is reduced from 9000/ml to less than 100/ml, or drinking-water standards, without chemical treatment. Careful studies have shown that a superconducting magnet of the size of the National Accelerator Laboratory's bubble-chamber magnet, described above, would have a flow capability sufficient to handle the entire 30×10^6-gal/day secondary water treatment from the city of Boston at a fraction of the capital cost of present systems.

3. *Fusion reactors*

It is estimated that by the year 2000 the peak electrical power generating capacity in the United States will increase from the present rate of 350,000 MW to 3,000,000 MW. Assuming that 6% of that has to be added or replaced yearly—currently the figure is 11%—it is then estimated that an additional 180,000 MW of power must be made available yearly during the first decade of 2000 (Powell, 1974). Half of that requirement is expected to be supplied by fusion-reactor power plants. A Tokamak-type fusion reactor proposed by the Oak Ridge National Laboratory, having a minor diameter of 11 m and a major diameter of 21 m with a toroidal field of 40 kOe, would deliver 2000 MW (Lubell *et al.*, 1972). Forty-five such plants, therefore, would be required yearly. The magnetic fields for such plants must be generated by a superconducting winding.

C. Hybrid Magnets

The term "hybrid magnet" has been coined to describe a magnet in which part of the field is generated by superconductors and part by non-superconductors. Such a system has the advantage of being able to produce higher fields than can be produced by superconductors alone. In addition, it can represent an economical method for generating very high fields for laboratories which already possess power supplies or refrigerators (Weggel and Montgomery, 1972).

The most compelling reason to consider hybrid systems is that super-

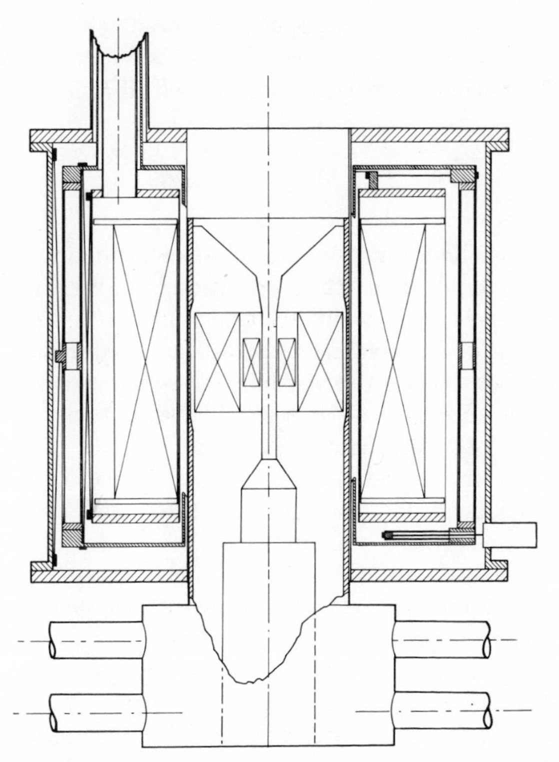

FIG. 7. Schematic of the 200-kOe hybrid magnet designed and constructed at the MIT Francis Bitter National Magnet Laboratory. A pair of water-cooled coils consuming 5 MW of dc power are operated inside a 14-in. room-temperature- bore 40-kOe superconducting coil.

conductors by themselves are limited by their upper critical fields. If one wants to produce a field above this level, it is clear that the superconductor cannot do it alone. If, on the other hand, we replace those sections of the winding which are exposed to a field too high for the superconductor by nonsuperconducting materials, we can increase the field to the limits of our ability to power or cool the dissipative inner sections. Needless to say, one must separate the two sections in a suitable cryostat. This cryostat must not only provide low-loss temperature isolation, but must withstand very large interaction forces as well.

The first hybrid magnet was designed at the Francis Bitter National Magnet Laboratory (Montgomery *et al.*, 1969b), and its magnet and cryostat arrangement are shown schematically in Fig. 7. Subsequently the magnet was constructed and tested, first the superconducting section alone (Leupold *et al.*, 1972), at the laboratory. The magnet has since been tested as a hybrid mode and generated a peak field of 200 kOe. A more powerful hybrid magnet, with a peak field of 250 kOe, has recently been operated in the Soviet Union (Cheremnyk *et al.*, 1973).

III. Magnet Design

The single most important feature of superconducting magnets is the negligible power required for their operation, making them particularly suitable for high-field, large-volume magnets. Power, an important parameter in the design of resistive magnets, will not concern us, and we shall focus our attention on the relationships between magnetic field, current density, magnetic and thermal stresses, field distributions, and the required amount and type of superconducting material.

In this section on magnet design, we shall first discuss the design of simple solenoidal magnets (Montgomery, 1969) and multipole magnets (Iwasa and Weggel, 1972), assuming the superconductor to be a perfect material limited only to a specified finite current density. After establishing some basic relationships for field, current density, volume, and energy, we shall examine the characteristics of superconductors in detail. Third, in this section, we shall discuss field distributions in simple solenoidal magnets. Finally, we shall discuss magnetic and thermal stresses in superconducting magnets, a most important consideration in the design of high-field, large-volume magnets.

A. Current Density, Field, Volume, and Energy

1. *Simple solenoidal, uniform-current-density magnets*

a. Current density and field. As a starting point, we take the field of the elemental circular loop shown in Fig. 8. The field H (Oe) on the axis of this loop, z (cm) from the loop, can be written as

$$H_z(z, 0) = \frac{\pi I}{5} \frac{a^2}{(a^2 + z^2)^{3/2}} \tag{1}$$

where I (A) is the loop current, and a (cm) is the radius of the loop. The central field ($z = 0$) of such a loop, H_0, is

$$H_0 = \pi I / 5a \tag{2}$$

Although Eq. (2) is derived for a single loop, a quick estimate of the central field of thin magnets such as pancakes can be done with Eq. (2) with a equal to the mean radius of a pancake.

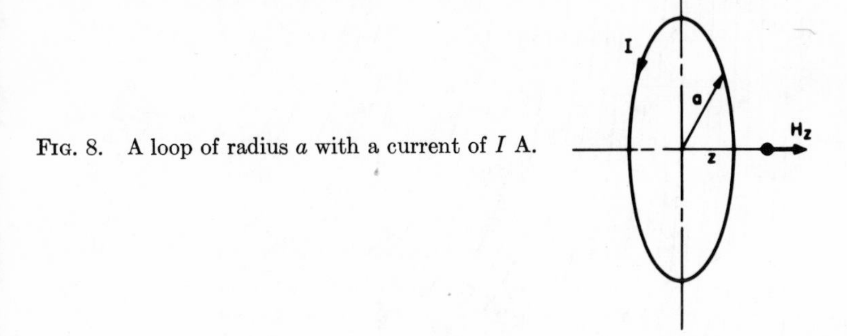

FIG. 8. A loop of radius a with a current of I A.

In general, a coil of finite length and finite radial build can be thought of as an aggregate of many single loops like the one considered above: to calculate the field, one need simply sum the contributions at z from each element loop. We then obtain the central field of a magnet having a total number of ampere turns, NI, an i.d. of $2a_1$, an o.d. of $2a_2$, and a length of $2b$:

$$H_0 = \frac{NI}{a_1} \frac{1}{2\beta(\alpha - 1)} F(\alpha, \beta) \tag{3}$$

where

$$F(\alpha, \beta) = \frac{4\pi\beta}{10} \ln\left[\frac{\alpha + (\alpha^2 + \beta^2)^{1/2}}{1 + (1 + \beta^2)^{1/2}}\right] \tag{4}$$

$$F(\alpha, \beta) = \frac{4\pi\beta}{10}\left(\sinh^{-1}\frac{\alpha}{\beta} - \sinh^{-1}\frac{1}{\beta}\right) \tag{5}$$

and $\alpha = a_2/a_1$, $\beta = b/a_1$. $F(\alpha, \beta)$ is plotted in Fig. 9. There are also tables available for $F(\alpha, \beta)$ (Montgomery, 1969).

Sometimes it is more conventional to work with current density than with ampere turns, because, as we shall see later, a superconductor is more suitably classified by its current density. We note that the cross section of a coil is given by

$$\text{overall cross section} = 2b(a_2 - a_1) = 2a_1^2\beta(\alpha - 1) \tag{6}$$

Thus,

$$NI/2a_1\beta(\alpha - 1) = \lambda j_c a_1 \tag{7}$$

where j_c is the intrinstic current density of the superconducting material, and λ is the fraction of cross section occupied by the superconductor. With

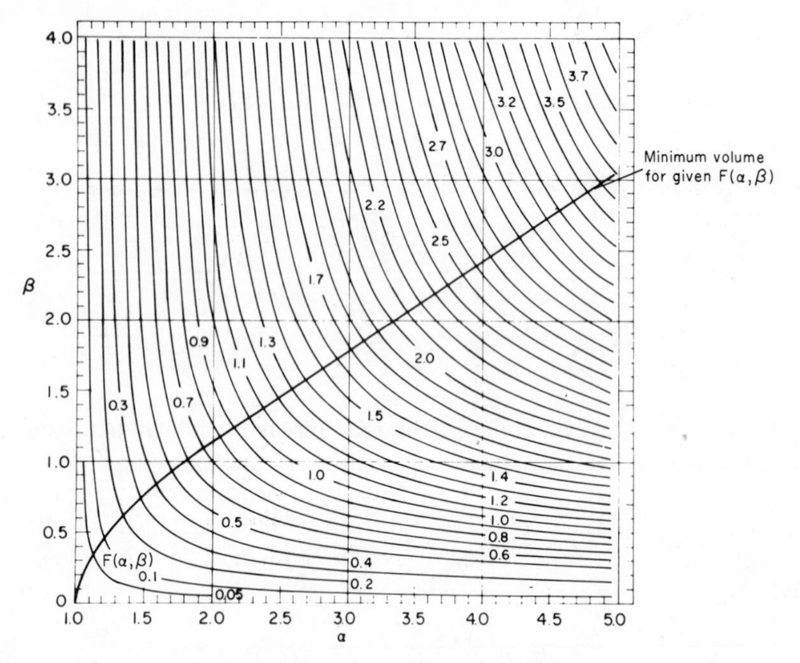

FIG. 9. Lines of constant $F(\alpha, \beta)$ plotted against α and β. The line passing through the values of α and β yielding the minimum volume for any given $F(\alpha, \beta)$ is also shown.

Eq. (7) substituted into Eq. (3), the central field is written as

$$H_0 = \lambda j_c a_1 F(\alpha, \beta) \tag{8}$$

For limiting values of α and β, Eqs. (3) and (8) can be simplified to obtain approximate values of H_0 quickly:

(1) $(\alpha/\beta)^2 \ll 1$, i.e., "long" coils:

$$H_0 = \pi NI/5a_1\beta = \pi NI/5b \tag{9}$$

$$H_0 = \frac{2\pi}{5} \lambda j_c a_1(\alpha - 1) \tag{10}$$

(2) $\beta \to 0$, i.e., "thin" coils:

$$H_0 = \pi NI \ln \alpha / 5a_1(\alpha - 1) \tag{11}$$

$$H_0 = \frac{2\pi}{5} \lambda j_c a_1 \beta \ln \alpha \tag{12}$$

(3) $\beta \to 0$ and $\alpha \to 1$, i.e., "loop" coils: As $\alpha \to 1$, $\ln \alpha \to \alpha - 1$, Eqs.

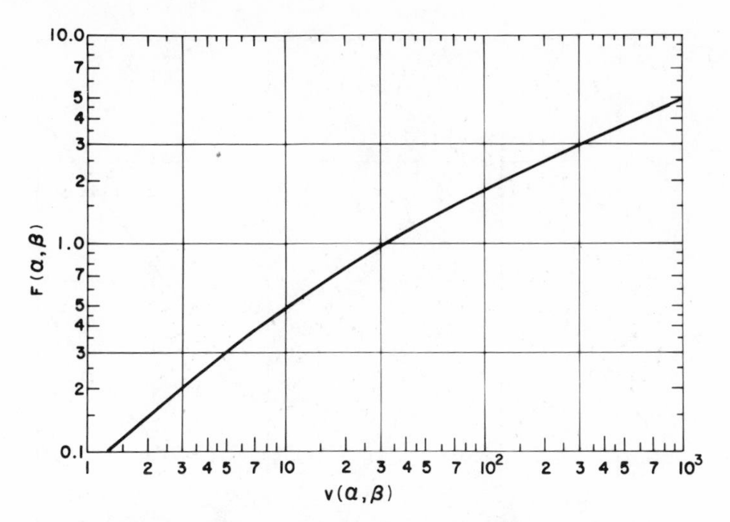

FIG. 10. The minimum volume factor necessary to achieve a given value of $F(\alpha, \beta)$ as calculated from the minimum-volume points in Fig. 2.

(11) and (12) become:

$$H_0 = \pi NI/5a_1 \tag{13}$$

$$H_0 = \frac{2\pi}{5} \lambda j_c a_1 \beta (\alpha - 1) \tag{14}$$

b. Current density and minimum conductor volume. The volume of a coil envelope can be written as

$$V = a_1^3 v(\alpha, \beta) \tag{15}$$

where

$$v(\alpha, \beta) = 2\pi\beta(\alpha^2 - 1) \tag{16}$$

The volume of conductor within the envelope is simply

$$V_c = \lambda V = \lambda 2\pi a_1^3 \beta(\alpha^2 - 1) \tag{17}$$

The length of conductor within the envelope is the coil colume multiplied by the turns per square centimeter, n:

$$l = n2\pi a_1^3 \beta(\alpha^2 - 1) = na_1^3 v(\alpha, \beta) \tag{18}$$

If we calculate $v(\alpha, \beta)$ along any constant $F(\alpha, \beta)$ line, we find that there is a unique point of minimum $v(\alpha, \beta)$, i.e., a shape of coil of the given $F(\alpha, \beta)$ which uses the minimum amount of material for given a_1 and constant F. The minimum volume factor necessary to achieve a given value

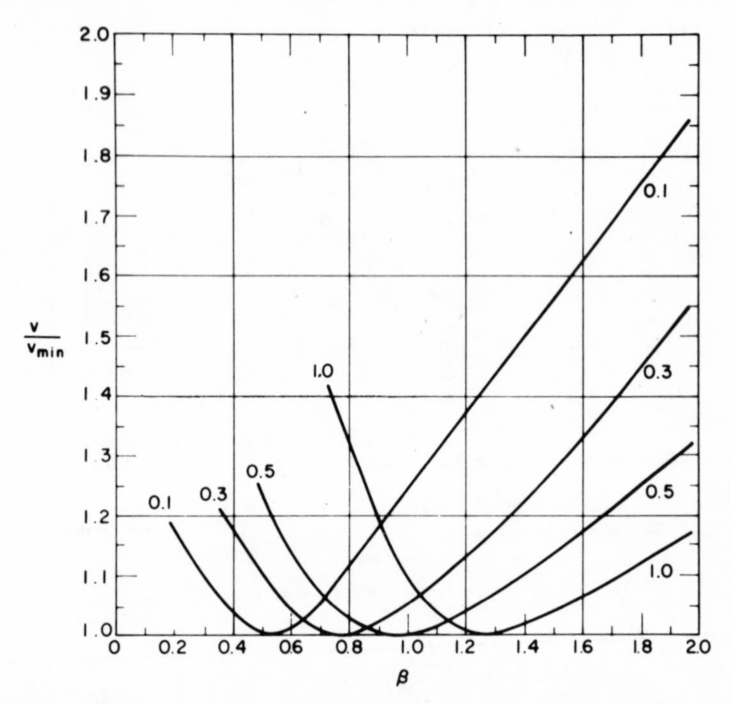

FIG. 11a. The increase in coil volume which results from a deviation away from the minimum-volume point for a given value of $F(\alpha, \beta)$. Lines of constant $F(\alpha, \beta)$ are shown plotted versus β in the range $0 < \beta < 2$. The α for any given β is chosen to achieve just the desired $F(\alpha, \beta)$.

of $F(\alpha, \beta)$, as calculated from the minimum volume points shown in Fig. 9, is plotted in Fig. 10.

When it is necessary to deviate from the minimum-volume point, more material is required for a given field. How much more is indicated by Figs. 11a,b, where the ratio of volumes has been plotted versus β. We note that, for increases in length in the neighborhood of the length necessary for minimum volume, the coil volume increases much more slowly than length. For example, a coil of $F = 4$ with a length 50% greater than the optimum uses only 12% more wire, an increase easily offset by the smaller cryostat and higher field uniformity. (We return to field uniformity later.)

There is another point that must be considered in choosing the coil shape, particularly for low-F-number coils; i.e., the maximum field seen by the superconductor is higher than the central field. If the critical current density is sensitive to the magnetic field, then a coil of greater length, smaller build, and lower maximum field at the winding will require less volume than the simple geometrical minimum shape. This can be explored

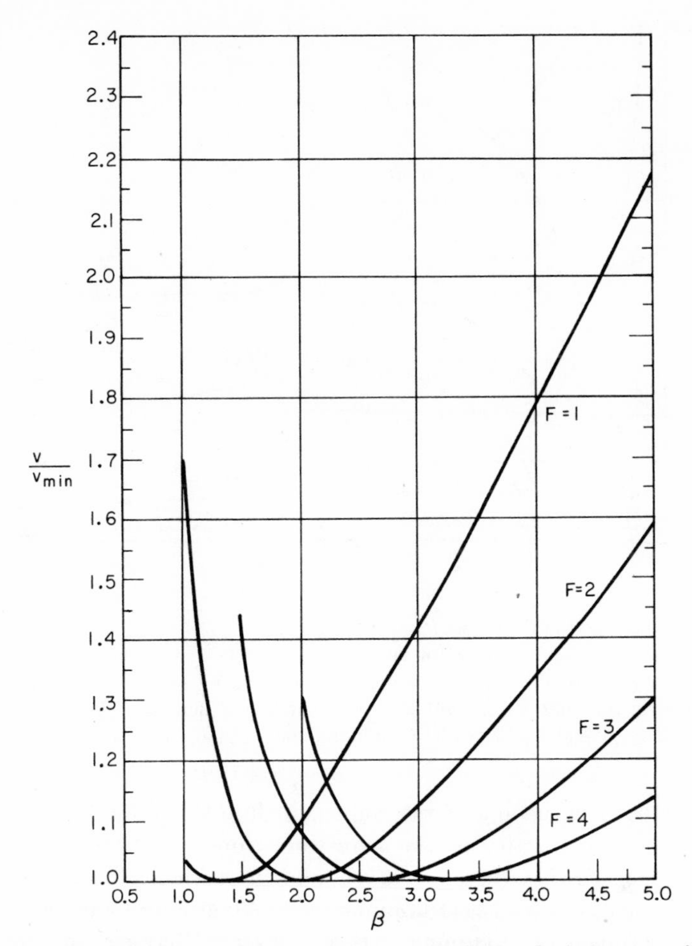

FIG. 11b. The increase in coil volume which results from a deviation away from the minimum-volume point for a given value of $F(\alpha, \beta)$. Lines of constant $F(\alpha, \beta)$ are shown plotted versus β in the range of $0.5 < \beta < 5$. The α for any given β is chosen to achieve just the desired $F(\alpha, \beta)$.

by the use of Fig. 12, in which the ratio of maximum to central field has been plotted for coils of various shapes. A useful, although approximate, expression for the maximum field in terms of the central field is

$$H_{\max}/H_0 = [1 + (0.64/4^{\beta})] \tag{19}$$

This consideration of the field at the winding only influences the choice of volume for coils of relatively low F number. An $F = 2$ coil of minimum geometrical volume is so long ($\beta = 2.0$) that the field ratio is less than

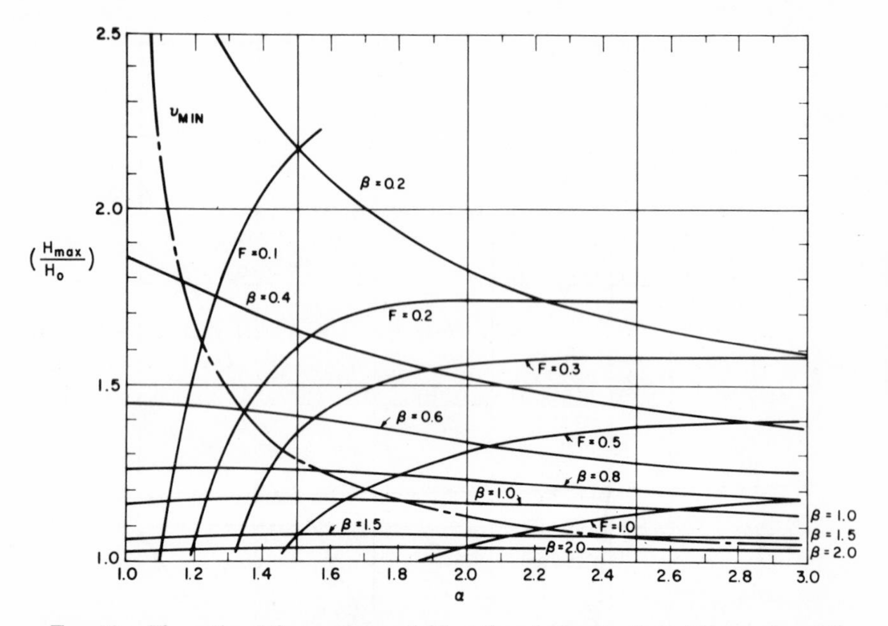

FIG. 12. The ratio of the maximum field at the winding to that at the center of the coil, H_{max}/H_0, as a function of α for various values of β. Lines of constant $F(\alpha\ \ \beta)$ are also shown and are determined by the value of α necessary to achieve the given value of $F(\alpha, \beta)$ along the lines of constant β shown. The line of minimum volume shown was determined by the value of α yielding minimum volume along the $F(\alpha, \beta)$ curves shown.

1.05. Further lengthening of the coil can reduce this only 5% at most, a reduction rapidly negated by the increasingly unfavorable shape. It should also be remarked that there are few applications where coil shapes can be chosen strictly on the basis of minimum volume; usually some requirement of the experiment or environment predominates. However, the graphs in Figs. 11 and 12 show the sacrifice we must make to satisfy a given experimental requirement. One requirement often encountered is that of uniformity.

c. *Coil volume and uniformity.* As we shall discuss later, the field around the center point of a uniform-current-density coil can be written as a power series, which along the z axis has the form

$$H = H_0[1 + E_2(\alpha, \beta)(z/a_1)^2 + E_4(\alpha, \beta)(z/a_1)^4 + \cdots] \tag{20}$$

The $E_n(\alpha, \beta)$ coefficients are functions of α and β, and hence, for any given z/a_1, the uniformity of the field can also be plotted as a function of α and β. For a simple solenoid, the $E_2(\alpha, B)$ coefficient is negative, and the field therefore decreases for increasing $|z|$. A sphere of radius z/a_1 will bound a region whose uniformity is comparable to that given by Eq. (20), because the inhomogeneity associated with any individual term is greatest in the z direction.

Figures 13a,b give, as a function of α and β, the uniformity over a sphere of radius $z/a_1 = 0.707$ as a percent deviation from the central field. The minimum-volume curve and a number of $F(\alpha, \beta)$ curves are also given. It can be seen that the minimum-volume coil shapes yield rather poor homogeneities for low-F-number coils, but they show a progressive improvement at higher F numbers.

Figures 13 and 11 can now be used to illustrate the increase in coil volume that must accompany a given required uniformity. For example, if a coil must have an $F = 2$, the minimum volume would result when $\beta = 2$, but the uniformity (over the sphere of radius $z/a_1 = 0.707$) would be only about 5%. This can be improved to 0.5% by an increase in coil length to $\beta = 4.4$, the volume increase being 45%. An $F = 4$ coil, however, already has nearly a 1% uniformity at the minimum volume, and an 0.5% uniformity can be obtained for an increase in volume of only 12%.

It can be seen from Eq. (20) that, if z/a_1 is much less than 1, the terms in the series rapidly diminish. Even for z/a_1 approaching 1, moreover, the diminishing magnitude of the E_n coefficients with increasing n continues to make E_2 the predominant term. Thus, the approximate uniformity for any value of z'/a_1 is the exact uniformity at z/a_1 multiplied by $(z'/z)^2$. If a sphere of radius $z'/a_1 = 0.5$ has a uniformity of 0.5%, then, over a sphere of radius $z'/a_1 = 1$, it will be approximately 2%. The scaling by (z'/z) provides a rapid estimate of the uniformity-versus-volume sacrifices necessary to meet a given requirement.

d. Conductor volume and current density. We have so far been concerned principally with the effects of shape or field uniformity on material or coil volume, while maintaining a constant value of current density. Often of greater interest, however, is the change in volume with changes in current density. For example, can a material of higher current density but greater cost reduce the required material enough to offset its increased cost? Or how will the material requirement depend on the overall current density? These questions often arise in considering stabilized conductors.

We can always answer these questions for any specific conditions by the straightforward calculation of the required F's at the two current densities, choosing an operating point along each F line to satisfy any constraints on coil volume, field uniformity, or some physical dimensions, and then calculating the volumes required for the two cases. It is more difficult, however, to form a completely general picture.

As we shall see later in connection with the stability of conductors, a

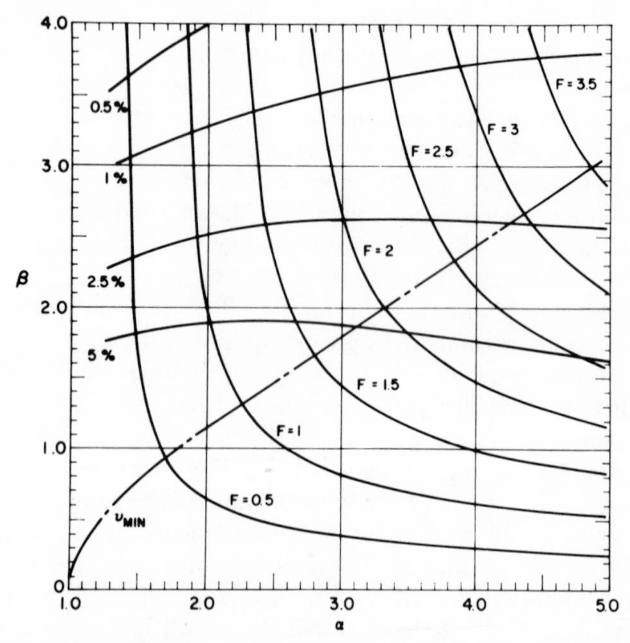

FIG. 13a. Field uniformity as calculated from Eq. (20) over a sphere of normalized radius $z/a_1 = 0.707$ in percent deviation from the central field. Also shown are lines of constant $F(\alpha, \beta)$ and minimum-volume curve.

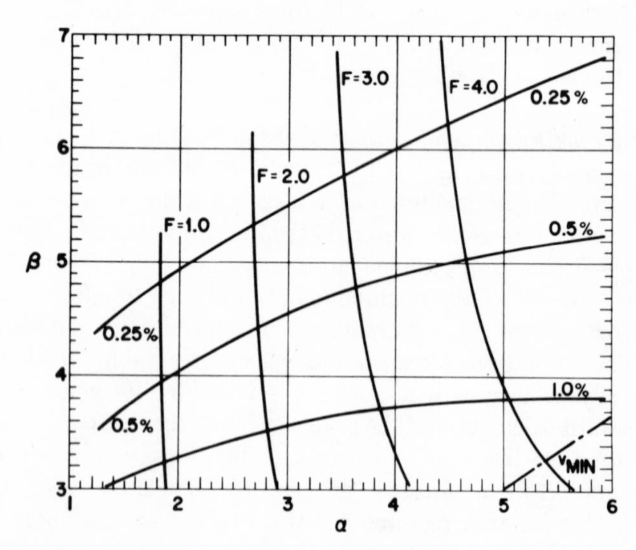

FIG. 13b. The uniformity of the central field as in Fig. 13a for the higher ranges of β.

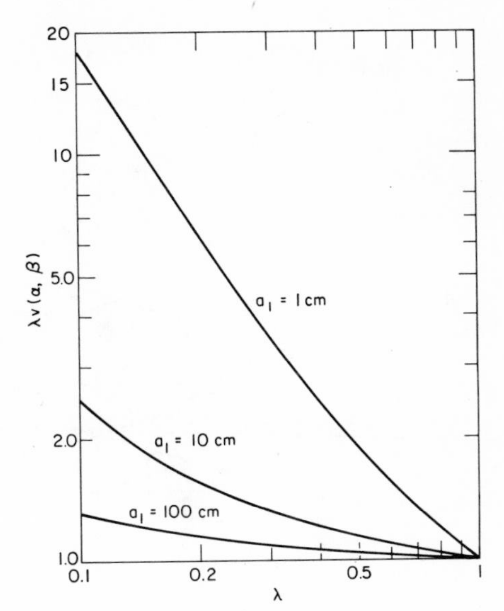

FIG. 14. The amount of superconducting material necessary versus the space factor λ for three solenoids with bore diameters of 1, 10, and 100 cm.

superconducting material alone can seldom be used to make a reliable conductor; it must generally be embedded in a matrix of copper or other normal conductor, and, depending on the amount of normal conductor, the overall current density of the conductor, λj_c, varies considerably. In particular, we wish to examine how the material requirement varies with λ for a given field intensity H and current density j_c. This we do using the minimum-geometrical-volume solution.

In Fig. 14 are plotted relative values of $\lambda v(\alpha, \beta)$, the superconducting material requirements normalized to that for $\lambda = 1$, as a function of λ for $a_1 = 1$, 10, and 100 cm. From Fig. 14, we can conclude that the larger the coil bore the more slowly will the material requirement increase with a decrease in λ, making it possible to operate large magnets economically at smaller overall current densities.

e. Magnetic energy and inductance calculations. Experience has shown that the overall current density achievable in large superconducting magnets is correlated with the stored magnetic energy of the magnets, as demonstrated in Section III.B.2. There we present Lubell's plot of data obtained from all of the known large superconducting magnets presently in operation, under construction, or in the design stage, relating the overall current density and the stored magnetic energy. Once the stored magnetic energy of a magnet is estimated, a sensible and dependable overall current density for the magnet can be estimated from the data. Thus, the stored

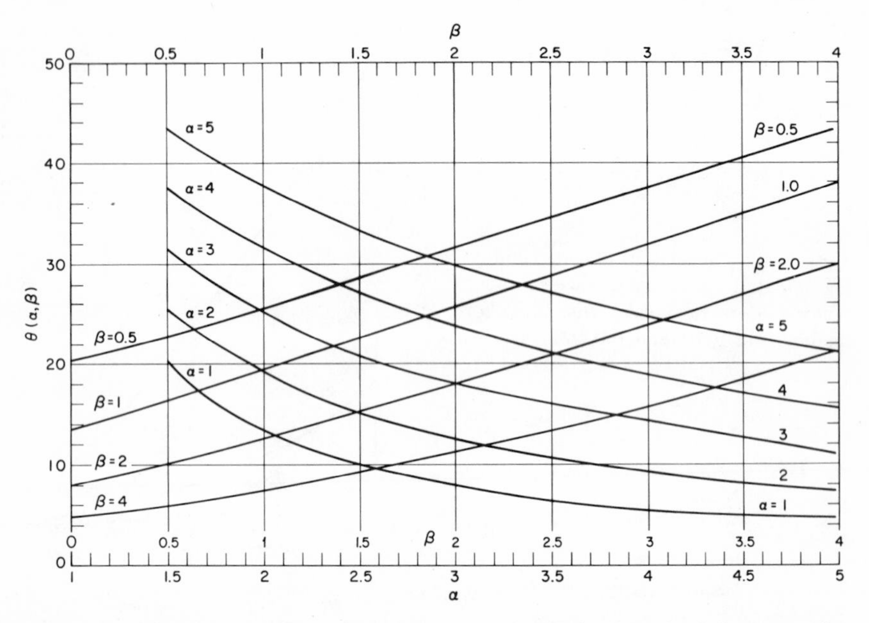

FIG. 15. The geometry-dependent factor $\theta(\alpha, \beta)$ which gives the inductance of a uniform-current-density coil as in Eq. (24). The factor $\theta(\alpha, \beta)$ is given for lines of constant α plotted versus β and lines of constant β plotted versus α.

magnetic energy is a very important parameter in the design of a superconducting magnet.

A general expression for the magnetic energy of a circuit energized with a current I is

$$E_M = \tfrac{1}{2}LI^2 \tag{21}$$

where L is the inductance of the circuit in henrys and E_M is the magnetic energy in joules. By definition, it is also equal to

$$E_M = \mu/2 \int_{\text{all space}} H^2\, dv \tag{22}$$

when all space is filled with a homogeneous, isotropic medium of permeability μ and H is the magnetic field produced by I. The magnetic energy, therefore, is spatial integral of the energy density $\mu H^2/2$. Since H is proportional to I with its proportionality constant, depending only on geometry, we see that the inductance of a magnet is a purely geometric and, through μ, medium-dependent factor:

$$L = \mu/I^2 \int_{\text{all space}} H^2\, dv \tag{23}$$

Here we present an inductance calculation for simple solenoidal magnets. Since we need not estimate the magnetic energy accurately (perhaps to within 20% would be adequate), only very simple approaches are given (Montgomery, 1969).

The inductance of a simple solenoidal, uniform-current-density air-core magnet of $2a_1$ i.d., $2\alpha a_1$ o.d., and $2\beta a_1$ length, and with N number of turns, is given:

$$L = a_1 N^2 \theta(\alpha, \beta) \text{ henry} \tag{24}$$

where a_1 is in centimeters. The inductance factor $\theta(\alpha, \beta)$ is plotted in Fig. 15.

2. *Circular-aperture multipole magnets*

The multipole magnets, particularly dipole and quadrupole, constitute a most important class of magnets. A majority of superconducting magnets, particularly those being built in the immediate future, are of this class: dipoles and quadrupoles in accelerators, and dipoles in MHD generators and in superconducting alternators and related machines.

The field generated by these magnets ideally is identical in all planes through the di- or quadrapolar axis; i.e., the field is completely specified in two dimensions and is always approximated as such. The aperture of these magnets, although not restricted to any particular shape, is usually circular or elliptical or, only occasionally, rectangular. We discuss here only the circular-aperture case, neglecting end effects but including the the effects of an iron shield (Iwasa and Weggel, 1972).

In a region where no current exists, the magnetic field H is the gradient of a potential ϕ which satisfies Laplace's equation, namely,

$$\mathbf{H} = -\nabla\phi \tag{25}$$

$$\nabla^2\phi = 0 \tag{26}$$

In two-dimensional polar coordinates, shown in Fig. 16, one class of ϕ is

$$\phi_n = \cos n\theta / r^n \qquad r > R \tag{27}$$

$$\phi_n = r^n \cos n\theta \qquad r < R \tag{28}$$

where n is an integer. At $r = R$, appropriate current filaments of infinitesimal thickness are distributed to give the above potentials. The field produced by a dipole corresponds to the case when $n = 1$ and that by a quadrupole to $n = 2$.

Real magnets differ from the above ideal cases in two important ways: (1) the thickness of current filaments at $r = R$ is not infinitesimal; and (2) the continuous variation of current density with θ required to produce

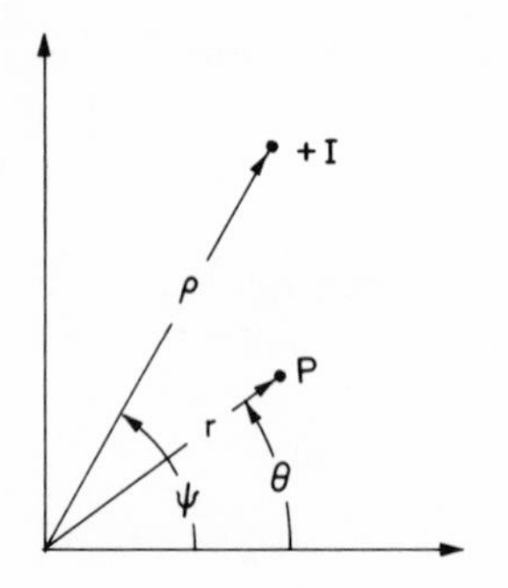

FIG. 16. Vector potential of line current $I(\rho, \psi)$ at point $p(r, \theta)$.

the above potential is not realizable in practice. Because of these and other reasons, the actual magnet always has components belonging to higher harmonics—an undesirable consequence. One of the central design problems in the construction of these multipole magnets, therefore, is to choose a proper winding scheme to eliminate as many of the higher harmonics as possible. Although idealized, two such winding schemes for the dipole and quadrupole are discussed, and solutions are presented.

We begin first with a general approach in solving the magnetic field in a multipole magnet (Asner, 1969; Green, 1971). This is essential because idealized schemes are useful only as an overall design approach—several compromising factors must be taken into account, and it becomes necessary eventually to compute the magnetic field for a current distribution which differs from any idealized case.

a. Vector potentials. The magnetic field H is given by the relationship $H = (\nabla \times A)/\mu_0$ where A is the vector potential. The vector potential A at a point $P(r, \theta)$ by a line current $+I$ at (ρ, ψ) (see Fig. 16) is

$$A = \frac{\mu_0 I}{4\pi} \ln[\rho^2 + r^2 - 2\rho r \cos(\psi - \theta)] \tag{29}$$

which for $r < \rho$ can be expressed as

$$A = \frac{\mu_0 I}{2\pi} \ln \rho - \frac{\mu_0 I}{2\pi} \sum_{n=1}^{\infty} \frac{1}{n} \left(\frac{r}{\rho}\right)^n \cos n\theta \tag{30}$$

and for $r > \rho$ as

$$A = \frac{\mu_0 I}{2\pi} \ln \rho + \frac{\mu_0 I}{2\pi} \ln \left|\frac{r}{\rho}\right| - \frac{\mu_0 I}{2\pi} \sum_{n=1}^{\infty} \frac{1}{n} \left(\frac{\rho}{r}\right)^n \cos n\theta \tag{31}$$

The expression for the vector potential and magnetic field generated by a current of any spatial distribution can be obtained by superposition.

The dipole magnet has a current distribution symmetric about the x axis and antisymmetric about the y axis. That is, for a line current $+I$ placed at $(\rho, 0 < \psi < \pi/2)$ along the z axis, there must exist a line current $+I$ at $(\rho, -\psi)$, a line current $-I$ at $(\rho, \pi + \psi)$, and finally a line current $-I$ at $(\zeta, \pi - \psi)$. Consequently, in computing distribution, we must consider these four line currents simultaneously, namely,

$$\begin{aligned} A_{\mathrm{D}} = (I\mu_0/4\pi)\{&\ln[\rho^2 + r^2 - 2\rho r\cos(\psi - \theta)] \\ &+ \ln[\rho^2 + r^2 - 2\rho r\cos(-\psi - \theta)] \\ &- \ln[\rho^2 + r^2 - 2\rho r\cos(\pi + \psi - \theta)] \\ &- \ln[\rho^2 + r^2 - 2\rho r\cos(\pi - \psi - \theta)]\} \end{aligned} \tag{32}$$

which reduces to

$$A_{\mathrm{D}} = -\frac{2I\mu_0}{\pi}\sum_{n=1}^{\infty}\frac{1}{(2n-1)}\left(\frac{r}{\rho}\right)^{2n-1}\cos(2n-1)\psi\cos(2n-1)\theta \qquad r < \rho \tag{33}$$

$$A_{\mathrm{D}} = -\frac{2I\mu_0}{\pi}\sum_{n=1}^{\infty}\frac{1}{(2n-1)}\left(\frac{\rho}{r}\right)^{2n-1}\cos(2n-1)\psi\cos(2n-1)\theta \qquad r > \rho \tag{34}$$

The quadrupole magnet, on the other hand, has a current distribution symmetric about $x = \pm y$ lines and antisymmetric about the x and y axes. That is, for a positive line current along the z axis placed in the first quadrant, a negative, positive, and negative line currents must be similarly placed, respectively, in the second, third, and fourth quadrants. The sum of the potentials due to these line currents gives the desired expression for the vector potential of a quadrupole-magnet current distribution:

$$A_{\mathrm{Q}} = -\frac{I\mu_0}{\pi}\sum_{n=1}^{\infty}\frac{1}{n}\left(\frac{r}{\rho}\right)^{2n}\sin 2n\psi\sin 2n\theta \qquad r < \rho \tag{35}$$

$$A_{\mathrm{Q}} = -\frac{I\mu_0}{\pi}\sum_{n=1}^{\infty}\frac{1}{n}\left(\frac{\rho}{r}\right)^{2n}\sin 2n\psi\sin 2n\theta \qquad r > \rho \tag{36}$$

b. Magnetic fields. The magnetic field can be obtained readily once the vector potential is known:

$$\mathbf{H} = \frac{1}{\mu_0}\nabla \times \mathbf{A} \tag{37}$$

Inserting Eqs. (33)–(36) into Eq. (37), we obtain expressions for the magnetic fields corresponding to the dipole and quadrupole consisting of four current lines.

For the dipole they are, expressed separately for the r and θ components, for $r < \rho$,

$$H_r = \frac{1}{r}\frac{\partial A_{\mathrm{D}}}{\partial \theta} = \frac{2I}{\pi}\sum_{n=1}^{\infty}\frac{r^{2n-2}}{\rho^{2n-1}}\cos(2n-1)\psi\sin(2n-1)\theta \tag{38}$$

$$H_\theta = -\frac{\partial A_{\mathrm{D}}}{\partial r} = \frac{2I}{\pi}\sum_{n=1}^{\infty}\frac{r^{2n-1}}{\rho^{2n-1}}\cos(2n-1)\psi\cos(2n-1)\theta \tag{39}$$

and for $r > \rho$,

$$\tilde{H}_{r,\theta} = \pm\frac{2I}{\pi}\sum_{n=1}^{\infty}\frac{\rho^{2n-1}}{r^{2n}}\cos(2n-1)\psi\begin{Bmatrix}\sin(2n-1)\theta\\ \\ \cos(2n-1)\theta\end{Bmatrix} \tag{40}$$

where the upper of the braced terms gives the r component, and the lower term gives the θ component of the field.

Similarly for the quadrupole, they are, for $r < \rho$,

$$H_{r,\theta} = \mp\frac{2I}{\pi}\sum_{n=1}^{\infty}\frac{r^{2n-1}}{\rho^{2n}}\sin 2n\psi\begin{Bmatrix}\cos 2n\theta\\ \\ \sin 2n\theta\end{Bmatrix} \tag{41}$$

and for $r > \rho$,

$$\tilde{H}_{r,\theta} = -\frac{2I}{\pi}\sum_{n=1}^{\infty}\frac{\rho^{2n}}{r^{2n+1}}\sin 2n\psi\begin{Bmatrix}\cos 2n\theta\\ \\ \sin 2n\theta\end{Bmatrix} \tag{42}$$

In the quadrupole magnet, one is interested also in the field gradient for $r < \rho$, the gradients along two important radii, $\theta = 0$ and $\theta = \pi/4$, being

$$g_r \equiv \frac{\partial H_r}{\partial r} \quad \text{at} \quad \theta = 0 \tag{43}$$

and

$$g_\theta \equiv \frac{\partial H_\theta}{\partial r} \quad \text{at} \quad \theta = \frac{\pi}{4} \tag{44}$$

Combining Eqs. (41)–(44), we have expressions for the field gradients. They are, for $r < \rho$,

$$g_r = -\frac{2I}{\pi}\sum_{n=1}^{\infty}\frac{(2n-1)}{\rho^{2n}}r^{2n-2}\sin 2n\psi \tag{45}$$

$$g_\theta = \frac{2I}{\pi}\sum_{n=1}^{\infty}\frac{(2n-1)}{\rho^{2n}}r^{2n-2}\sin 2n\psi\sin\frac{\pi}{2}n \tag{46}$$

c. Dipoles. The magnetic field expressions given by Eqs. (38)–(40) are those of a dipole energized with four current lines. For any other dipole-like current distribution, all we need to do to obtain proper field

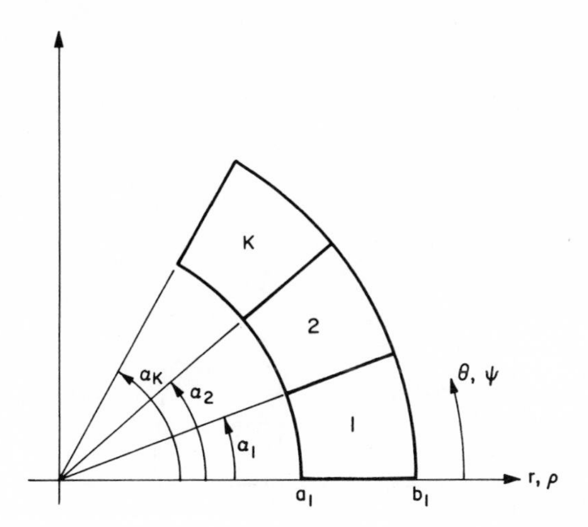

FIG. 17. Current-block arrangement for the first quadrant of the constant-winding-thickness dipole.

expressions is to replace I in Eqs. (38)–(46) by $j(\rho, \psi)\rho\, d\rho\, d\psi$ and integrate over one quadrant of the ρ and ψ space where $j(\rho, \psi)$ is nonzero. For a dipole, the ideal current distribution, the one for which only the fundamental component results, is $j \sim \cos\psi$. In practice, current-carrying blocks, within each of which the current density is uniform, are placed around the aperture to approximate the ideal current distribution. Two schemes are considered here. In the first, shown in Fig. 17, the current density differs from one block to the next, the winding thickness being identical; in the other, shown in Fig. 18, the winding thickness differs from one block to the next, the current densities being identical.

In the first scheme, if K represents the number of blocks in one quadrant, the harmonics corresponding to $n = 2$ through $n = 2K$ can be eliminated by a proper adjustment of the current density and location of the blocks. In the second scheme, as we shall see, except for the case $K = 1$, which is identical to the first scheme, K must be an even number, and the highest harmonic which can be eliminated corresponds to $n = 3K/2$. In either scheme, if n^* represents the first nonzero harmonic term, the r and θ components of the magnetic field in such a dipole magnet can always be expressed as a sum of the fundamental and a power series of nonvanishing harmonics. Thus, for $r < a_1$, where a_1 is the innermost radius of the winding,

$$\frac{H_{r,\theta}}{H_0} = D_1(K, c_1)\begin{Bmatrix}\sin\theta \\ \\ \cos\theta\end{Bmatrix} + \sum_{n=n^*}^{\infty} D_n(K, n, c_1)\left(\frac{r}{a_1}\right)^{2n-2}\begin{Bmatrix}\sin(2n-1)\theta \\ \\ \cos(2n-1)\theta\end{Bmatrix} \tag{47}$$

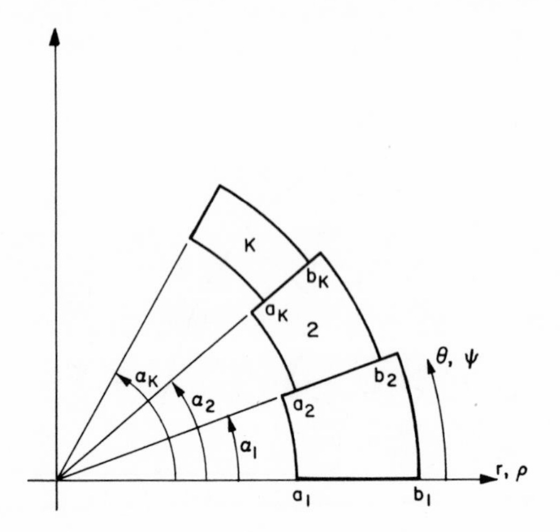

FIG. 18. Current-block arrangement for the first quadrant of the variable-winding-thickness dipole.

and for $r > b_1$, where b_1 is the outermost radius of the winding,

$$\frac{\tilde{H}_{r,\theta}}{H_0} = \tilde{D}_1(K, c_1)\left(\frac{b_1}{r}\right)^2 \begin{Bmatrix}\sin\theta \\ \\ \cos\theta\end{Bmatrix} + \sum_{n=n*}^{\infty} \tilde{D}_n(K, n, c_1)\left(\frac{b_1}{r}\right)^{2n} \begin{Bmatrix}\sin(2n-1)\theta \\ \\ \cos(2n-1)\theta\end{Bmatrix} \tag{48}$$

The normalizing field H_0 (Oe) is given by

$$H_0 = 0.4\pi a_1(c_1 - 1)\, j_1 \tag{49}$$

where j_1 (A/cm²) is the current density in the first block, and $c_1 = b_1/a_1$.

i. CONSTANT WINDING THICKNESS. For a constant winding thickness, it can be shown that

$$D_n(K, n, c_1) = \frac{2(c_1^{3-2n} - 1)}{(2n-1)(3-2n)(c_1 - 1)} \times \left\{\frac{1}{j_1}\sum_{k=1}^{K} j_k[\sin(2n-1)\alpha_k - \sin(2n-1)\alpha_{k-1}]\right\} \tag{50}$$

$$\tilde{D}_n(K, n, c_1) = \frac{2(c_1^{2n+1} - 1)}{c_1^{2n}(2n-1)(2n+1)(c_1 - 1)} \times \left\{\frac{1}{j_1}\sum_{k=1}^{K} j_k[\sin(2n-1)\alpha_k - \sin(2n-1)\alpha_{k-1}]\right\} \tag{51}$$

TABLE I

PARAMETERS AND COEFFICIENTS FOR CONSTANT-WINDING-THICKNESS DIPOLE

K = 1	K = 2	K = 3	K = 4
$\alpha_1 = 60°$	$\alpha_1 = 36°$; $\alpha_2 = 72°$	$\alpha_1 = 25\,5/7°$; $\alpha_2 = 51\,3/7°$; $\alpha_3 = 77\,1/7°$	$\alpha_1 = 20°$; $\alpha_2 = 40°$; $\alpha_3 = 60°$; $\alpha_4 = 80°$
	$j_2/j_1 = 0.6180$	$j_2/j_1 = 0.8019$; $j_3/j_1 = 0.4450$	$j_2/j_1 = 0.8794$; $j_3/j_1 = 0.6527$; $j_4/j_1 = 0.3473$

D_1	K = 1 1.7320			K = 2 1.6258			K = 3 0.7994			K = 4 0.7940		
c_1	$\tilde{D}_1$	$-D_3 \times 10$	$-\tilde{D}_3 \times 10$	$\tilde{D}_1$	$-D_5 \times 10$	$-\tilde{D}_5 \times 10$	$\tilde{D}_1$	$-D_7 \times 10^2$	$-\tilde{D}_7 \times 10^2$	$\tilde{D}_1$	$-D_9 \times 10^2$	$-\tilde{D}_9 \times 10^2$
1.0	1.7320	3.4640	3.4640	1.6258	1.8064	1.8064	1.5988	12.300	12.300	1.5880	9.3420	9.3420
1.2	1.4594	2.4320	2.1400	1.3698	0.9302	0.8528	1.3472	4.8380	4.6000	1.3380	2.9120	2.8580
1.4	1.2844	1.8348	1.5678	1.2056	0.5840	0.5606	1.1856	2.7260	2.8560	1.1776	1.5470	1.7180
1.6	1.1638	1.4546	1.2706	1.0924	0.4140	0.4354	1.0742	1.8528	2.1840	1.0670	1.0370	1.3110
1.8	1.0764	1.2958	1.0952	1.0102	0.3174	0.3688	0.9936	1.3954	1.8446	0.9868	0.7784	1.1062
2.0	1.0104	1.0104	1.0104	0.9820	0.2560	0.3282	0.9326	1.1176	1.6398	0.9264	0.6228	0.9834
2.2	0.9590	0.8718	0.9036	0.9002	0.2142	0.3010	0.8854	0.9316	1.5032	0.8794	0.5190	0.9014
2.4	0.9182	0.7652	0.8464	0.8618	0.1839	0.2816	0.8476	0.7986	1.4056	0.8418	0.4448	0.8428
2.6	0.8848	0.6806	0.8032	0.8306	0.1611	0.2608	0.8168	0.6968	1.3324	0.8112	0.3892	0.7990
2.8	0.8592	0.6122	0.7692	0.8046	0.1433	0.2554	0.7912	0.6212	1.2754	0.7860	0.3460	0.7646
3.0	0.8340	0.5560	0.7420	0.7828	0.1290	0.2464	0.7698	0.5590	1.2298	0.7646	0.3114	0.7374

$$D_n(K, n, C_1) = \frac{(C_1^{3-2n} - 1)}{(2n-1)(3-2n)(C_1-1)} \frac{1}{j_1} \sum_{k=1}^{K} j_k \left(\{\sin(2n-1)\alpha_k - \sin(2n-1)\alpha_{k-1}\}\right)$$

$$\tilde{D}_n(K, n, C_1) = \frac{(C_1^{2n+1} - 1)}{C_1^{2n}(2n-1)(2n+1)(C_1-1)} \frac{1}{j_1} \sum_{k=1}^{K} j_k \left(\{\sin(2n-1)\alpha_k - \sin(2n-1)\alpha_{k-1}\}\right)$$

Both j_k, the current density in the kth block ($k = 1$ to K), and α_k (with $\alpha_0 = 0$) have been solved by Beth (1969). D_n, $\tilde{D}_n$, j_k, and α_k are given in Table I. For this winding scheme, the field within the winding is relatively simple. We consider the dominant ($n = 1$) term only:

$$\frac{H_{r,\theta}}{H_0} = \left[\left(\frac{b_1 - r}{b_1 - a_1}\right) D_1 \pm \left(\frac{r^3 - a_1^3}{b_1^3 - a_1^3}\right)\left(\frac{b_1}{r}\right)^2 \tilde{D}_1\right] \begin{Bmatrix} \sin\theta \\ \cos\theta \end{Bmatrix} \tag{52}$$

The maximum occurs at $r = a_1$.

ii. VARIABLE WINDING THICKNESS. Often a winding of variable thickness is more convenient and economical of superconductor, because it maintains the same current density throughout the winding. Other apparently simpler schemes which do this, such as those shown in Fig. 19, eliminate only half as many harmonics per block of current. It is desirable to eliminate harmonics in the region $r > b_1$ as well as in the region $r < a_1$ because, as we shall see later when discussing iron shielding, the field in region $r > b_1$ determines the harmonics of the induced field generated by

TABLE II

PARAMETERS AND COEFFICIENTS FOR VARIABLE-WINDING-THICKNESS DIPOLE

K = 2

c_1	a_2/a_1	b_2/a_1	$\alpha_1(0)$	$\alpha_2(0)$	D_1	$\tilde{D}_1$	$-D_4 \times 10^2$	$-\tilde{D}_4 \times 10^2$
1.2	1.039	1.163	36	72	1.6276	1.3714	0.5390	0.4648
1.4	1.077	1.329	36	72	1.6346	1.2120	1.1606	0.9296
1.6	1.113	1.399	36	72	1.6436	1.1038	1.5290	1.1884
1.8	1.145	1.671	36	72	1.6530	1.0262	1.7004	1.3160
2.0	1.175	1.845	36	72	1.6624	0.9680	1.7530	1.3722
2.2	1.202	2.021	36	72	1.6716	0.9228	1.7398	1.3906
2.4	1.226	2.199	36	72	1.6800	0.8870	1.6920	1.3890
2.6	1.248	2.377	36	72	1.6880	0.8576	1.6276	1.3774
2.8	1.268	2.556	36	72	1.6954	0.8332	1.5566	1.3610
3.0	1.286	2.735	36	72	1.7020	0.8128	1.4846	1.3426

K = 4

c_1	a_2/a_1	b_2/a_1	a_3/a_1	b_3/a_1	a_4/a_1	b_4/a_1	$\alpha_1(0)$	$\alpha_2(0)$	$\alpha_3(0)$	$\alpha_4(0)$	D_1	$\tilde{D}_1$	$-D_7 \times 10^3$	$-\tilde{D}_7 \times 10^4$
1.2	1.012	1.189	1.035	1.167	1.066	1.136	19.98	39.97	59.98	79.99	1.5908	1.3404	0.159	0.153
1.4	1.024	1.378	1.069	1.336	1.133	1.277	19.92	39.87	59.87	79.94	1.1858	1.1858	0.743	0.474
1.6	1.034	1.569	1.102	1.508	1.199	1.420	19.93	39.94	60.04	80.07	1.6104	1.0812	1.440	0.403
1.8	1.043	1.762	1.130	1.684	1.261	1.567	19.82	39.75	59.85	79.98	1.6216	1.0060	1.909	0.203
2.0	1.050	1.934	1.156	1.861	1.320	1.718	19.73	39.62	59.73	79.95	1.6328	0.9498	2.010	1.070
2.2	1.057	2.148	1.179	2.040	1.375	1.871	19.64	39.49	59.63	79.92	1.6434	0.9060	2.086	2.022
2.4	1.062	2.341	1.199	2.219	1.427	2.026	19.57	39.98	59.55	79.91	1.6534	0.8712	2.086	2.956
2.6	1.067	2.535	1.218	2.400	1.477	2.182	19.50	39.29	59.49	79.91	1.6630	0.8428	2.046	3.824
2.8	1.071	2.729	1.234	2.581	1.524	2.339	19.44	39.21	59.46	79.94	1.6716	0.8192	1.988	4.616
3.0	1.075	2.924	1.249	2.762	1.568	2.497	19.40	39.15	55.45	79.99	1.6798	0.7994	1.924	5.340

$$D_n(K, n, C_1) = \frac{1}{(2n-1)(2-2n)(C_1-1)} \sum_{k=1}^{K} \left[\left(\frac{b_k}{a_1}\right)^{3-2n} - \left(\frac{a_k}{a_1}\right)^{3-2n}\right] \{\sin(2n-1)\alpha_k - \sin(2n-1)\alpha_{k-1}\}$$

$$D_n(K, n, C_1) = \frac{1}{C_1^{2n}(2n-1)(2n+1)(C_1-1)} \sum_{k=1}^{K} \left[\left(\frac{b_k}{a_1}\right)^{2n+1} - \left(\frac{a_k}{a_1}\right)^{2n+1}\right] \{\sin(2n-1)\alpha_k - \sin(2n-1)\alpha_{k-1}\}$$

the iron shield. For this scheme, it can be shown that

$$D_n(K, n, c_1) = \frac{2}{(2n-1)(3-2n)(c_1-1)} \sum_{k=1}^{K} \left[\left(\frac{b_k}{a_1}\right)^{3-2n} - \left(\frac{a_k}{a_1}\right)^{3-2n}\right] \times [\sin(2n-1)\alpha_k - \sin(2n-1)\alpha_{k-1}] \tag{53}$$

$$\tilde{D}_n(K, n, c_1) = \frac{2}{c_1^{2n}(2n-1)(2n+1)(c_1-1)} \sum_{k=1}^{K} \left[\left(\frac{b_k}{a_1}\right)^{2n+1} - \left(\frac{a_k}{a_1}\right)^{2n+1}\right] \times [\sin(2n-1)\alpha_k - \sin(2n-1)\alpha_{k-1}] \tag{54}$$

a_k, b_k, α_k, D_n, and $\tilde{D}_n$ are given in Table II.

d. Quadrupoles. A procedure entirely similar to that used for the dipole is applicable here. Again, we examine two winding schemes analogous to those examined for the dipole and obtain appropriate expressions for the field. As before, the magnetic field for the r and θ components can be expressed as a sum of the fundamental and a power series of harmonics.

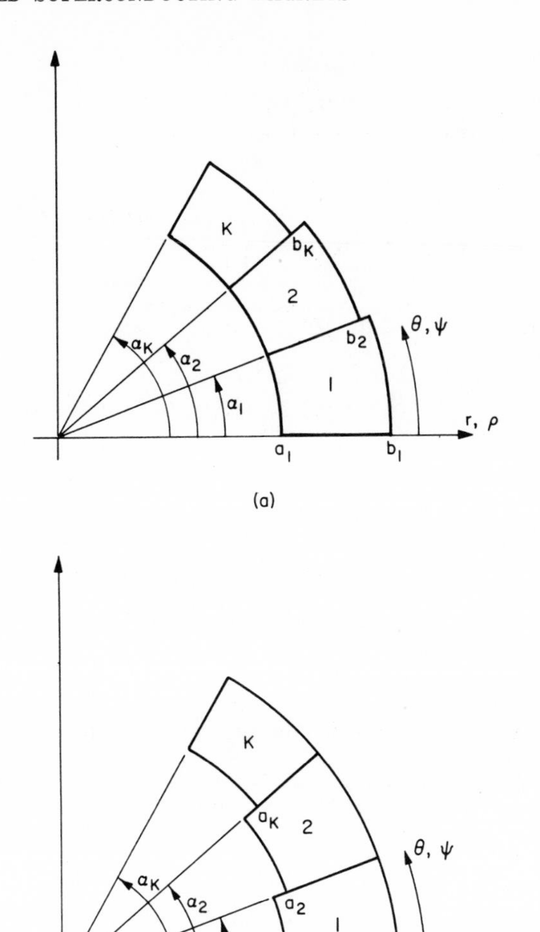

FIG. 19. Two other possible current-block arrangements for the first quadrant of the dipole.

That is, for $r < a_1$,

$$\frac{H_{r,\theta}}{H_0} = \pm Q_1(K, c_1) \left(\frac{r}{a_1}\right) \begin{Bmatrix} \cos 2\theta \\ \sin 2\theta \end{Bmatrix}$$

$$\pm \sum_{n=n*}^{\infty} Q_n(K, n, c_1) \left(\frac{r}{a_1}\right)^{4n-3} \begin{Bmatrix} \cos(4n - 2)\theta \\ \sin(4n - 2)\theta \end{Bmatrix} \tag{55}$$

and for $r > b_1$,

$$\frac{\tilde{H}_{r,\theta}}{H_0} = +\tilde{Q}_1(K, c_1)\left(\frac{b_1}{r}\right)^3 \begin{Bmatrix}\cos 2\theta \\ \sin 2\theta\end{Bmatrix} + \sum_{n=n*}^{\infty} \tilde{Q}_n(K, n, c_1)\left(\frac{b_1}{r}\right)^{4n-1} \begin{Bmatrix}\cos(4n-2)\theta \\ \sin(4n-2)\theta\end{Bmatrix} \tag{56}$$

The field gradients, for $r < a_1$, are given by

$$g_r(\theta = 0) = -\frac{H_0}{a_1}\left[Q_1(K, c_1) + \sum_{n=n*}^{\infty} Q_n(K, n, c_1)(4n-3)\left(\frac{r}{a_1}\right)^{4n-4}\right] \tag{57}$$

$$g_\theta\left(\theta = \frac{\pi}{4}\right) = \frac{H_0}{a_1}\left[Q_1(K, c_1) + \sum_{n=n*}^{\infty} (-1)^{n+1} Q_n(K, n, c_1)(4n-3)\left(\frac{r}{a_1}\right)^{4n-4}\right] \tag{58}$$

where the normalizing field H_0 is given by Eq. (49). Note that the first block, corresponding to $k = 1$, is now the one adjacent to the 45° line (see Figs. 20 and 21).

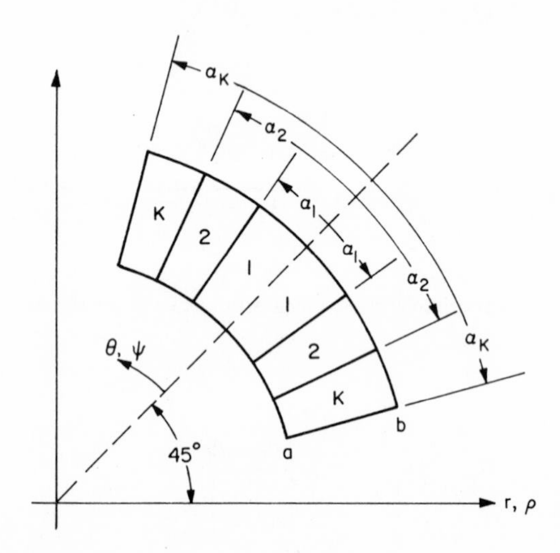

FIG. 20. Current-block arrangement for the first quadrant of the constant-winding-thickness quadrupole.

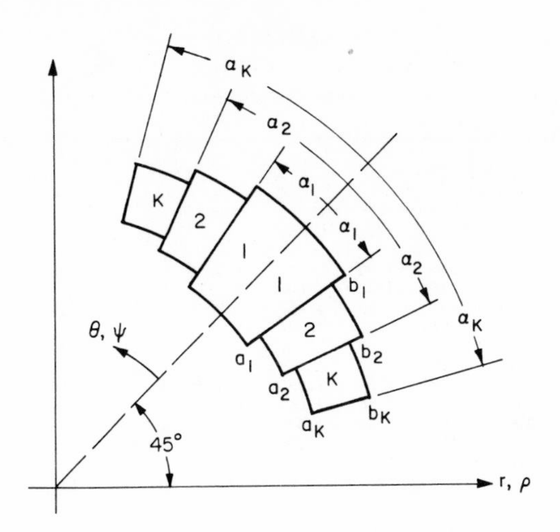

FIG. 21. Current-block arrangement for the first quadrant of the variable-winding-thickness quadrupole.

i. CONSTANT WINDING THICKNESS. For this scheme, shown in Fig. 20, it can be shown that

$$Q_1(K, c_1) = \frac{2}{c_1 - 1} \ln c_1 \left[\frac{1}{j_1} \sum_{k=1}^{K} j_k(\sin 2\alpha_k - \sin 2\alpha_{k-1}) \right] \tag{59}$$

$$Q_{n>1}(K, n, c_1) = \frac{(-1)^{n+1}(c_1^{4n-4} - 1)}{c_1^{4n-4}(2n-1)(2n-2)(c_1-1)} \times \left\{ \frac{1}{j_1} \sum_{k=1}^{K} j_k[\sin(4n-2)\alpha_k - \sin(4n-2)\alpha_{k-1}] \right\} \tag{60}$$

$$\tilde{Q}_n(K, n, c_1) = \frac{(-1)^{n+1}(c_1^{4n} - 1)}{c^{4n-1}(2n-1)(2n)} \times \left\{ \frac{1}{j_1} \sum_{k=1}^{K} j_k[\sin(4n-2)\alpha_k - \sin(4n-2)\alpha_{k-1}] \right\} \tag{61}$$

The required value of each α_k, now measured from the 45° line, is one-half that for the dipole of constant winding thickness; each j_k is unchanged. Q_1, Q_n, and $\tilde{Q}_n$ are tabulated in Table III. For this winding scheme, the

TABLE III
PARAMETERS AND COEFFICIENTS FOR
CONSTANT-WINDING-THICKNESS QUADRUPOLE

	K = 1; $\alpha_1 = 30°$; $j_1/j_1 = 1$						K = 2; $\alpha_1 = 18°$; $\alpha_2 = 36°$; $j_2/j_1 = 0.6180$					
c_1	Q_1	$\tilde{Q}_1$	$Q_3 \times 10$	$\tilde{Q}_3 \times 10$	(r/a_1)*	(H/H_c)†	Q_1	$\tilde{Q}_1$	$Q_5 \times 10^2$	$\tilde{Q}_5 \times 10^2$	(r/a_1)*	(H/H_c)†
1.0	1.7320	1.7320	3.4640	3.4640	1.00	1.000	1.6258	1.6258	13.500	13.500	1.00	1.000
1.2	1.5790	1.3452	1.6616	1.5378	1.00	1.000	1.4820	1.2626	5.3400	5.2780	1.00	1.000
1.4	1.4570	1.1210	1.0092	0.9926	1.00	1.000	1.3676	1.0522	2.8100	3.1580	1.06	1.013
1.6	1.3568	0.9786	0.7048	0.7670	1.00	1.000	1.2736	0.9184	1.8806	2.4080	1.12	1.034
1.8	1.2726	0.8814	0.5364	0.6490	1.07	1.005	1.1946	0.8274	1.4112	2.0320	1.17	1.054
2.0	1.2006	0.8120	0.4314	0.5772	1.14	1.019	1.1270	1.1270	1.1290	1.8064	1.23	1.075
2.2	1.1380	0.7600	0.3602	0.5292	1.20	1.035	1.0682	0.7134	0.9408	1.6560	1.29	1.096
2.4	1.0832	0.7200	0.3090	0.4948	1.26	1.053	1.0166	0.6758	0.8064	1.5484	1.35	1.117
2.6	1.0342	0.6882	0.2706	0.4690	1.33	1.071	0.9710	0.6460	0.7056	1.4678	1.14	1.138
2.8	0.9908	0.6626	0.2405	0.4490	1.39	1.090	0.9300	0.6220	0.6272	1.4050	1.48	1.159
3.0	0.9514	0.6416	0.2164	0.4330	1.46	1.110	0.8930	0.6022	0.5646	1.3548	1.55	1.180

	K = 3; $\alpha_1 = 12\ 6/7°$; $\alpha_2 = 25\ 5/7°$; $\alpha_3 = 38\ 4/7°$; $j_2/j_1 = 0.8019$; $j_3/j_1 = 0.4450$						K = 4; $\alpha_1 = 10°$; $\alpha_2 = 20°$; $\alpha_3 = 30°$; $\alpha_4 = 40°$; $j_2/j_1 = 0.8794$; $j_3/j_1 = 0.6527$; $j_4/j_1 = 0.3473$					
c_1	Q_1	$\tilde{Q}_1$	$Q_7 \times 10^2$	$\tilde{Q}_7 \times 10^2$	(r/a_1)*	(H/H_c)†	Q_1	$\tilde{Q}_1$	$Q_9 \times 10^2$	$\tilde{Q}_9 \times 10^2$	(r/a_1)*	(H/H_c)†
1.0	1.5988	1.5988	12.300	12.300	1.00	1.000	1.5880	9.5880	9.3400	9.3400	1.00	1.000
1.2	1.4576	1.2476	2.5300	2.6200	1.03	1.007	1.4476	1.2334	1.4554	1.5558	1.04	1.013
1.4	1.3450	1.0348	1.2808	1.5372	1.09	1.029	1.3358	1.0278	0.7298	0.9082	1.09	1.036
1.6	1.2524	0.9032	0.8540	1.1714	1.14	1.051	1.2440	0.8972	0.4866	0.6920	1.14	1.058
1.8	1.1748	0.8136	0.6406	0.9882	1.19	1.072	1.1668	0.8082	0.3650	0.5838	1.20	1.080
2.0	1.1082	0.7494	0.5124	0.8784	1.25	1.094	1.1008	0.7444	0.2920	0.5190	1.25	1.101
2.2	1.0506	0.7016	0.4270	0.8052	1.31	1.115	1.0434	0.6968	0.2432	0.4758	1.31	1.123
2.4	0.9998	0.6646	0.3660	0.7530	1.37	1.136	0.9930	0.6600	0.2086	0.4448	1.38	1.145
2.6	0.9548	0.6354	0.3202	0.7138	1.43	1.158	0.9484	0.6310	0.1825	0.4216	1.44	1.166
2.8	0.9146	0.6117	0.2846	0.6832	1.50	1.179	0.9084	0.6076	0.1622	0.4036	1.51	1.188
3.0	0.8782	0.5922	0.2562	0.6588	1.57	1.201	0.8724	0.5882	0.1460	0.3892	1.58	1.210

$$Q_1(K, C_1) = \frac{1}{(C_1 - 1)} \ln C_1 \frac{1}{j_1} \sum_{k=1}^{K} j_k (\sin 2\alpha_k - \sin 2\alpha_{k-1})$$

$$Q_{n>1}(K, n, C_1) = \frac{(-1)\ (C_1^{4n-4} - 1)}{(C_1 - 1)(2n-1)(4n-4)\, C_1^{4n-4}} \frac{1}{j_1} \sum_{k=1}^{K} \{\sin(4n-2)\alpha_k - \sin(4n-2)\alpha_{k-1}\}$$

$$\tilde{Q}_n(K, n, C_1) = \frac{(-1)\ (C_1^{4n} - 1)}{(2n-1)(4n)(C_1 - 1)\, C^{4n-1}} \frac{1}{j_1} \sum_{k=1}^{K} \{\sin(4n-2)\alpha_k - \sin(4n-2)\alpha_{k-1}\}$$

*Location of maximum field. $H_c = H_0 Q_1$.

†Maximum field.

dominant term of the magnetic field within the winding is

$$\frac{H_{r,\theta}}{H_0} = \pm Q_1 \left(\frac{\ln b_1/r}{\ln c_1}\right)\left(\frac{r}{a_1}\right)\begin{Bmatrix}\cos 2\theta \\ \\ \sin 2\theta\end{Bmatrix} + \tilde{Q}_1 \left(\frac{r^4 - a_1^4}{b_1^4 - a_1^4}\right)\left(\frac{b_1}{r}\right)^3 \begin{Bmatrix}\cos 2\theta \\ \\ \sin 2\theta\end{Bmatrix} \qquad (62)$$

which is maximum for $r > a_1$. The location and value of this maximum are included in Table III.

ii. VARIABLE WINDING THICKNESS. For this scheme, shown in Fig. 21, we have

$$Q_1(K, c_1) = \frac{2}{c_1 - 1}\left[\frac{1}{j_1}\sum_{k=1}^{K} \ln\left(\frac{b_k}{a_k}\right)(\sin 2\alpha_k - \sin 2\alpha_{k-1})\right] \tag{63}$$

$$Q_{n>1}(K, n, c_1) = \frac{(-1)^n}{(c_1 - 1)(2n - 1)(2n - 2)}\left\{\frac{1}{j_1}\sum_{k=1}^{K}\left[\left(\frac{b_k}{a_1}\right)^{4-4n} - \left(\frac{a_k}{a_1}\right)^{4-4n}\right]\right.$$

$$\left.\times\left[\sin(4n - 2)\alpha_k - \sin(4n - 2)\alpha_{k-1}\right]\right\} \tag{64}$$

$$\tilde{Q}_n(K, n, c_1) = \frac{(-1)^n}{c_1^{4n-1}(2n - 1)(2n)(c_1 - 1)}\left\{\frac{1}{j_1}\sum_{k=1}^{K}\left[\left(\frac{b_k}{a_1}\right)^{4n} - \left(\frac{a_k}{a_1}\right)^{4n}\right]\right.$$

$$\left.\times\left[\sin(4n - 2)\alpha_k - \sin(4n - 2)\alpha_{k-1}\right]\right\} \tag{65}$$

Appropriate values of a_k, b_k, α_k are tabulated together with Q_1, Q_n, and $\tilde{Q}_n$ in Table IV.

e. Iron shielding for circular-aperture dipoles and quadrupoles. It is usually desirable to minimize the intense fields immediately outside multipole magnets. This can be done either with reverse windings, which decrease the useful field, or with an iron shield, which increases it. The iron shield also makes the internal field less sensitive to its external environment. Here we discuss iron shields for circular-aperture multipoles in general, and derive expressions for the total magnetic field for the circular-aperture dipoles and quadrupoles analyzed above. We assume the magnetic permeability of the iron to be infinite and calculate the thickness of the shield necessary to make the assumption valid.

Given a circular field winding confined between $r = a_1$ and $r = b_1 > a_1$, let the resultant magnetic fields be $H(r/a_1, \theta)$ for $r < a_1$ and $\tilde{H}(b_1/r, \theta)$ for $r > b_1$. Suppose we add a circular-aperture iron shield of inside radius $b_s \geq b_1$. Since the permeability is assumed infinite, the tangential component of the total field must vanish at $r = b_s$. That is,

$$H_\theta(b_1/b_s, \theta) + H_{s\theta}(r = b_s, \theta) = 0 \tag{66}$$

where $H_{s\theta}(r, \theta)$ is the θ component of the additional magnetic field due to the presence of the shield and derivable from a potential in the manner of Eqs. (25) and (26). Since $H(r/a_1, \theta)$ and $\tilde{H}(b_1/r, \theta)$ themselves satisfy Eqs. (25) and (26), it is reasonable to assume $H_s(r, \theta)$ is either of the form $H(r/a_1, \theta)$ or $\tilde{H}(b_1/r, \theta)$. Because $H_s(r, \theta)$ must be finite for $r \leq b_s$, it fol-

TABLE IV
PARAMETERS AND COEFFICIENTS FOR
VARIABLE-WINDING-THICKNESS QUADRUPOLE

K = 2

c_1	a_2/a_1	b_2/a_1	$\alpha_2^{(0)}$	$\alpha_2^{(0)}$	Q_1	$\tilde{Q}_1$	$Q_4 \times 10^2$	$-\tilde{Q}_4 \times 10^2$
1.2	1.037	1.164	18	36	1.4918	1.2708	1.1863	1.0570
1.4	1.067	1.337	18	36	1.3975	1.0743	1.6537	1.4591
1.6	1.088	1.516	18	36	1.3223	0.9508	1.5880	1.4511
1.8	1.102	1.700	18	36	1.2576	0.8659	1.4024	1.3529
2.0	1.110	1.886	18	36	1.1998	0.8037	1.2195	1.2528
2.2	1.116	2.073	18	36	1.1476	0.7563	1.0650	1.1691
2.4	1.119	2.261	18	36	1.1000	0.7190	0.9389	1.1023
2.6	1.121	2.449	18	36	1.0567	0.6891	0.8363	1.0492
2.8	1.123	2.637	18	36	1.0170	0.6646	0.7523	1.0064
3.0	1.124	2.825	18	36	0.9805	0.6443	0.6826	0.9716

K = 4

c_1	a_2/a_1	b_2/a_1	a_3/a_1	b_3/a_1	a_4/a_1	b_4/a_1	$\alpha_1^{(0)}$	$\alpha_2^{(0)}$	$\alpha_3^{(0)}$	$\alpha_4^{(0)}$	Q_1	$\tilde{Q}_1$	$-Q_7 \times 10^3$	$\tilde{Q}_7 \times 10^4$
1.2	1.011	1.189	1.034	1.168	1.065	1.138	9.97	19.95	29.94	39.97	1.4601	1.2437	0.6838	2.6894
1.4	1.020	1.381	1.060	1.343	1.123	1.285	9.89	19.83	29.83	39.91	1.3719	1.0541	1.5177	5.0271
1.6	1.025	1.576	1.078	1.525	1.166	1.444	9.78	19.63	29.59	39.76	1.3019	0.9350	1.5721	4.1790
1.8	1.028	1.771	1.089	1.711	1.196	1.612	9.70	19.47	29.41	39.63	1.2421	0.8533	1.4064	2.9763
2.0	1.030	1.967	1.096	1.898	1.216	1.785	9.64	19.37	29.28	39.54	1.1890	0.7937	1.2309	2.1341
2.2	1.031	2.164	1.101	2.087	1.229	1.961	9.60	19.30	29.21	39.49	1.1407	0.7482	1.0819	1.5910
2.4	1.032	2.360	1.104	2.276	1.238	2.138	9.58	19.26	29.17	39.46	1.0964	0.7124	0.9597	1.2340
2.6	1.033	2.557	1.106	2.466	1.244	2.316	9.57	19.24	29.14	39.44	1.0556	0.6835	0.8592	0.9921
2.8	1.033	2.753	1.108	2.655	1.248	2.494	9.56	19.22	29.12	39.43	1.0179	0.6598	0.7760	0.8235
3.0	1.033	2.950	1.109	2.845	1.251	2.672	9.55	19.21	29.11	39.42	0.9831	0.6399	0.7064	0.7033

$$Q_1(K, C_1) = \frac{(-1)^n}{(C_1 - 1)} \frac{1}{j_1} \sum_{k=1}^{K} \ln\left(\frac{b_k}{a_k}\right)\{\sin 2\alpha_k - \sin 2\alpha_{k-1}\}$$

$$Q_{n>1}(K, n, C_1) = \frac{(-1)^n}{(2n-1)(2n-2)} \frac{1}{j_1} \sum_{k=1}^{K} \left[\left(\frac{b_k}{a_1}\right)^{4-4n} - \left(\frac{a_k}{a_1}\right)^{4-4n}\right]\{\sin(4n-2)\alpha_k - \sin(4n-2)\alpha_{k-1}\}$$

$$\tilde{Q}_n(K, n, C_1) = \frac{(-1)^{n+1}}{(C_1 - 1)(2n-1)(2n)C_1^{4n-1}} \frac{1}{j_1} \sum_{k=1}^{K} \left[\left(\frac{b_k}{a_1}\right)^{4n} - \left(\frac{a_k}{a_1}\right)^{4n}\right]\{\sin(4n-2)\alpha_k - \sin(4n-2)\alpha_{k-1}\}$$

lows that $H_s(r, \theta)$ is of the form $H(r/a_1, \theta)$ with a_1 replaced by b_s. Thus,

$$H_s(r, \theta) = C_s H(r/b_s, \theta) \tag{67}$$

where C_s is a constant determined solely by the boundary condition given by Eq. (66). It is clear, therefore, that any harmonic present in $H(b_1/r, \theta)$ will appear in $H_s(r, \theta)$, indicating the importance of eliminating harmonics for $r > b_1$ as well as $r < a_1$.

General expressions for the total magnetic fields for $r < a_1$ and $r > b_1$ are, for $r < a_1$,

$$H_T(r, \theta) = H(r/a_1, \theta) + C_s H(r/b_s, \theta) \tag{68}$$

and, for $b_1 < r < b_s$,

$$\tilde{H}_T(r, \theta) = \tilde{H}(b_1/r, \theta) + C_s H(r/b_s, \theta) \tag{69}$$

i. DIPOLES. Applying Eqs. (66) and (67), we have expressions for the r and θ components of $H_s(r, \theta)$ as follows:

$$\left(\frac{H_s}{H_0}\right)_{r,\theta} = \tilde{D}_1 \left(\frac{b_1}{b_s}\right)^2 \begin{Bmatrix} \sin\theta \\ \cos\theta \end{Bmatrix} + \sum_{n=n*}^{\infty} \tilde{D}_n \left(\frac{b_1}{b_s}\right)^{2n} \left(\frac{r}{b_s}\right)^{2n-2} \begin{Bmatrix} \sin(2n-1)\theta \\ \cos(2n-1)\theta \end{Bmatrix} \tag{70}$$

Thus, from Eqs. (68) and (69), we have expressions for the total field, $H_T(r, \theta)$, for $r \leq a_1$:

$$\left(\frac{H_T}{H_0}\right)_{r,\theta} = \left[D_1 + \tilde{D}_1 \left(\frac{b_1}{b_s}\right)^2\right] \begin{Bmatrix} \sin\theta \\ \cos\theta \end{Bmatrix} + \sum_{n=n*}^{\infty} \left[D_n + \tilde{D}_n \frac{(a_1 b_1)^{2n}}{(b_s)^{4n}} \left(\frac{b_s}{a_1}\right)^2\right] \left(\frac{r}{a_1}\right)^{2n-2} \begin{Bmatrix} \sin(2n-1)\theta \\ \cos(2n-1)\theta \end{Bmatrix} \tag{71}$$

and for $b_1 \leq r \leq b_s$:

$$\left(\frac{\tilde{H}_T}{H_0}\right)_{r,\theta} = \pm\tilde{D}_1 \left[\left(\frac{b_1}{r}\right)^2 \pm \left(\frac{b_1}{b_s}\right)^2\right] \begin{Bmatrix} \sin\theta \\ \cos\theta \end{Bmatrix} + \sum_{n=n*}^{\infty} \tilde{D}_n \left[\left(\frac{b_1}{r}\right)^{2n} \pm \frac{(a_1 b_1)^{2n}}{b_s^{4n}} \left(\frac{b_s}{a_1}\right)^2 \left(\frac{r}{a_1}\right)^{2n-2}\right] \begin{Bmatrix} \sin(2n-1)\theta \\ \cos(2n-1)\theta \end{Bmatrix} \tag{72}$$

ii. QUADRUPOLES. Similarly for the quadrupoles, the total magnetic field is, for $r \leq a_1$,

$$\left(\frac{H_T}{H_0}\right)_{r,\theta} = \pm\left[Q_1 + \tilde{Q}_1 \frac{a_1 b_1^3}{b_s^4}\right] \left(\frac{r}{a_1}\right) \begin{Bmatrix} \cos 2\theta \\ \sin 2\theta \end{Bmatrix} \pm \sum_{n=n*}^{\infty} \left[Q_n + \tilde{Q}_n \frac{(a_1 b_1)^{4n}}{b_s^{8n}} \frac{b_s^4}{(a_1^3 b_1)}\right] \left(\frac{r}{a_1}\right)^{4n-3} \begin{Bmatrix} \cos(4n-2)\theta \\ \sin(4n-2)\theta \end{Bmatrix} \tag{73}$$

and, for $b_1 \leq r \leq b_s$,

$$\left(\frac{H_T}{H_0}\right)_{r,\theta} = +\tilde{Q}_1 \left[\left(\frac{b_1}{r}\right)^3 \pm \frac{a_1 b_1^3}{b_s^4} \left(\frac{r}{a_1}\right)\right] \begin{Bmatrix} \cos 2\theta \\ \sin 2\theta \end{Bmatrix} + \sum_{n=n*}^{\infty} \tilde{Q}_n \left[\left(\frac{b_1}{r}\right)^{4n-1} \pm \frac{(a_1 b_1)^{4n} b_s^4}{b_s^{8n}(a_1^3 b_1)} \left(\frac{r}{a_1}\right)^{4n-3}\right] \begin{Bmatrix} \cos(4n-2)\theta \\ \sin(4n-2)\theta \end{Bmatrix} \tag{74}$$

iii. THICKNESS OF THE IRON SHIELD. The magnetic induction within an iron shield should not exceed 16,000 G or so. Its thickness W (cm), therefore, should satisfy the relationship

$$16{,}000 \geq b_s/W \int_0^{\pi/2} [H_T(r = b_s)]_r \, d\theta \tag{75}$$

for the dipole and

$$16{,}000 \geq b_s/W \int_0^{\pi/4} [H_T(r = b_s)]_r \, d\theta \tag{76}$$

for the quadrupole. Combining Eqs. (72) and (75) for the dipole,

$$\left(\frac{W}{b_s}\right) \geq \frac{2H_0\tilde{D}_1}{16{,}000}\left(\frac{b_1}{b_s}\right)^2 \tag{77}$$

and Eqs. (74) and (76) for the quadrupole,

$$\left(\frac{W}{b_s}\right) \geq \frac{H_0\tilde{Q}_1}{16{,}000}\left(\frac{b_1}{b_s}\right)^3 \tag{78}$$

f. Magnetic energy. We give here approximate expressions for the magnetic energy storage of the dipole and quadrupole of constant winding thickness with a circular-aperture iron shield located at $r = b_s$, $b_1 \leq b_s < \infty$. Since we are concerned only with an approximate value for the energy storage, the results should suffice for the dipole and quadrupole of variable winding thickness.

i. DIPOLES. The magnetic energy per unit length, E_M (J/m), is

$$\frac{E_M}{E_0} = \left\{1 + \frac{(c_1^2 + c_1 + 1)^2}{9c_1^2}\left[1 + 2\left(\frac{b_1}{b_s}\right)^2\right]\right\} \equiv e_D\left(c_1, \frac{b_1}{b_s}\right) \tag{79}$$

where

$$E_0 = (0.8) \times 10^{-6}[a_1^2(c_1 - 1)j_1D_1]^2 \quad \text{J/m} \tag{80}$$

We have neglected the magnetic energy stored in the iron shield; since the thickness of the shield is chosen to keep the iron from saturating, the energy stored in the shield is indeed small. For the case with no shield, the (b_1/b_s) term becomes zero. $e_D(c_1, b_1/b_s)$ is given in Fig. 22.

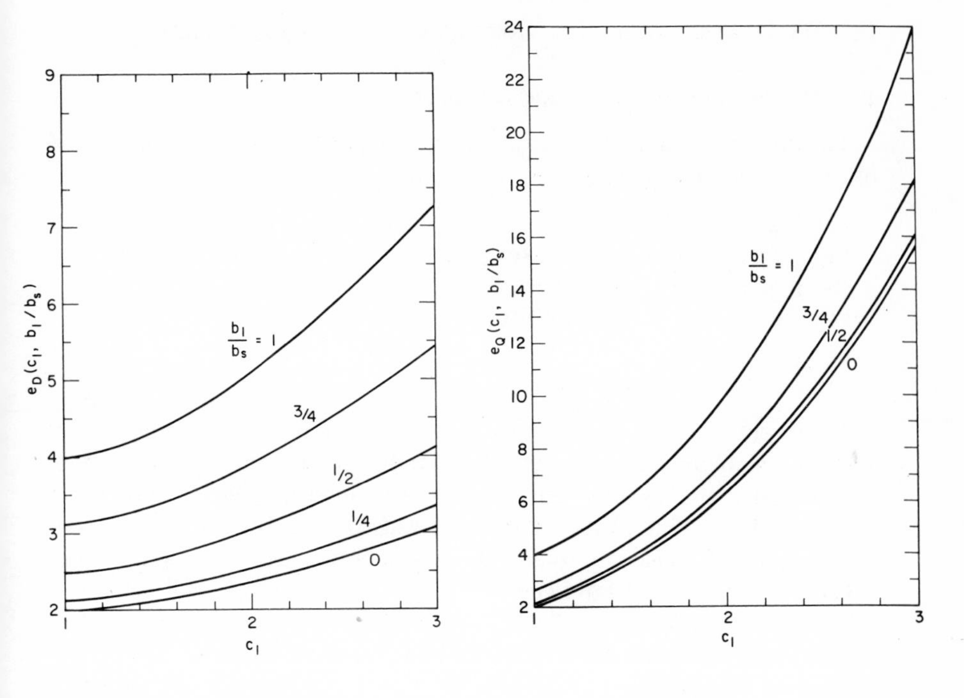

FIG. 22 FIG. 23

FIG. 22. The geometry-dependent factor $e_D(c_1, b_1/b_s)$ which gives the energy of a constant-winding-thickness dipole of the type shown in Fig. 17 and as in Eq. (79). The factor $e_D(c_1, b_1/b_s)$ is given for lines of constant b_1/b_s plotted versus c_1.

FIG. 23. The geometry-dependent factor $e_Q(c_1, b_1/b_s)$ which gives the energy of a constant-winding-thickness quadrupole of the type shown in Fig. 21 and as in Eq. (81). The factor $e_Q(c_1, b_1/b_s)$ is given for lines of constant b_1/b_s plotted versus c_1.

ii. QUADRUPOLES. Similarly, the stored energy for a quadrupole of constant winding thickness is

$$\frac{E_M}{E_0} = \left\{\left[\frac{c_1^4 - 1}{4(\ln c_1)^2} - \frac{1}{\ln c_1}\right] + \left(\frac{b_1}{b_s}\right)^4 \frac{(c_1^4 - 1)^2}{8c_1^4(\ln c_1)^2}\right\} \equiv e_Q\left(c_1, \frac{b_1}{b_s}\right) \tag{81}$$

where

$$E_0 = (0.4) \times 10^{-6}[a_1^2(c_1 - 1)j_1Q_1]^2 \quad \text{J/m} \tag{82}$$

$e_Q(c_1, b_1/b_s)$ is plotted in Fig. 23.

B. Materials, Conductor Configurations, and Stability

1. *High-field, high-current superconductors*

The properties a material should have to be practical for the construction of superconducting magnets are

(1) high upper critical field, H_{c2};
(2) high critical current density, J_c;
(3) high critical temperature, T_c;
(4) good strength and ductility;
(5) high stability;
(6) availability in quantity.

At present, three materials come closest to meeting these requirements: alloys of niobium and titanium, Nb–Ti; the intermetallic compound of niobium and tin, Nb_3Sn; and, finally, the intermetallic compound of vanadium and gallium, V_3Ga. These materials are largely complementary—Nb–Ti is used for fields up to 95 kOe, Nb_3Sn between 80 and 150 kOe, and V_3Ga between 100 and 180 kOe.

The high critical current density of these materials results because they exhibit large "pinning forces" located at "pinning centers." The field-dependent pinning force balances the Lorentz force, $J \times B$, resulting from the current density, J, and the ambient magnetic field, B. For any given external field B there exists a "critical current density" J_c for which the Lorentz force just equals the pinning force. Beyond J_c the material becomes resistive. The metallurgical "defects" of the structure, both mechanical and chemical as well as grain boundaries, are believed to provide pinning sites. To improve the current-carrying capacity of a material, therefore, defects are introduced deliberately during the processing. The most common methods are by cold-working and heat treatment. Cold-working introduces mechanical defects such as dislocations, whereas heat treatment controls the size of grains and phase precipitates. Consequently, J_c is strongly influenced by the metallurgical history of the material.

a. Niobium-titanium composite superconductors. Alloys of Nb and Ti containing approximately 50 wt% of each element have proved to be the most successful in meeting the above requirements (Berlincourt and Hake, 1963), and several Nb–Ti alloys of similar performance characteristics are now available commercially (Critchlow *et al.*, 1971). A typical short-sample critical current density versus magnetic field for Nb–Ti is shown in Fig. 24 as curve A. Although its upper critical field is near 120 kOe, the alloy typically has been used only for magnets designed to generate fields of no more than 90 kOe. [Although it is not impossible to generate a mag-

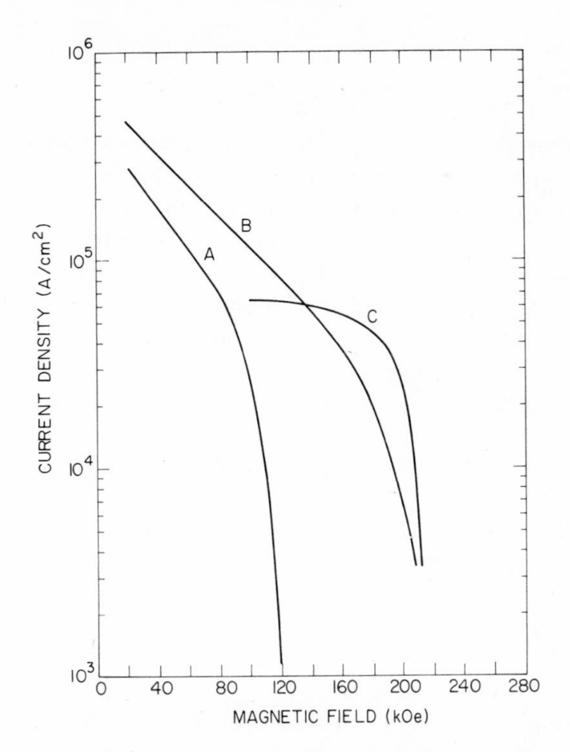

FIG. 24. Short sample critical-current density plotted as a function of magnetic field. Curve A for Nb–Ti alloy; curve B for diffusion-processed Nb_3Sn; curve C for diffusion-processed V_3Ga. For curves B and C the areas used for computing the critical current densities are those occupied by the superconductor and the substrate.

netic field as high as 100 kOe at 4.2K (Coffey *et al.*, 1965), a standard procedure currently practiced when attempting to generate fields up to 100 kOe is to operate the magnet at 2K.] Nb_3Sn is more economical at fields higher than 90 kOe.

The performance shown in curve A of Fig. 24 indicates that, if one wishes to build a magnet capable of generating a field of 60 kOe at an operating current of 1000 A, for example, the required conductor cross section is about 10^{-2} cm², or a diameter of 1 mm. In practice, however, such a large cross section in a single-core configuration is never used for magnets of any size. Instabilities inherent in large-size superconductors make it impossible to operate such a magnet successfully. Furthermore, the alloy, no matter how small its cross section, is rarely used alone: for stability it is always cladded with a high-conductivity normal metal. The whole question of stability will be discussed in the next section.

b. Niobium–tin tapes and composite wires. Until recently, when V_3Ga became available, the intermetallic compound Nb_3Sn was the only widely available high-field, high-current superconductor outside of Nb–Ti alloys. There now appear to be three methods to manufacture Nb_3Sn tapes suitable for magnet construction (Echarri and Spadoni, 1971).

One is the so-called vapor-deposition method, in which a mixture of hydrogen and gaseous chlorides of niobium and tin is reacted and a very thin layer ($\approx 5\ \mu M$) of Nb_3Sn is deposited on a thin, flexible stainless-steel tape. Another is the so-called diffusion method, in which a diffusion layer of Nb_3Sn is produced on a thin tape of niobium substrate after it has been run through a bath of molten tin. In the so-called plasma-spray method, a mixture of niobium and tin powder is dispensed into an argon plasma, and a mixture of molten niobium and tin is then sprayed on to a copper or copper/stainless-steel tape. The tape is subsequently heat-treated at 800°C to form a layer of Nb_3Sn, as thick as 125 μ, on the tape. The principal advantage of this method, aside from a possible reduction in the production cost, is its ability to form a layer of Nb_3Sn on a variety of complex and large surfaces, making it ideal in the preparation of surfaces for magnetic shielding.

The main drawback of Nb_3Sn is that it is hard and brittle. This has made it extremely difficult to wind large-cross-section Nb_3Sn wires into a coil, and only a small quantity of wire has ever been produced (Hechler *et al.*, 1969). Recent work with multifilament Nb_3Sn (Crow and Suenaga, 1972) appears to overcome this difficulty, but at present tape is the standard configuration; the mechanical difficulties are minimized by keeping the superconductor in a thin layer at the neutral bending axis (Benz, 1968). To increase its mechanical strength and stability, additional tapes of stainless steel and copper are usually soldered to the superconductor to make a composite tape.

A typical critical current density versus magnetic field for a diffusion-processed tape is shown in Fig. 24 as curve B. In computing the current density, we have taken a cross section including the niobium substrate as well as the Nb_3Sn. In practice, the useful current density is less, owing to the additional stainless steel and copper tapes. Again, as in the case of Nb–Ti alloys, the material is not ordinarily used as a magnet material up to its upper critical field of 220 kG; the maximum field produced by Nb_3Sn has been limited to 150 kOe by economic considerations.

c. Vanadium–gallium tapes and composite conductors. Until 1966, when a modified technique of processing was developed (Tachikawa and Tanaka, 1966), the intermetallic compound V_3Ga was known only as a high-field superconductor with a small current-carrying capacity. In both the old and new processes, a thin layer of V_3Ga is produced by diffusion in much the same way that Nb_3Sn is produced in the diffusion method. A vanadium substrate passes through a molten bath of gallium before it is reacted at an elevated temperature. It is this temperature which is the key to the differences between the old and new processes. The old process requires a temperature above 1200°C; in the new process, a temperature near 700°C suffices because copper is used as a catalyst. The lower heat temperature limits grain growth, resulting in a higher grain boundary density. As grain boundaries act as pinning centers, this results in a higher pinning center density and an improved current-carrying capacity.

A typical performance characteristic of V_3Ga tape is shown in Fig. 24 as curve C (Tachikawa and Iwasa, 1970). As in the case of Nb_3Sn, both the substrate and superconductor cross sections are used for the computation of the critical current density. From Fig. 24 we note that Nb_3Sn and V_3Ga are complementary—for fields up to 140 kOe, Nb_3Sn is preferred, and above that V_3Ga is useful up to about 180 kOe.

Although V_3Ga and Nb_3Sn are both considered to be brittle and have equally poor mechanical properties, V_3Ga is more likely to be ductile than Nb_3Sn. This speculation is based upon a comparison of their melting temperatures; greater ductility is generally associated with lower melting temperature. Among intermetallic compounds, V_3Ga is first to be fabricated successfully into a multifilament configuration (Furuto *et al.*, 1973).

d. Other superconductors. Superconductors with high transition temperatures and critical fields are always regarded as promising materials.† Several candidates, all compounds or alloys based on niobium or vanadium, are listed in Table V. However, the current-carrying capacities of these materials are generally still poor or have not yet been measured because their undesirable mechanical properties have prevented easy fabrication in the form of tape or wire. Recent data on Nb–Al–Ge compound, although still limited, are promising (Lohberg *et al.*, 1973).

e. Temperature dependence of J_c. Curves of relative critical-current density versus temperature at constant external fields for Nb–Ti, Nb_3Sn, and V_3Ga are shown, respectively, in Figs. 25 (Hampshire *et al.*, 1969) 26 (Crow and Suenaga, 1972; Aron and Ahlgren, 1968; Haller and Belanger, 1971), and 27 (Crow and Suenaga, 1972). As the majority of data for J_c versus H are given only at 4.2K, we must combine Figs. 25–27 with J_c

† Suenaga and Sampson (1972), Lohberg *et al.* (1973), Gavaler (1973), and Foner *et al.* (1974).

TABLE V

TRANSITION TEMPERATURES AND CRITICAL FIELDS OF SOME PROMISING SUPERCONDUCTORS[a]

Material	T_c (K)	H_{c2} (kOe)[b]	Reference
Nb–Ge	22.3	~370	Gavaler (1973); Foner *et al.* (1974)
$Al_{0.153}Ge_{0.057}Nb_{0.79}$	20.7	410	Foner *et al.* (1970)
$Al_{0.75}Ge_{0.25}Nb_3$[c]	18.5	420	Foner *et al.* (1970)
$Nb_{0.68}Ga_{0.32}$	20.0	330	Foner *et al.* (1972)
Nb_3Al[c]	18.5	295	Kohr *et al.* (1972)
V_3Si[c]	10.9	235	Suenaga and Sampson (1972)
V–Hf–Nb	10.4	257	Inoue and Tachikawa (1972)
V–Hf–Zr[c]	10.1	240	Inoue and Tachikawa (1972)
V–Hf–Ta	10.0	261	Inoue and Tachikawa (1972)

[a] In zero magnetic field at 4.2K.

[b] Pulsed-field data.

[c] Limited data of J_c versus H available (Kohr *et al.*, 1972; Inoue *et al.*, 1971; Lohberg *et al.*, 1973).

versus H data at 4.2K such as those shown in Fig. 24 and can then predict approximately the performance of a material for a broad range of temperature.

For Nb–Ti in fields above 24 kOe, we can use a "linear-equation" approximation, namely,

$$J_c(T, H) = J_{c0}(T_0, H) \frac{T_c(H) - T}{T_c(H) - T_0} \tag{83}$$

where $T_c(H)$ is the critical temperature of the material at a given external field. As indicated by the dotted lines in Fig. 25, it is probably valid to extend the linear curves down to the vicinity of 2K.

In Nb_3Sn the relationship between J_c and T is more complicated, as is evident from Fig. 26 for diffusion-processed tape, diffusion-processed wire, and vapor-deposit-processed tape. There is no direct relationship in the ordinate of Fig. 26 among the three samples. That is, the diffusion-processed tape (solid curves) does not imply, for example, 8.5/7 times more capacity than the vapor-deposit-processed tape (dotted curves) at $T = 5$K and $H = 10$ kOe.

In V_3Ga, J_c again decreases linearly with temperature, but only to within a degree or so of the critical temperature, where the decrease dimin-

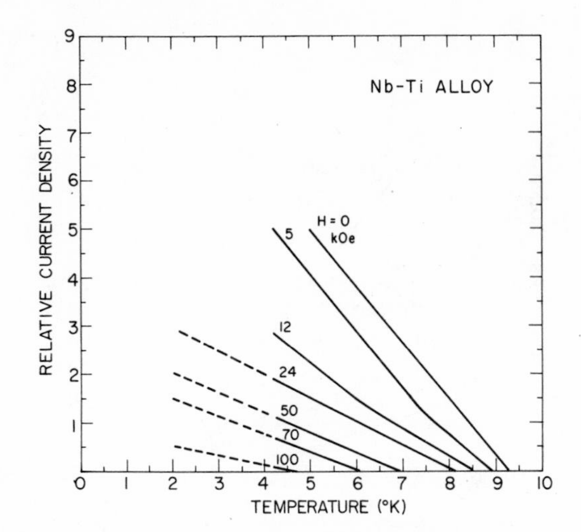

FIG. 25. Relative critical current density versus temperature at constant magnetic field for Nb–Ti composite.

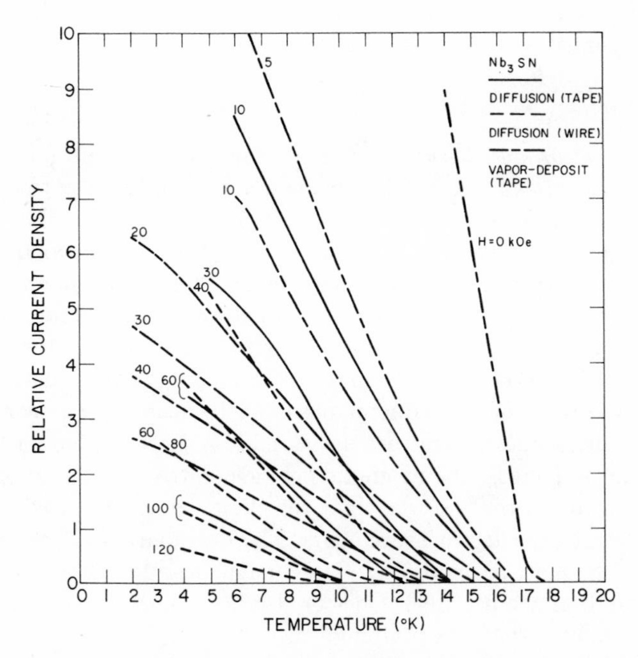

FIG. 26. Relative critical current density versus temperature at constant magnetic field for Nb_3Sn conductors. (Comparison not valid from one conductor to another.)

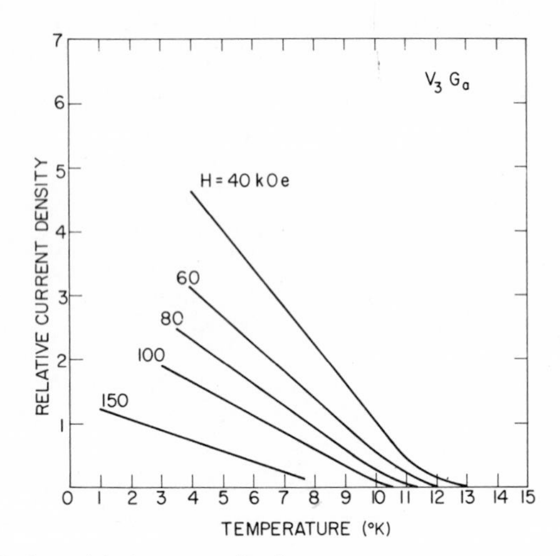

FIG. 27. Relative critical current density versus temperature at constant magnetic field for V_3Ga wire.

ishes. If Eq. (83) is to be used, therefore, T_c must be replaced by an appropriate temperature a degree or so lower than T_c.

2. *Stabilization*

a. Basic ideas and theory. An important requirement for superconducting magnets, as for any other system, is reliability. If potential instabilities exist in a system, reliable operation is not possible unless they are suppressed or circumvented. High-current, high-field superconductors, as we shall see shortly, can be inherently unstable. Thus, when superconducting magnets were first operated in the early 1960s, without any form of stabilization, the result was catastrophic (Lubell *et al.*, 1963; Kim *et al.*, 1963; Schrader *et al.*, 1964). Now commonly known as the degradation effect, the critical current of a conductor observed in magnets was consistently and drastically reduced from that measured in a short sample for the same magnetic field. That is, the magnetic field generated by such magnets was sometimes unpredictably lower than the designed value based on short-sample current capacity measurements. The degradation effect was found generally to be most pronounced in large magnets and high-current-density conductors(Berlincourt, 1963; Chester, 1967).

It is now agreed that flux jumping is usually responsible for the degradation effect. Flux jumping is a phenomenon in which a superconducting shielding current, set up in the conductor by an external magnetic field, decays suddenly, in the time scale of milliseconds, and the magnetic flux penetrates into the conductor. The jump is usually triggered by disturbances such as temperature and magnetic field fluctuations within the conductor. That is, the flux jump is a thermal runaway process: the local generation of heat not only exceeds the removal of heat, but is intensified as the process advances. Heat is generated by the movement of flux lines and is removed by conduction. By examining the diffusion processes of flux lines and heat, we can see that condition is inherently unstable in high-current, high-field superconductors (Wilson *et al.*, 1970).

The diffusion of magnetic flux lines in a material is governed by the following equation:

$$\nabla^2 H = \frac{1}{D_{\mathrm{M}}} \frac{\partial H}{\partial t} \tag{84}$$

where H (Oe) is the magnetic field and D_{M} (cm²/sec) is the magnetic diffusivity of the material. D_{M} is proportional to ρ (Ω cm), the electrical resistivity:

$$D_{\mathrm{M}} = (10^9/4\pi)\rho \tag{85}$$

The propagation speed of magnetic flux lines within a finite region bounded by radius R (cm) is proportional to D_{M}/R.

In a similar manner, the diffusion of heat is governed by

$$\nabla^2 T = \frac{1}{D_{\mathrm{H}}} \frac{\partial T}{\partial t} \tag{86}$$

where T (K) is the temperature and D_{H} (cm²/sec) is the thermal diffusivity. D_{H} is given by the ratio of k (W/cm K), the thermal conductivity, and C (J/cm³ K), the heat capacity:

$$D_{\mathrm{H}} = k/C \tag{87}$$

Similarly, the propagation speed of heat within the same finite region is proportional to D_{H}/R.

Since the movement of flux lines generates heat and the diffusion of heat controls its removal, a thermal runaway condition would prevail if

$$D_{\mathrm{M}} > D_{\mathrm{H}} \tag{88}$$

The order-of-magnitude values of D_{M} and D_{H} for a typical high-current, high-field superconductor, based on values of ρ, k, and C listed in Table VI, are 10^3–10^4 cm²/sec and 1 cm²/sec, respectively. The corresponding values of D_{M} and D_{H} for pure conductors such as copper, also listed in Table VI, are 1 cm²/sec and 10^3–10^4 cm²/sec, respectively. These values indicate that the superconductors can be unstable and suggest that some form of stabilization might be possible with the use of copper.

TABLE VI
PROPERTIES OF ELEMENTS AND ALLOYS AT 4.2K

Material	Density (g/cm³)	Thermal capacity (J/cm³ K)	Thermal conductivity (W/cm K)
Cu(99.99% an)	8.95	8.95×10^{-4}	2.5
Al(99.996%)	2.7	14.3×10^{-4}	3.3
Nb_3Sn	8.91[a]	1.1×10^{-3}	
	5.4[b]	1.49×10^{-3} [c]	4×10^{-4}
Nb 50 wt% Ti	6.	3.6×10^{-3}	1.2×10^{-3}
V_3Ga	6.64[a]	3.2×10^{-3} [c]	—

[a] Theoretical (using x-ray-determined lattice parameters and atomic weight).
[b] See Brechna (1969).
[c] Normal-state value (Heiniger *et al.*, 1966).

Historically copper was incorporated to limit quench voltages in the manufacture of superconducting wires and tapes, before any detailed investigation of the stability of the superconductor was underway. Because its electrical conductivity is much better than that of normal-state superconductors, copper was plated or bonded to the surface of wires and tapes to form composite conductors soon afrer the degradation effect was discovered.

The copper provides a less resistive path for the transport current and, in so doing, practically eliminates Joule heating within the normal-state superconductor. If a sufficient amount of copper is provided, Joule heating within the copper will be less than the heat which can be removed to liquid helium from its surface (Stekly and Zar, 1965). Under such a condition, the temperature of the copper will remain near 4.2K, and the adjoining section of the superconductor, in the normal state and thus at its critical temperature, will in time return to the superconducting state by the transfer of excess heat to the copper. Unfortunately, several times more copper than superconductor typically is required; the overall current density is then several times less than that in the superconductor itself.

The Joule heating generated per unit length of conductor is Q_J (W/cm) $=$ $(\rho I_t^2)/A_{Cu}$, where ρ (Ω cm) is the resistivity of the copper, I_t the transport current, and A_{Cu} (cm²) the cross section of the copper. The heat removed into liquid helium from the copper unit length is Q_f (W/cm) $= q_f p$, where q_f (W/cm²) is the heat flux rate per unit area and p (cm) the perimeter of the copper cross section. In order for the copper to remain at 4.2K, we require

$$Q_J \leq Q_f \tag{89}$$

or

$$\rho I_t^2 / A_{Cu} \leq q_f p \tag{90}$$

and

$$p = g(A_{Cu} + A_s)^{1/2} \tag{91}$$

where A_s (cm²) is the cross section of the superconductor and g is a proportionality constant. For a circular cross section, the value of g is $2(\pi)^{1/2}$; for a square it is 4. Combining Eqs. (89)–(91), we obtain

$$\rho I_t^2 / A_{Cu} \leq g q_f (A_{Cu} + A_s)^{1/2} \tag{92}$$

This equation gives the necessary minimum A_{Cu} in terms of ρ, I_t, q_f, A_s, and g. Once A_{Cu} is solved, we can compute the overall current density:

$$J_{\text{overall}} = \frac{I_t}{A_{Cu} + A_s} = \frac{A_s J_c}{A_{Cu} + A_s} = \frac{J_c}{1 + (A_{Cu}/A_s)} \tag{93}$$

where J_c is the short-sample critical current density of the superconductor. I_t as given by $A_s J_c$ is thus the maximum stable transport current. The extent of the reduction in current density necessary to achieve stability is evident from the following numerical illustration.

We can solve for A_{Cu}/A_s from Eq. (92):

$$(A_{Cu}/A_s)^3 + (A_{Cu}/A_s) = (\rho^2 J_c^3 / g^2 q_f^2) I_t \tag{94}$$

where $I_t = J_c A_s$. The dependence of A_{Cu}/A_s on the quantities appearing in the right-hand side of Eq. (94) makes sense physically. Typical values of A_{Cu}/A_s for various values of I_t are given in Table VII when $\rho = 2 \times 10^{-8}$ Ω cm, $J_c = 10^5$ A/cm², $g = 5$, $q_f = 0.1$ and 1 W/cm². The heat flux of

TABLE VII
RATIO OF COPPER TO SUPERCONDUCTOR

Transport current (A)	$(A_{Cu}/A_s)^a$ with	
	$q_f = 0.1$ W/cm²	$q_f = 1$ W/cm²
50	4.2	0.6
200	6.8	1.3
500	9.3	1.8
2000	14.7	3
5000	20	4.2

[a] $\rho = 2 \times 10^{-8}$ Ω cm; $J_c = 10^5$ A/cm²; $g = 5$ [see Eq. (94)].

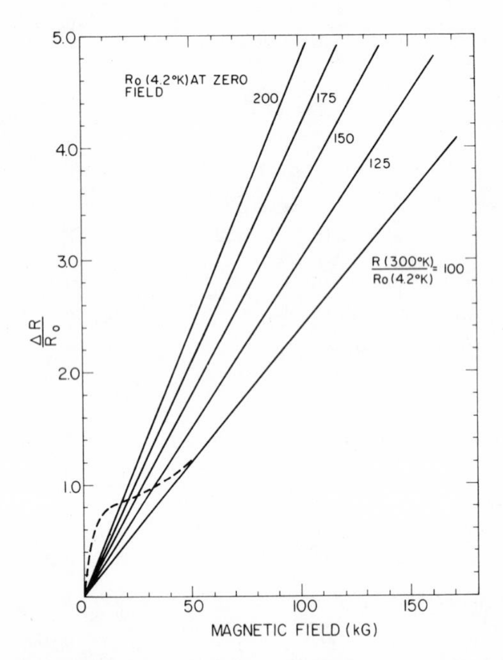

FIG. 28. Fractional change in resistivities of copper (——) and aluminum (- - -) versus magnetic field at 4.2K. Lines of constant "resistance ratio" are shown for copper. Metals 99.995% purity annealed; ρ(4.2K $= 3.7 \times 10^{-9}\ \Omega$ cm.

0.1 W/cm² corresponds to heat transfer by film boiling, whereas 1 W/cm² corresponds to that by nucleate boiling (Smith, 1969).

It has been verified experimentally that a large release of heat by a single flux occurring locally can not only drive the conductor normal at the location but can also drive the temperature of the surrounding normal metal high enough to force heat transfer there by film boiling (Iwasa and Montgomery, 1971). Such large flux jumps were mainly responsible for the failure of many early superconducting magnets. The conductor size and configuration dependence of magnetization is treated in the next section. In spite of the remarkable progress of recent years in the area of conductor design to reduce or eliminate the effect of flux jumping, the larger magnets are still designed conservatively, and the film-boiling heat-transfer rate is chosen to compute the minimum required value of A_{Cu}/A_s. Of course, A_{Cu}/A_s must be computed using a value for ρ which takes into account magnetoresistance (see Fig. 28) (Benz, 1969).

Despite this reduction in the overall current density, this form of stabilization has contributed enormously to the development of superconducting magnet technology, because it has insured the reliable and predictable operation of magnets. Many large-scale magnets have, in fact, been built and operated successfully (Wong *et al.*, 1968; Stekly *et al.*, 1968), using this stabilization technique. Some of them have already been discussed. Although the reduction in the overall current density might seem a definite drawback from the economic point of view, the cost of large magnets is, in fact, insensitive to the overall current density, as was pointed out earlier in Section II.4.

Of course, the complete elimination of flux jumps is always desirable. Moreover, a high overall-current-density conductor is definitely required in some cases, and the presence of a large volume of high-conductive normal metal is undesirable in other cases. In the next section, we discuss the intrinsic stabilization technique.

b. Magnetization and stability criteria. Shielding or magnetization currents are induced in a superconductor to oppose the penetration of flux lines, when it is subject to an external field. Since the shielding current density must be limited to the critical current density J_c, the complete shielding of flux lines is not possible, and flux lines penetrate into the superconductor according to Maxwell's equation, $\nabla \times H = (4\pi/10)J_c$, where H is expressed in Oe, J_c in A/cm², and lengths in cm (Bean, 1962; Kim *et al.*, 1963). If J_c is assumed constant with respect to magnetic field, the profile of the magnetic field within a simple slab of width $2a$ would be one of those shown in Fig. 29, depending on the magnitude of the external field H_a with respect to the J_c and the slab width. In all cases, magnetiza-

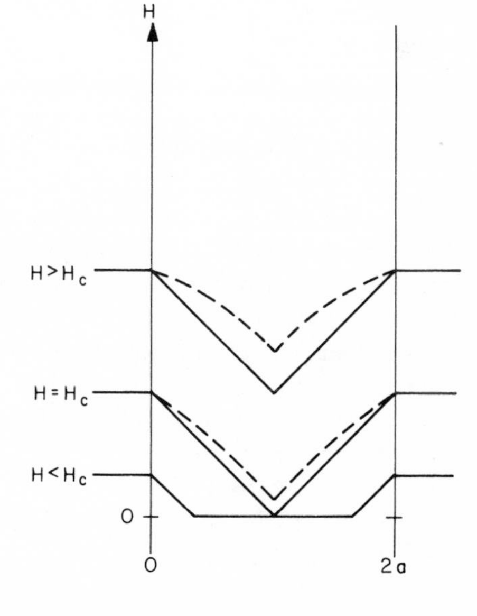

FIG. 29. Magnetic field distributions inside a hard superconducting slab of width $2a$. Solid lines for the case J_c = const and dotted lines for the case $J_c = J_c(H)$.

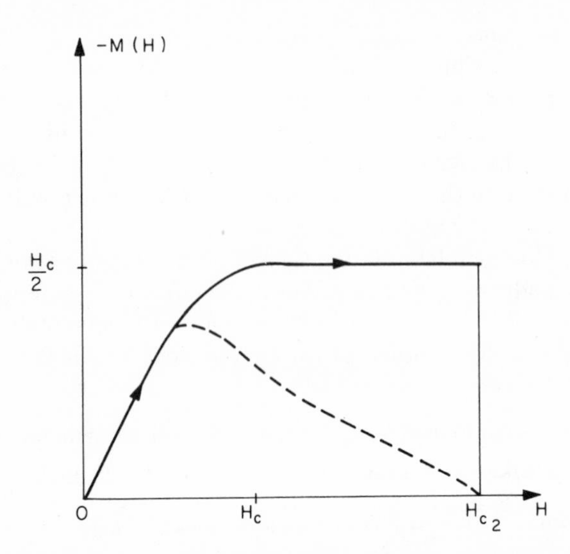

FIG. 30. Magnetization of a slab as a function of ambient field. The solid-line magnetization is based on J_c = const and the dotted-line magnetization on $J_c = J_c(H)$.

tion is simply the average magnetic induction within the superconductor. Thus for a one-dimensional case such as this slab, it is given by

$$M = \frac{1}{(2a)} \int_0^{2a} [H(x) - H_0]\, dx \tag{95}$$

From Fig. 29 and Eq. (95), we can compute $M(H_a)$ for the slab:

$$-M(H_a) = H_a - \frac{H_a^2}{2H_c} \qquad H_a \leq H_c = \frac{4\pi}{10} J_c a \tag{96}$$

$$-M(H_a) = H_c/2 \qquad H_c \leq H_a < H_{c2} \tag{97}$$

$M(H_a)$ is plotted in Fig. 30. The dotted curves in Fig. 30 schematically illustrate the more realistic case in which J_c decreases with magnetic field and $J_c = 0$ at $H_a = H_{c2}$ (Yasukochi *et al.*, 1964; Fietz *et al.*, 1964).

Once magnetization is known, we can compute the heat released within the material during a partial or total flux jump. The heat released is equal to the difference between the initial and final magnetic energies stored. The magnetic energy stored, e_M (J/cm^3) corresponding to magnetization M is given by

$$e_M = (10^{-7}/6\pi) M^2 \tag{98}$$

Thus the total energy released as heat is

$$\Delta e_M = -(10^{-7}/6\pi)[M_f^2 - M_i^2] = -(10^{-7}/6\pi)\int_{M_i}^{M_f} M\,dM \quad (99)$$

where M_i and M_f are, respectively, the initial and final magnetization of the material. For $H_a > H_c$, in which $-M = H_c/2$ [Eq. (97)], Eq. (99) becomes

$$\Delta e_M = -(4\pi/3)10^{-9}a^2 \int_{J_{ci}}^{J_{cf}} J_c\,dJ_c \quad (100)$$

where J_{ci} and J_{cf} are the critical current densities of the material corresponding, respectively, to the initial and final states of the material. The rise in temperature of the material corresponding to a release of the above magnetization energy is given by

$$\Delta e_M = \int_{T_i}^{T_f} C(T)\,dT \quad (101)$$

where $C(T)$ (J/cm^3 K) is the heat capacity of the material. T_i and T_f are, respectively, the initial and final temperatures of the material. The equality sign nearly holds because $D_M \gg D_T$ for the superconductor, and very little of the released heat escapes, i.e., the condition is adiabatic. Differentiating both sides with respect to temperature and evaluating at $T = T_i$, we obtain

$$\left(\frac{4\pi}{3}\right)10^{-9}\frac{a^2 J_{ci}^2}{T_e} = C(T_i) \quad (102)$$

where T_e is defined:

$$T_e = -\left(\frac{1}{J_{ci}}\right)\left(\frac{\partial J_c}{\partial T}\right)_{T=T_i} \quad (103)$$

A thermal runaway condition or a flux jump can be prevented if we require

$$\frac{4\pi}{3}10^{-9}\frac{a^2 J_{ci}^2}{T_e} < C(T_i) \quad (104)$$

Since $C(T_i)$, J_{ci}, and T_e are fixed for a given material, the only variable is its dimension, a. Thus, to stabilize the material against flux jumps, we must restrict the size of the conductor such that it satisfies the following conditions:

$$a < J_{ci}^{-1}[(3/4\pi)10^9 C(T_i)T_e]^{1/2} \quad (105)$$

The condition derived above is commonly known as the adiabatic stability criterion (Hancox, 1965). In deriving the above expression, we have implicitly assumed that the heat released by the decay of magnetization is uniformly distributed. This assumption clearly is violated here because the process is supposed to be adiabatic ($D_M \gg D_T$): heat cannot be distributed uniformly in the time scale of a flux jump. Nevertheless, the above expression comes within 10% of the correct one (Wipf, 1967; Swartz and Bean, 1968):

$$a < J_{ci}^{-1}[(\pi/16)10^9 C(T_i) T_e]^{1/2} \tag{106}$$

A typical numerical value of a may be obtained for Nb–Ti. From Table VI, $C(T_i)$ is about 3×10^{-3} J/cm³ K at $T_i = 4.2$K, and T_e is 4.3K. Then, for $J_c = 10^5$ A/cm³, $a = 0.016$ cm, which corresponds to a maximum transport current of 20 A. If any magnet is to be designed for a transport current greater than 20 A, and if the adiabatic stability criterion is to be adopted, then more than one such "filamentary" conductor must be used in parallel. In practice, the conductors are bonded together in a copper matrix.

In the presence of a highly conductive normal metal such as copper, an entirely different means of stabilization may be achieved (Wilson *et al.*, 1970; Hart, 1969). A basic contribution of such metal is to slow down the motion of flux lines, resulting in a slower rate of heat generation within the superconductor, making possible its removal by conduction. That is, within the composite of superconductor and copper, the condition becomes more isothermal than adiabatic. This stability technique is known as the dynamic stability criterion. An approximate expression similar to Eq. (105) is derived below.

In this stability criterion, Eq. (101) is modified:

$$k_s \nabla^2 T = (\partial/\partial t)\,(\Delta e_M) \tag{107}$$

where k_s (W/cm K) is the thermal conductivity of the superconductor. Here we have implicitly assumed than an isothermal condition exists and neglected the heat capacity term. For this order-of-magnitude calculation, the left-hand side of Eq. (107) may be approximated as $k_s\,\Delta T/a^2$, and, inserting Eq. (100) into the right-hand side of Eq. (107), we have

$$k_s \frac{\Delta T}{a^2} = -\left(\frac{4\pi}{3}\right) 10^{-9} a^2 \frac{\partial}{\partial t} \int_{J_{ci}}^{J_{cf}} J_c\, dJ_c \tag{108}$$

Differentiating both sides of Eq. (108) with temperature and evaluating at $T = T_i$, we have

$$\frac{k_s}{a^2} = \left(\frac{4\pi}{3}\right) 10^{-9} a^2 \frac{\partial}{\partial t} \left(\frac{J_{ci}^2}{T_e}\right) \tag{109}$$

For our present purpose, it is reasonable to express

$$\frac{\partial}{\partial t}\left(\frac{J_{ci}^2}{T_e}\right) \approx \frac{1}{\Delta t}\frac{J_{ci}^2}{T_e} \tag{110}$$

where Δt is the effective time scale for magnetic flux diffusion across the copper matrix of dimension a; i.e., $\Delta t \sim t_M$, the diffusion time constant. We are here assuming the sizes of the superconductor and copper to be about the same. In this case, then,

$$\Delta t \sim \tau_M = \left(\frac{4}{\pi^2}\right)\frac{a^2}{D_M} = \frac{16 \cdot 10^{-9}}{\pi}\frac{a^2}{\rho} \tag{111}$$

where ρ (Ω cm) is the electrical resistivity of copper. Inserting Eqs. (110) and (111) into Eq. (109), we have

$$\frac{k_s}{a^2} \approx \frac{\pi^2}{12}\frac{\rho J_{ci}^2}{T_e} \tag{112}$$

Again for stability, we require

$$k_s > \frac{\pi^2}{12}\frac{a^2 \rho J_{ci}^2}{T_e} \tag{113}$$

And the required restriction for the size of the superconductor becomes

$$a < J_{ci}^{-1}\left(\frac{12}{\pi^2} k_s \frac{T_e}{\rho}\right)^{1/2} \tag{114}$$

A typical value of a for Nb–Ti is about 0.005 cm, with $k_s = 10^{-3}$ W/cm K and $\rho = 2 \times 10^{-8}$ Ω cm for copper. Although it is not explicit in Eq. (114), we can increase this value a little further by making copper occupy a greater fraction of the total volume. Such an effort, however, has limited value because, when the critical size given by the dynamic stability criterion exceeds that given by the adiabatic stability criterion, it is doubtful that isothermal conditions can be maintained within such a filament (Wilson *et al.*, 1970).

The two analyses treated above represent two extremes. In reality, conductors, wires or tapes, are always in the form of composites of superconductor and normal metal, and the condition of flux diffusion falls in between the two limits. This means, for example, that not only the heat capacity of the superconductor but also that of the normal metal must be taken into account in the adiabatic stability criterion (Wilson *et al.*, 1970). In fact, there have been a number of attempts, resulting in qualified success, to stabilize magnets with the use of normal metals such as lead and mercury, both of which have a large heat capacity (Williams, 1967; Hale and Williams, 1968; Schrader, 1969). Similarly, the thermal conductivity of the normal metal cannot be neglected entirely in the dynamic stability condition (Hart, 1969). It is evident that, in any case, a decrease of J_c should help. As the size depends inversely on J_c, the total transport current also depends inversely on J_c. The conductor is generally much more stable in higher fields, where J_c is smaller.

In practice, the upper limit for the size of a filament is generally taken to be that given by Eq. (106), which is much too small for practical magnets if a single filament were to be used. Therefore, practical conductors are composed of many filaments bonded together in a copper matrix.

c. Multifilament composite conductors. As we have seen in the previous section, if the size of a superconductor is reduced to meet the basic stability requirement, a conductor consisting of only one such filament of superconductor becomes virtually useless as a magnet material—its current-carrying capacity is about 20 A, whereas for most applications current-carrying capacity ranges from 50 to 2000 A. A logical solution to this difficulty was the development of multifilament composite conductors in which many superconducting filaments are embedded in a matrix of normal metal, usually copper. The current-carrying capacity of such a conductor then is greater than that of a single-filament conductor by the number of filaments. The critical question, however, is if stability requirements can still be met in the multifilament composite.

It is apparent that, if each individual filament acts independently of every other, then the stability requirement can be met when the size of each filament is made less than the critical value. Since filaments are in a matrix of normal metal, usually copper, it is clear that they are independent in dc but not normally in ac. In this section, we develop a simplified analysis of the behavior of multifilaments in a matrix of normal metal subject to a time-varying external magnetic field. Based on the analysis, we obtain necessary conditions for multifilament composite conductors to be stable in a given ac condition.

We begin with the simplest geometry (Fig. 31), consisting of two superconducting slabs, each of thickness d (cm) and length $2l$ (cm), separated by a normal material of thickness w (cm) and length $2l$ (cm) and of electrical resistivity of ρ (Ω cm), subject to a time-varying magnetic field $\dot{H}$ (Oe/sec), normal to w and l (Wilson *et al.*, 1970).

An electric field $E(x)$ induced within the normal material by $\dot{H}$ at $x < l$ from the origin is given by

$$E(x) = 10^{-8} \times \dot{H} \tag{115}$$

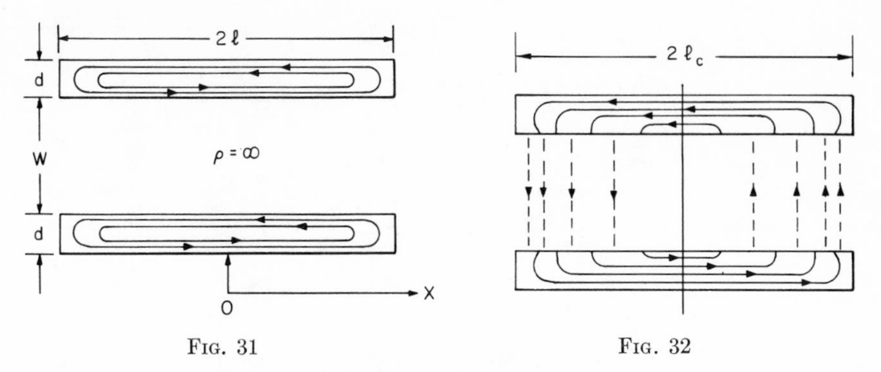

FIG. 31 FIG. 32

FIG. 31. Two parallel superconducting slabs separated by a perfect insulator.
FIG. 32. Two parallel superconducting slabs separated by resistive material. The conductor's length is just equal to the critical length given in Eq. (119).

where E is in V/cm. If the electrical resistance of the material is infinite, then there will be no current flowing from one slab to the other, and the induced current will be confined within each slab, as illustrated in Fig. 31. Obviously the two slabs are acting independently in this case.

Next we consider the case in which the electrical resistivity of the material is finite (Fig. 32). For a "slow" time-varying field, the electric field in the material is still given by Eq. (115), but associated with $E(x)$ will be a flow of current, which between the origin and location $x < l$ away is given, per unit depth, by

$$I_t(x) = \frac{1}{\rho} \int_0^x E(x)\, dx \tag{116}$$

Inserting Eq. (115) into Eq. (116), we have

$$I_t(x) = \frac{10^{-8}}{2\rho} x^2 \dot{H} \tag{117}$$

At $x = l_c$, I_t is equal to the critical current of the slab. Thus,

$$J_c d = \frac{10^{-8}}{2\rho} l_c^2 \dot{H} \tag{118}$$

or

$$l_c^2 = 2\rho J_c d 10^8 / \dot{H} \tag{119}$$

where J_c (A/cm^2) is the critical current density of the slab. That is, l_c (cm) is just sufficient to transport the total current induced in one slab to the other, as shown in Fig. 32. A typical value of l_c for Nb–Ti filaments is about 10 cm if $J_c = 10^5$ A/cm^2, $d = 0.016$ cm [$\approx a$ given in Eq. (24)],

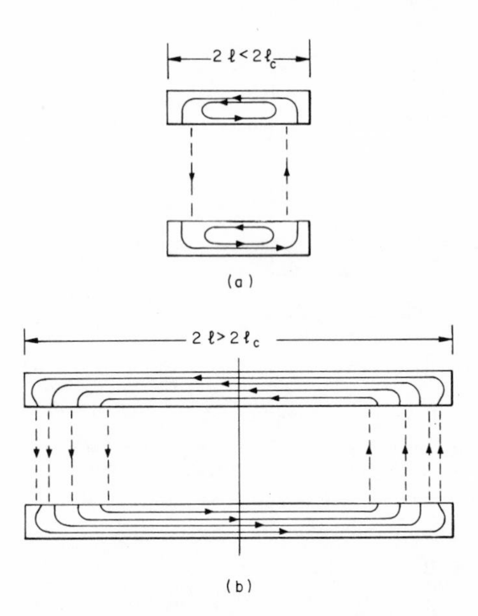

FIG. 33. Distribution of current in a length $2l$ of the Fig. 32 composite: (a) for $l < l_c$, (b) for $l > l_c$.

$\rho = 2 \times 10^{-8}\ \Omega$ cm, and $\dot{H} = 100$ Oe/sec. From Eq. (119), we note that l_c depends particularly on ρ and $\dot{H}$; if the design is in accord with Eq. (106), then we know $J_c d$ is independent of J_c. Therefore, the slabs act more independently as ρ is increased and $\dot{H}$ decreased, the limit of both cases we have already discussed.

For any other choice of l, $l < l_c$ or $l > l_c$, with ρ and $\dot{H}$ kept constant, the distribution of the induced current will look as shown in Fig. 33a or 33b. Thus for $l < l_c$ (Fig. 33a), only a fraction of the induced current in one slab commutates to the other slab, and the rest is confined within the slab, a behavior which approaches the case $\rho = \infty$ (Fig. 31) in the limit of $l/l_c \rightarrow 0$. That is, by choosing l much shorter than l_c, we can make the two slabs act indepndently even in the presence of a time-varying magnetic field. On the other hand, for $l > l_c$ the entire induced current in one slab commutates to the other slab, and, as seen from Fig. 33b, the time-varying magnetic field is completely excluded from the region $(l - l_c) < x < (l - l_c)$; the entire conductor in this region, including the normal metal, acts as one superconductor of thickness $\omega + 2d$, violating the size requirement.

As computed above, a typical value of l_c is 10 cm, clearly much shorter than wire lengths of practical interest. If a composite wire is twisted along

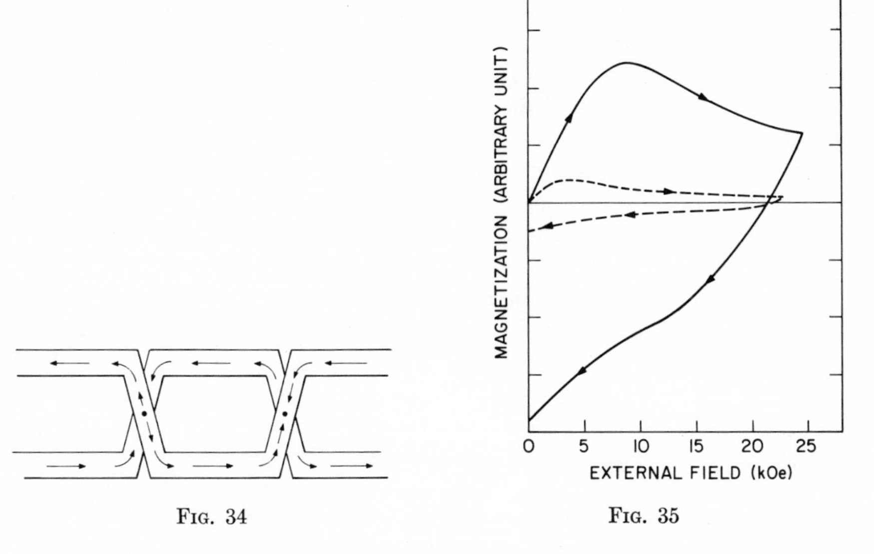

FIG. 34 FIG. 35

FIG. 34. Distribution of current in a composite conductor when two superconductors are twisted.

FIG. 35. Magnetization traces for two otherwise identical multifilament Nb–Ti composite conductors: nontwisted (solid curve) and twisted (dotted curve).

its length, however, the effective length of the wire is reduced to its pitch length. This can be seen as follows.

As illustrated in Fig. 34 for a two-slab geometry, the direction of the induced currents in the slabs alternates from one pitch to the next, forcing the induced currents to be zero at each crossing. That is, as far as the induced currents are concerned, the slabs can be cut off at each crossing without any change in the distribution of the induced currents. Consequently, each pitch can be considered separately, and it is evident that the size requirement can be met when the pitch length is made less than l_c.

In practice, conductors normally contain more than two filaments, and the above analysis must be modified accordingly. A more accurate expression of l_c, first derived by Wilson *et al.* (1970), is

$$l_c^2 = \frac{2\rho J_c d 10^8}{\dot{H}} \lambda \cdot \left(\frac{w}{w + d}\right) \tag{120}$$

where λ is the fraction of wire occupied by the superconductor and w is the distance between filaments.

Figure 35 shows the magnetization curves of two multifilament Nb–Ti composite conductors, identical in every respect except that in one conductor (the solid curve) the filaments are not twisted, whereas in the other (dotted curve) they are (Iwasa and Montgomery, 1971). The conductors, containing 121 filaments of about 10^{-2} cm diam, have cross-sectional dimensions of 0.32×0.13 cm, with the narrow side facing the external magnetic field. These magnetization curves indicate that the filaments are not acting independently in the untwisted conductor, whereas in the twisted conductor they are. Let us confirm this by two independent quantitative means.

From Eqs. (96) and (97), we know that magnetization is proportional to the "size" of the conductor facing the external field. If filaments act independently, the correct size is the diameter of the filament itself (10^{-2} cm); if they do not, the correct size is approximately the distance between the filaments furthest apart (0.1 cm). Consequently, the magnetization of the untwisted conductor should be greater than that of the twisted conductor by a factor of about 10; indeed this is the case.

The pitch length of the twisted conductor is 1.25 cm; let us see if the critical length given by Eq. (120) is reasonable. Taking the following values appropriate for these conductors, $\rho = 10^{-8}\ \Omega$ cm, $J_c = 4 \times 10^5$ A/cm^2 (at 10 kOe), $d = 10^{-2}$ cm, $\lambda = \frac{1}{4}$, and $w \approx d$, we obtain $l_c = 7.5$ cm for $\dot{H} = 100$ Oe/sec. As these magnetization curves were taken in field-sweep rates near 100 Oe/sec, Eq. (120) seems consistent with the experimental results. Since l_c depends on $\dot{H}$, it follows that the magnetization of twisted conductors depends on $\dot{H}$ (Iwasa, 1969; McInturff, 1969; Dahl *et al.*, 1969).

In Fig. 36 is shown the cross section of a multifilament Nb–Ti composite conductor of 1 mm in overall diameter containing extremely fine filaments (Popley *et al.*, 1972). The conductor, especially developed for use in fast pulse magnets, consists of 55 strands, each strand containing 241 5-μM filaments embedded in a copper matrix. The copper matrix is in turn subdivided with sheets and fins of Cu–Ni to minimize eddy currents. The extreme smallness of filament size is not so much to satisfy the stability criterion of Eq. (106) but to keep ac losses as low as possible.

d. AC losses in superconductors. All electrical devices are subject to ac conditions in one way or another: even a dc magnet encounters ac transients while it is being charged or discharged. Moreover, a majority of electrical devices now in use are operated on ac.

The high-current, high-field superconductors of interest in most of this book are, unfortunately, not lossless in ac applications. This has, of course, caused a great many disappointments as well as delay in the utilization of superconductor in ac devices. Succinctly (but from a somewhat oversim-

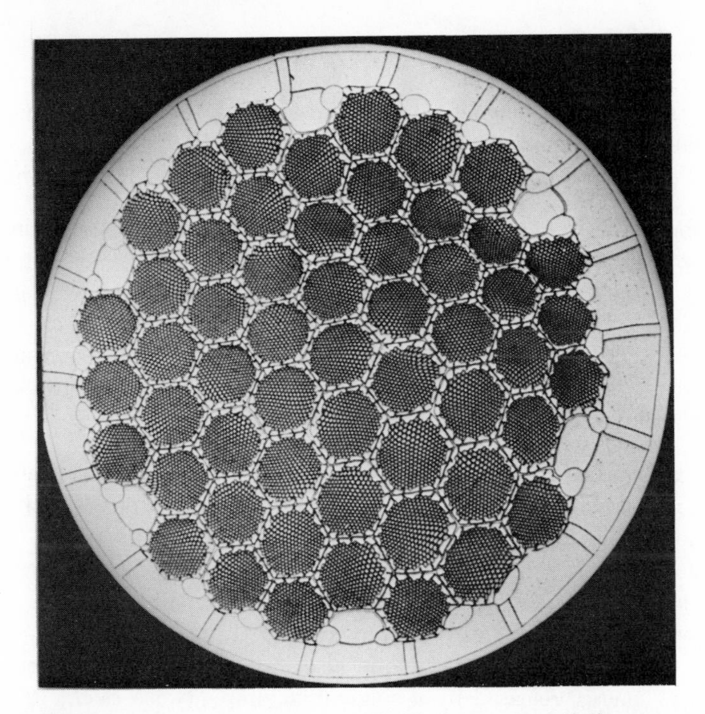

FIG. 36. Cross section of a multifilament Nb–Ti of 1-mm overall diameter, consisting of 13,255 5-μM filaments.

plified viewpoint), these superconductors are lossy in ac because of the presence of normal-state regions in them in a magnetic field above a few kilo-oersteds. Any movement of flux, created by a time-varying external field and/or transport current, consequently generates heat in the normal-state regions, and the conductors as a whole appear lossy. With the advent of multifilament conductors discussed above and of cables and braids presented next, however, the prospect for superconductors in ac devices has increased greatly.

There are three basic equations with which one can calculate the energy losses (Wipf, 1969): (1) the area of the magnetization loop, E (J/cm^3) = $(10^{-7}/4\pi)\oint H\ dB$, (2) the Poynting vector, P (W) = $(10/4\pi)\int_{\text{surf}}(\mathbf{E} \times \mathbf{H})\ ds$, and (3) the Joule heating, P (W) = VI. Using the first approach, we can make the following observations:

(1) As the area of magnetization is a function of magnetic field, the energy losses critically depend on the peak magnetic field at the surface of the conductor.

(2) As the transport current creates a surface magnetic field, the energy

losses due to ac transport currents can be predicted from those due to ac external fields, or vice versa.

(3) As the area of magnetization decreases with the size of conductor, it is desirable to use small-size conductors to reduce the losses.

(4) Power is proportional to the area of magnetization times frequency.

Presently the most active areas of application are for magnets in high-energy physics, where frequencies are less than 1 Hz, and in generators and motors, where the conductors are essentially in a dc field with a small unbalanced field whose fundamental frequency is 120 Hz.

Here we first present some simple experimental techniques to measure the energy losses, followed by some data on wires. The energy losses in the magnets below frequencies of 1 Hz are then presented.

i. Experimental techniques. There are three simple techniques used to measure energy losses: (1) the calorimetric method, (2) the magnetization method, and (3) the ohmic method.

In the calorimetric method, the amount of helium boiled off due to the energy losses alone is measured. The method is normally used in a steady-state condition, enabling one to convert directly the helium boil-off rate to the energy losses. The technique also has been used to measure the amount of energy released by a single-flux jump (Iwasa *et al.*, 1970). In this modified scheme, one measures a pressure pulse created by the flux jump rather than the amount of helium boil-off (it being normally too small to measure); pressure pulses can be calibrated with known energy pulses.

As mentioned above, the area under a full cycle of the magnetization curve is the loss, and in the magnetization method the magnetization as a function of field is measured. Since the loss per cycle is independent of frequency, it can be performed under a slowly varying field. In addition to the loss, moreover, the technique reveals other important information about the conductor, e.g., flux jump activities and the dependence of the critical current density on the field, making this technique the most fundamental of the three.

In the ohmic method, the time integral over one cycle of the product of the current and voltage across the specimen is measured. For all practical purposes, the method is restricted to those cases involving time-varying transport currents.

ii. Loss data on conductors. We present loss data to illustrate the essential characteristics for the loss curve and also to give typical numerical values.

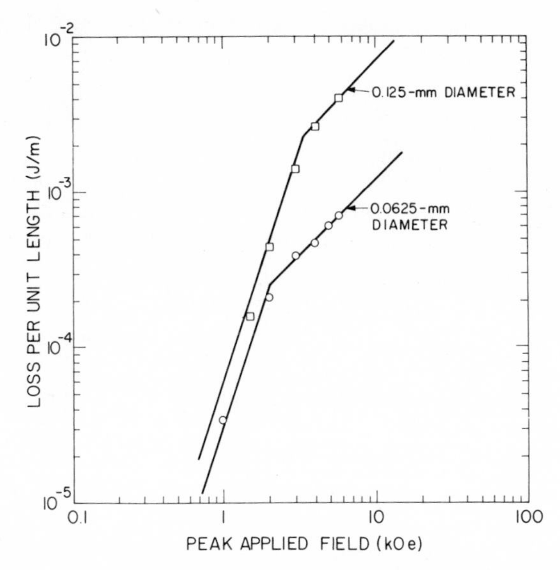

FIG. 37. AC losses in Nb–Ti wires as a function of peak applied field.

Figure 37 presents data for two Nb–Ti wires of 0.0625- and 0.125-mm diam (DiSalvo, 1966). The losses per unit length of wire (J/m) are plotted in the figures as a function of peak magnetic field.

From Fig. 37 we can make the following observations, which are quite general:

(1) For fields less than a certain field [H_c in Eq. (96)], the losses increase as the third power of the field.

(2) For fields above H_c they increase linearly with the field.

(3) The losses are proportional to the size of the conductor for fields below H_c.

(4) Above H_c, they are proportional to the second to third power of the size.

Although not obvious from Fig. 37, simple computations based on the magnetization curves given by Eqs. (96) and (97) reveal further that

(5) Below H_c, the losses are inversely proportional to the critical current density, J_c.

(6) Above H_c, they are proportional to J_c.

Taking the 0.125-mm-diam Nb–Ti wire as an example, we find that the loss in a small magnet wound with this conductor is formidable when operated at 60 Hz—a magnet consisting of a conductor 10,000 m long would dissipate when subject to an average peak field of 10 kOe, 5 kW. No wonder we do not as yet have truly ac superconducting devices! When it is operated at 0.1 Hz, however, the losses are reduced to about 10 W, and the problem becomes more manageable.

iii. AC LOSSES IN MAGNETS. AC losses generated in magnets, particularly in pulsed magnets for use in high-energy physics, have been extensively investigated by a number of groups, notably those at the Rutherford Laboratory (Wilson *et al.*, 1970), Brookhaven National Laboratory (McInturff *et al.*, 1972), and Lawrence Berkeley Laboratory (Gilbert *et al.*, 1972).

The total power dissipation (W) in a magnet is (Wilson *et al.*, 1970)

$$W \cong \frac{V \cdot d \cdot J_0 \cdot H_0 \nu}{2 \cdot 10^8} \ln\left(\frac{H_{\max} + H_0}{H_1 + H_0}\right) \tag{121}$$

where V (cm^3) is the total volume of superconductor; d (cm) the filament diameter; and $H_{\max}$ (Oe) the maximum field; H_1 is a lower limit for the field, approximately equal to $\frac{4}{3}H_c$, where H_c is the field at which penetration of the filament is complete [Eq. (96)]; ν (Hz) is the frequency. J_0 and H_0 are parameters which define the field dependence of the critical current density:

$$J_c = J_0 H_0/(H_0 + H) \tag{122}$$

Typical values of J_0 and H_0 are 10^6 A/cm^2 and 10^4 Oe, respectively.

A typical beam-handling magnet, 1 m long and of 0.1-m diam, might contain 10^4 cm^3 of superconductor. With a filament diameter of 10^{-3} cm and $H_{\max}$ of 10^4 Oe, Eq. (121) gives a loss of approximately 100 W when operated at 1 Hz.

Equation (121) is applicable only as long as the filament twist pitch is less than the critical pitch length [l_c in Eq. (120)]; for frequencies beyond which $l > l_c$, filaments couple and the losses increase. The losses are then given by Eq. (121) with the filament diameter d replaced by the composite diameter D.

e. Cables and braids. Along with multifilament composite conductors, cables and braids have been developed to make superconductors useable in pulsed magnets. The need for such conductors becomes quite apparent once we examine numerical values. To keep the ac losses within a tolerable level, the size of each filament must be near or below 10^{-3} cm: the critical current of each filament is thus about 0.1 A. Since the pulsed magnets are

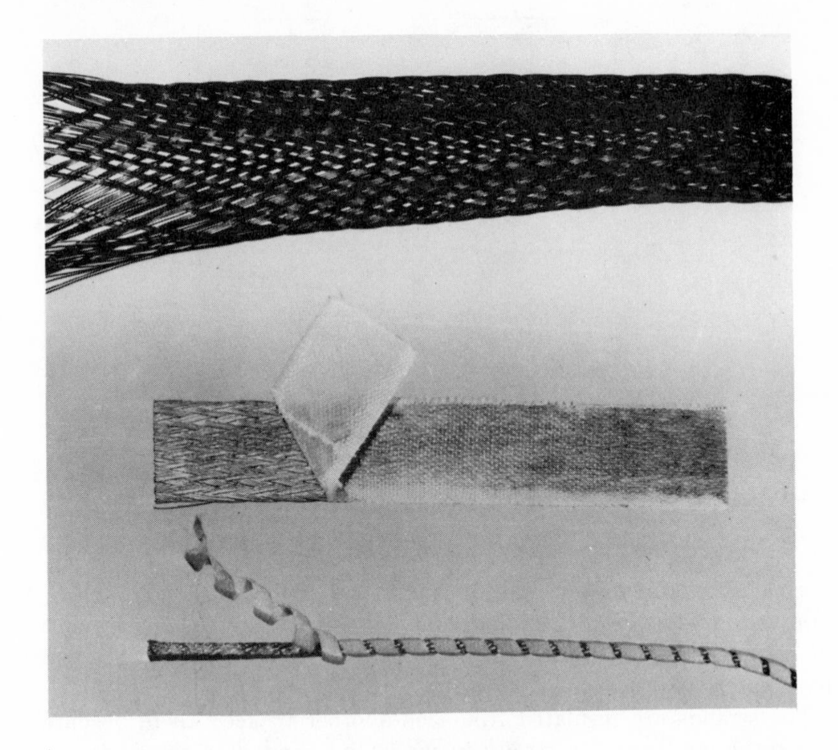

FIG. 38. Braids and cable used in synchrotron magnets. Top: Formvar braid, 1.58 cm wide; middle: Sn–Ag-filled braid, 1.4 cm wide, about 3000 A at kOe; bottom: 18 × 7 compacted cable, 0.29 × 0.19 cm, about 3000 A at 40 kOe.

normally energized with a current of thousands of amperes, the number of filaments for one conductor ranges from 10^4 to 10^5. The cables and braids are, therefore, practical solutions to the problem of assembling a great number of filaments in one conductor, while keeping them decoupled at frequencies up to 1 Hz. A typical cable consists of up to 100 strands containing up to 10^4 filaments. The diameter of a strand ranges from 0.01 to 0.1 cm. To achieve a high degree of field homogeneity, it is better to use cables or braids of fewer strands: the fewer the strands, the easier it will be to control the exact location of the conductor during and after winding. Each strand is bigger, however, resulting in a more rigid conductor.

The choice of normal metal to constitute the matrix in the strand depends on the following factors. A metal of high thermal conductivity is desirable to limit the temperature rise within the inner filaments. However, the high

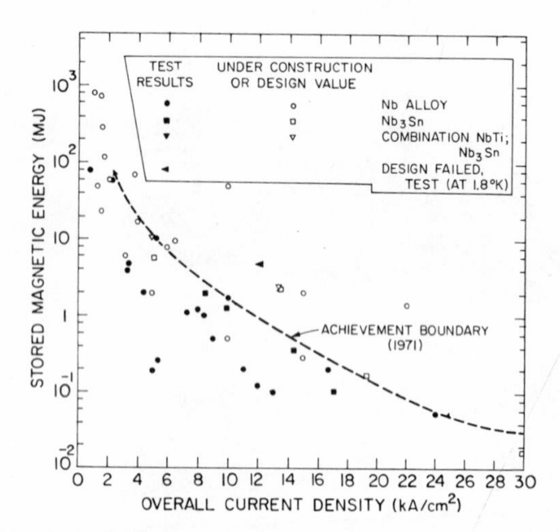

FIG. 39. Magnetic energy storage versus over-all current density for various large superconducting magnets presently in operation, under construction, and in the design stage.

thermal conductivity normally means a high electrical conductivity, requiring a higher rate of twist to decouple the filaments. To keep the critical pitch length long, alloys such as cupronickel have been tried as the matrix material.

The strands are insulated from one another to keep them decoupled. The insulation must be strong to withstand conductor compaction and abrasion caused by strand motion. Two types of materials have been widely used: (1) organic, and (2) metallic or intermetallic. The organic materials, being insulators, accomplish perfect decoupling. However, they not only inhibit the commutation of transport current from one strand to the next but also reduce heat transfer. An option for commutation is desirable for emergencies such as when one strand goes normal locally. The metallic or intermetallic materials, on the other hand, are much superior as additional "stabilizing" materials, although, being conductors, they make complete decoupling more difficult. In an attempt to increase the electrical resistivity of these materials, a variety of metallurgical processes have been introduced. One notable method is that by McInturff of Brookhaven (McInturff, 1972), in which individual strands are tinned (95% Sn–4% Ag) and heat treated for 240 hr at 325°C. The braid made of these strands is then filled with In–Tl eutectic at 275°C and subsequently heat-treated at 300°C for hours to tens of hours, depending on a desired thickness of Cu–Sn–Ag–In–Tl composition which is formed around each filament. Figure 38 shows a number of cables and braids prepared at the Brookhaven National Laboratory and Lawrence Berkeley Laboratory.

f. Overall current density and size of the magnet. Experience has shown us that the larger the magnet, the more unstable it is, forcing a lower overall current density. There is no accepted analytical explanation for this phenomenon; plausibly it must have something to do with the poor overall thermal environment associated with large size and with mechanical motions due to increased thermal and magnetic stresses. Lubell (1972) of the Oak Ridge National Laboratory has compiled data, obtained from all of the known large superconducting magnets presently in operation, under construction, or in the design stage, relating the overall current density and the stored magnetic energy. (The stored magnetic energy and the size of a magnet are closely related.) His data, shown in Fig. 39, are a useful guide in deciding the overall current density for a given magnet.

C. Field Analyses and Syntheses of Solenoidal Magnets

1. *Field along the axis*

The field H_z $(z, r = 0)$ along the axis of a coil of axially uniform excitation may be calculated from the simple superposition of the end fields from two subcoils into which the original solenoid has been partitioned (Montgomery, 1969). The schemes to subdivide a coil and subsequently to calculate the fields of the subdivisions are shown in Fig. 40 when z is inside the coil ends and when it is outside. By symmetry, the end field of any coil is half the central field of a coil of twice the length—hence the dotted line extensions indicated in Fig. 40. Thus, calculation of the field anywhere along the axis of a coil requires no more than the central field formula given in Section III.A.

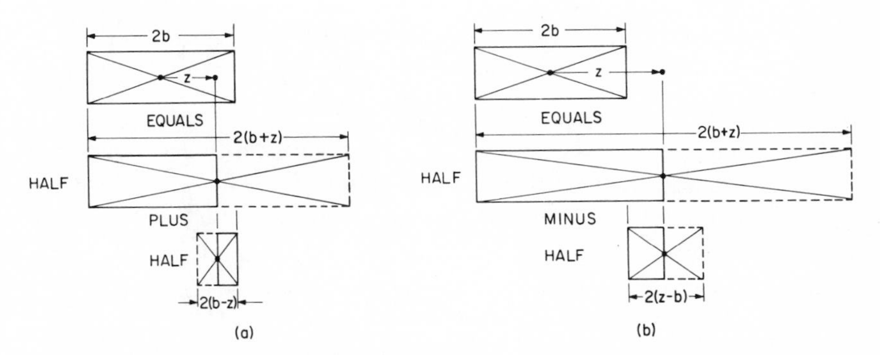

Fig. 40. Calculation of the axial field, $H_z(z, r = 0)$, at points along the axis of a coil by the superposition of appropriate fictitious coils: (a) when points are inside the coil ends, and (b) when they are outside.

For example, to determine the field at axial point z in a coil of length $2b$ as in Fig. 40a, we need only calculate the central field of the two superimposed coils shown and take half the value of their sum. We write this summation as follows:

$$H_z\left(\frac{z}{a_1}\right) = j\,\frac{\lambda a_1}{2}\left[F\left(\alpha, \beta + \frac{z}{a_1}\right) + F\left(\alpha, \beta - \frac{z}{a_1}\right)\right] \tag{123}$$

$$H_z\left(\frac{z}{a_1}\right) = H_z(0)\left[\frac{F(\alpha, \beta + z/a_1) + F(\alpha, \beta - z/a_1)}{2F(\alpha, \beta)}\right] \tag{124}$$

For $(z/a_1) \ll \beta$, points near the center of the coil, Eq. (124) can be approximated by

$$H_z\left(\frac{z}{a_1}\right) = H_z(0)\left\{1 - \frac{1}{2F(\alpha, \beta)}\,\frac{4\pi}{10\beta}\left[\frac{\alpha^3}{(\beta^2 + \alpha^2)^{3/2}} - \frac{1}{(\beta^2 + 1)^{3/2}}\right]\left(\frac{z}{a_1}\right)^2\right\} \tag{125}$$

That is, the field profile near the center of the coil is parabolic. (A more detailed discussion of field profile in the central zone is given below.) In the case $(\alpha/\beta) \ll 1$, Eq. (125) can be further reduced to a convenient form:

$$H_z\left(\frac{z}{a_1}\right) = H_z(0)\left\{1 - \left[\frac{\alpha^2 + \alpha + 1}{2\beta^4}\left(1 - \frac{4\alpha^2}{3\beta^2}\right)\right]\left(\frac{z}{a_1}\right)^2\right\} \tag{126}$$

When the field point lies beyond the end of the coil, $z > b$, we must use the relation

$$F[\alpha, \beta - (z/a_1)] = -F[\alpha, (z/a_1) - \beta] \tag{127}$$

For $(z/a_1) \gg \beta$, points far from the coil:

$$\lim_{(z/a_1)\to\infty}\left[F\left(\alpha, \frac{z}{a_1} + \beta\right) - F\left(\alpha, \frac{z}{a_1} - \beta\right)\right] = \frac{4\pi\beta}{10}\left(\frac{2}{3}\right)(\alpha^3 - 1)\left(\frac{a_1}{z}\right)^3 \tag{128}$$

That is, the field approaches that of a dipole.

The technique of superposition is equally applicable to any number of coaxial solenoids, provided that the proper set of coordinates for each is chosen. One typical case, a pair of coils separated by a gap, is illustrated in Fig. 41. The field is

$$\begin{aligned} H_z\left(\frac{z}{a_1}\right) = j\,\frac{\lambda a_1}{2}\Bigg[&F\left(\alpha, 2\beta + \frac{g + z}{a_1}\right) + F\left(\alpha, 2\beta + \frac{g - z}{a_1}\right) \\ &- F\left(\alpha, \frac{g + z}{a_1}\right) - F\left(\alpha, \frac{g - z}{a_1}\right)\Bigg] \end{aligned} \tag{129}$$

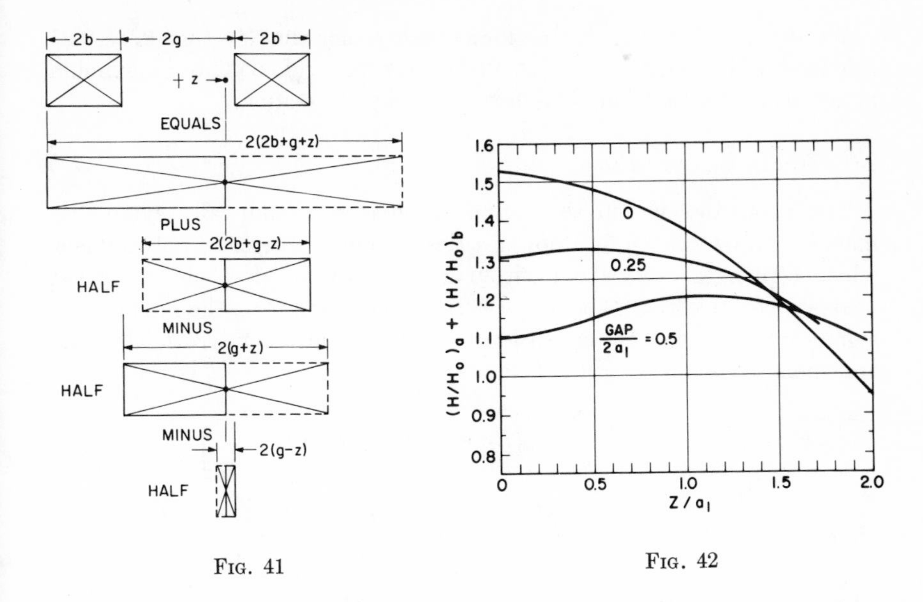

FIG. 41

FIG. 42

FIG. 41. Calculation of the axial field, $H_z(z, r = 0)$, at points along the axis of a pair of coils separated by a gap by the superposition of appropriate fictitious coils.

FIG. 42. Field profile of a pair of $\alpha = 3$, $\beta = 1$ coils for gaps of $\beta_g = 0, 0.25$, and 0.5. The field is normalized to the central field of an $\alpha = 3$, $\beta = 1$ coil.

The field at the center of the gap is simply

$$H_z(0) = j\lambda a_1 F_0(\alpha, \beta, \beta_g) \tag{130}$$

where

$$F_0(\alpha, \beta, \beta_g) = F(\alpha, \beta_t) - F(\alpha, \beta_g), \qquad \beta_t = 2\beta + \beta_g, \quad \beta_g = g/a_1 \tag{131}$$

We recognize Eq. (131) as the equivalent of a coil of length $4b + 2g$ from which a coil of length $2g$ has been subtracted.

To show a variety of field profiles obtainable with difierent values of β_g, we have plotted in Fig. 42 the field profiles of a pair of $\alpha = 3$, $\beta = 1$ coils separated by a gap of $\beta_g = 0$, 1, and $\frac{1}{2}$. The profile with zero gap corresponds to that of an $\alpha = 3$, $\beta = 2$ coil. We note that, when the coils are separated more than a certain amount, the field in the center becomes progressively smaller than the peak field. In some cases the peak field can become considerably larger than the central field. This means that high-field materials or reduced current densities must be used even for modest central fields. The peak field seen by the superconductor is, in fact, even higher than that along the axis.

We also note from Fig. 42 that for a spacing somewhat less than $\beta_g = 0.25$ the field will be extremely flat in the central region. This "maximum homogeneity" separation is discussed later in this section.

2. *Field in the central zone*

The magnetic field in the central zone of a cylindrically symmetric magnet can be written as a power series, involving Legendre polynomials, which converges everywhere within a sphere which does not reach the nearest magnet corner (Garrett, 1962). If the origin is on the midplane of symmetry of the coil, the expansion will contain only even terms. The axial and radial components of the field are

$$H_z(\rho, \theta) = H_0\left[1 + E_2\left(\frac{\rho}{a_1}\right)^2 P_2(u) + E_4\left(\frac{\rho}{a_1}\right)^4 P_4(u) + \cdots\right] \tag{132}$$

$$H_r(\rho, \theta) = -H_0\left[0 + E_2\left(\frac{\rho}{a_1}\right)^2 \frac{P_2'(u)}{3} + E_4\left(\frac{\rho}{a_1}\right)^4 \frac{P_4'(u)}{5} + \cdots\right] \tag{133}$$

where ρ and θ are conventional spherical coordinates of radius and polar angle, the Legendre polynomials $P_n(u)$ and their derivatives $P_n'(u)$ are given in Table VIII, and the E_n coefficients are defined from the standard formula for the coefficients in a Taylor series:

$$E_{2n} = \frac{1}{H_0}\frac{1}{(2n)!}\left[\frac{d^{2n}H_z(z, \theta = 0)}{dz^{2n}}\right]_{z=0} \tag{134}$$

To find these E_n coefficients, it is therefore necessary only to take derivatives of the field with respect to z (that is, along the axis) and to evaluate them at $z = 0$. The first few E_n coefficients for uniform-current-density coils are given in Table IX. They depend only on α and β.

The first three even E_n coefficients for the uniform-current-density case

TABLE VIII
TABLE OF EVEN-ORDER LEGENDRE POLYNOMIALS AND THEIR FIRST DERIVATIVE

$P_0(u)$	$= 1,\ u = \cos\theta,\ u' = \sin\theta$
$P_0'(u)$	$= 0$
$P_2(u)$	$= (1/2)(3u^2 - 1)$
$P_2'(u)$	$= (1/2)(6u)u'$
$P_4(u)$	$= (1/8)(35u^4 - 30u^2 + 3)$
$P_4'(u)$	$= (1/8)(140u^3 - 60u)u'$
$P_6(u)$	$= (1/16)(231u^6 - 315u^4 + 105u^2 - 5)$
$P_6'(u)$	$= (1/16)(1386u^5 - 1260u^3 + 210u)u'$
$P_8(u)$	$= (1/128)(6435u^8 - 12{,}012u^6 + 6930u^4 - 1260u^2 + 35)$
$P_8'(u)$	$= (1/128)(51{,}480u^7 - 72{,}072u^5 + 27{,}720u^3 - 2520u)u'$

TABLE IX

TABLE OF ERROR COEFFICIENTS FOR UNIFORM-CURRENT-DENSITY COILS

$$C_1 = \frac{1}{1+\beta^2}, \quad C_2 = \frac{\beta^2}{1+\beta^2}, \quad C_3 = \frac{\alpha^2}{\alpha^2+\beta^2}, \quad C_4 = \frac{\alpha^2}{\alpha^2+\beta^2}$$

$$F = \frac{4\pi}{10}\,\beta\left(\sinh^{-1}\frac{\alpha}{\beta} - \sinh^{-1}\frac{1}{\beta}\right)$$

$$FE_2(\alpha,\beta) = \frac{4\pi}{10}\,\frac{1}{2\beta}\,(C_1^{3/2} - C_3^{3/2})$$

$$FE_4(\alpha,\beta) = \frac{4\pi}{10}\,\frac{1}{24\beta^3}\,[C_1^{3/2}(2 + 3\,C_2 + 15\,C_2^2) - C_3^{3/2}(2 + 3\,C_4 + 15\,C_4^2)]$$

$$FE_6(\alpha,\beta) = \frac{4\pi}{10}\,\frac{1}{240\beta^5}\,[C_1^{3/2}(8 + 12\,C_2 + 15\,C_2^2 - 70\,C_2^3 + 315\,C_2^4) - C_3^{3/2}(8 + 12\,C_4 + 15\,C_4^3 + 315\,C_4^4)]$$

$$FE_8(\alpha,\beta) = \frac{4\pi}{10}\,\frac{1}{896\beta^7}\,[C_1^{3/2}(16 + 24\,C_2 + 30\,C_2^2 + 35\,C_2^3 + 315\,C_2^4 - 2079\,C_2^5 + 3003\,C_2^6) - C_3^{3/2}(16 + 24\,C_4 + 30\,C_4^2 + 35\,C_4^3 + 315\,C_4^4 - 2079\,C_4^5 + 3003\,C_4^6)]$$

for a wide range of α and β are tabulated in the form of $F(\alpha, \beta)E_n(\alpha, \beta)$ in the computer table given by Montgomery (1969). This is the more useful form for the design of the compensated homogeneous systems that we discuss next, but the E_n coefficients themselves can easily be separated out. Used in the tabulated form $F(\alpha, \beta)E_n(\alpha, \beta)$, the coefficients are used as in Eqs. (135) and (136); thus,

$$H_z(\rho, \theta) = j\lambda a_1\left[F(\alpha, \beta) + F(\alpha, \beta)E_2(\alpha, \beta)\left(\frac{\rho}{a_1}\right)^2 P_2(u) + \cdots\right] \tag{135}$$

$$H_r(\rho, \theta) = j\lambda a_1\left[F(\alpha, \beta)E_2(\alpha, \beta)\left(\frac{\rho}{a_1}\right)^2 \frac{P_2'(u)}{3} + \cdots\right] \tag{136}$$

For the axial field along either the z axis or the central plane, Eqs. (135) and (136) reduce to

$$H_z(z, 0) = H_0[1 + E_2(z/a_1)^2 + E_4(z/a_1)^4 + \cdots] \tag{137}$$

$$H_z(r, \pi/2) = H_0[1 - \tfrac{1}{2}E_2(r/a_1)^2 + \tfrac{3}{8}E_4(r/a_1)^4 - \cdots] \tag{138}$$

Near the center the E_2 term clearly predominates, and we see that the axial field deviation in the radial direction is only half as great (and of opposite sign) as in the axial direction. As E_2 for a simple solenoid is always negative, the field decreases along the z axis and increases in the radial direction, having the form of a saddle. We recognize that Eq. (125) is the leading term of Eq. (137).

3. *Synthesis of high-homogeneous-field coils*

Here we use superposition and the series expansion of the field to design solenoids with improved field homogeneity. We do this by "subtracting" from a large coil a smaller one chosen to cancel one or more terms of the series expansion of the original field. When this second coil has a current density equal to that of the main coil and is superimposed on the larger coil, it will appear as a "gap," "notch," or "void" in the main coil.

We have a great deal of freedom in choosing the parameters of the small coil as to its location, inner and outer radius, and relative current density. In general, the number of expansion terms of the main coil which can be canceled is set by the number of variables of the compensating coil. With one variable, we can cancel the first expansion term E_2, and with two, E_2 and E_4, etc. The three most common compensated-coil types are shown in Fig. 43. In the first (Fig. 43a), the superimposed negative coil produces a gap, and, because the main coil and the gap have the same α, the only

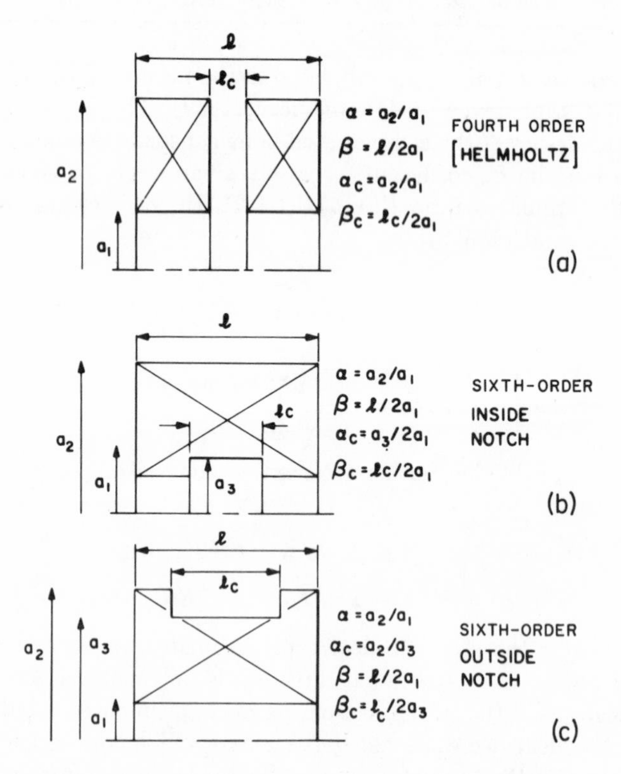

FIG. 43. Definition of parameters for coil systems compensated for increased field homogeneity.

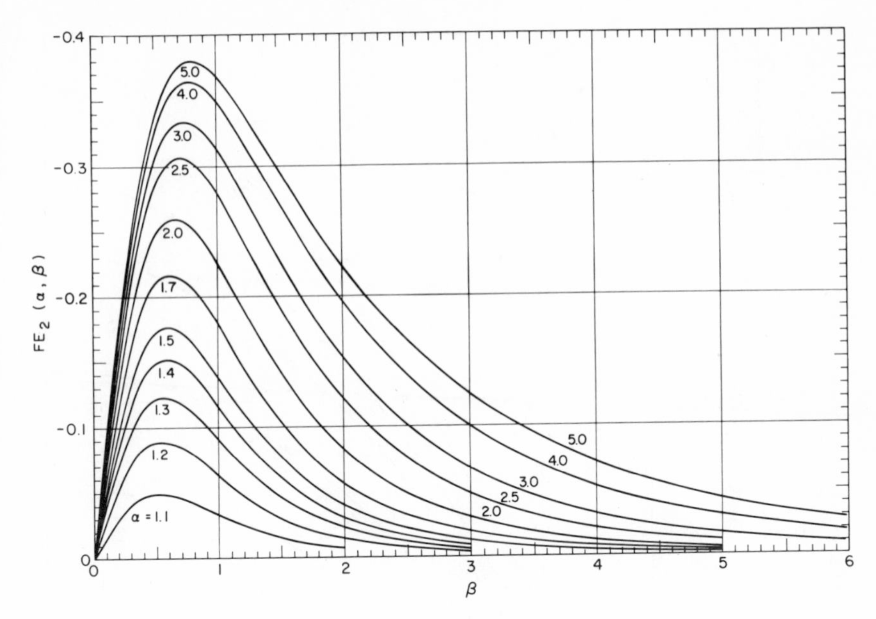

FIG. 44. The product of the $F(\alpha, \rho)$ factor and the "second-order error coefficient." Contours of constant α are plotted versus β.

variable is β_g. We can thus cancel the E_2 term in the field expansion by the proper choice of β_g. In the remaining two cases (Figs. 43b and 43c), we have placed a notch on either the inside or the outside, and, because we are free to choose both the α_c and β_c of the notch, we can cancel two terms in the expansion.

A certain price must always be paid for this type of cancelation, because the reverse-current coil subtracts from the main field, and it also usually increases the first uncanceled term. We first illustrate the use of this technique by graphical means, and we then present the computer solution of compensated systems.

a. Graphical solutions. Compensating coils for Helmholtz and inside-notched coils have the same inside diameter as the main coil, and we can write the on-axis field expansion for the combination as

$$H_z(z) = j\lambda a_1 \left\{F_0(\alpha, \beta) - F_c(\alpha_c, \beta_c) + [F_0E_2(\alpha, \beta) - F_cE_2(\alpha_c, \beta_c)] \left(\frac{z}{a_1}\right)^2 \right.$$

$$\left. + [F_0E_4(\alpha, \beta) - F_cE_4(\alpha_c, \beta_c)] \left(\frac{z}{a_1}\right)^4 + \cdots \right\} \qquad (139)$$

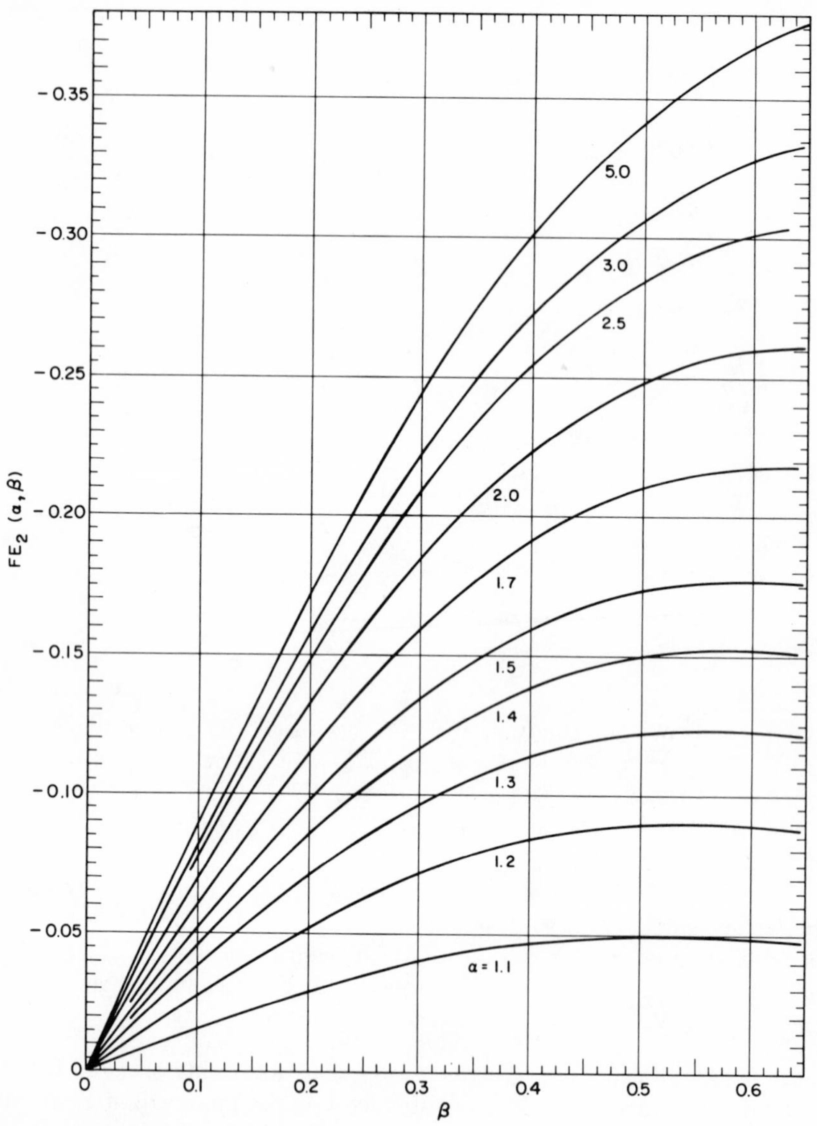

FIG. 45. Enlargement of Fig. 44 for small β. Lines of constant α are plotted versus β.

By making $F_c E_2(\alpha_c, \beta_c) = F_0 E_2(\alpha, \beta)$, we can cancel the original E_2 term and compensate the coil "to fourth order."

Figure 44 plots $F(\alpha, \beta)$ versus β for constant α. A main coil of $\alpha = 3$, $\beta = 2$, for example, has $EF_2(3, 2) = -0.153$. A Helmholtz design based on this main coil likewise requires $EF_2(\alpha', \beta') = -0.153$, with $\beta' = \beta_g \neq 2$.

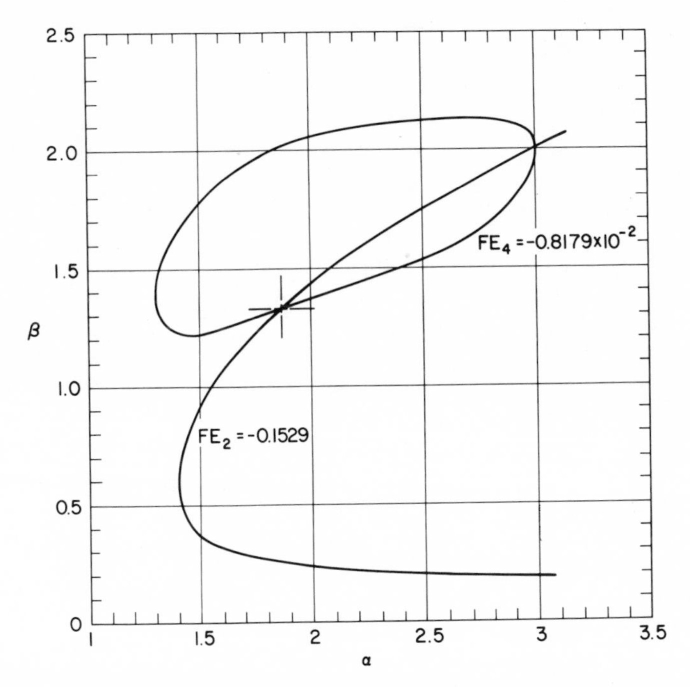

FIG. 46. Example of the method of graphical solution yielding an appropriate inside notch to cancel the second- and fourth-order error coefficients of an $\alpha = 3$, $\beta = 2$ coil. Contours of constant FE_4 and constant FE_2 which pass through the point $\alpha = 3$, $\beta = 2$ are shown, and their second intersection then yields the proper compensating coil of $\alpha_c = 1.87$, $\beta_c = 1.33$.

From Fig. 44, we find the solution to be $\beta_g \approx 0.2$ or, more precisely, $\beta_g = 0.192$ (Fig. 45). This is the gap that will yield the maximum central homogeneity.

If one is interested not in maximizing the homogeneity at the center, but rather in minimizing the mean-square deviation over some axial interval, the gap spacing should be slightly larger. The field profile then dips in the center, as in Fig. 48, but the length over which the homogeneity is maintained is increased. The effect can be explored by means of Eq. (139) with $(z/a_1)_{max}$ set equal to the end of the desired zone.

If our coil need not have a gap for access purposes, α_c need not equal α, and we can choose from an infinite family of combinations of α_c and β_c which will have the same FE_2 as our main coil. A locus of the (α_c, β_c) points that satisfy $FE_2 = -0.153$ is shown in Fig. 46 to be used in finding the compensating coil canceling two expansion terms.

We now require two variables. If we again choose the notch on the i.d.,

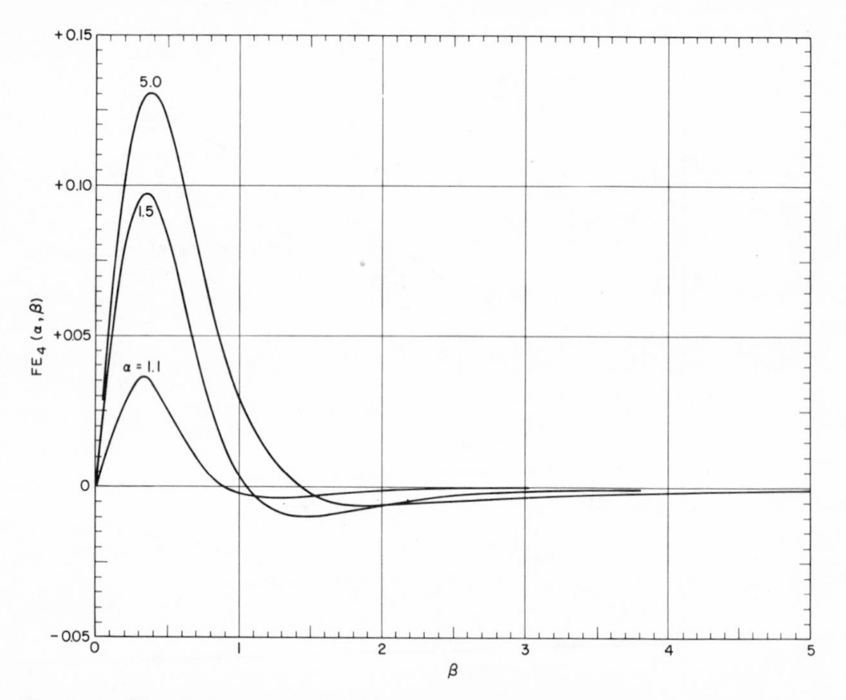

FIG. 47. The product of the $F(\alpha, \beta)$ and the "fourth-order error coefficient." A few contours of constant α are plotted versus β to show the general features of the functions. Note that near $\beta = 1$ the curves cross the axis, indicating simple coils with no E_4 term.

the two coils have the same i.d., and we can again use Eq. (139). The $FE_4(\alpha, \beta)$ term is plotted in Fig. 47 to show its general form, and a zone of particular usefulness is enlarged in Fig. 48.

To find a compensating coil that will cancel the E_4 term of the expansion for our $\alpha = 3$, $\beta = 2$ coil, we plot (Fig. 46) a second locus of (α_c, β_c) points, this time of FE_4, with $FE_4(\alpha_c, \beta_c) = -0.819 \times 10^{-2}$, the same as that for the main coil. The intersection of the two curves locates the compensating coil which will cancel both the E_2 and E_4 of the main coil: $\alpha_c = 1.87$ and $\beta_c = 1.33$. The central field of the compensated coil will be

$$H_0 = j\lambda a_1(F_0 - F_c) = j\lambda a_1(1.045) \tag{140}$$

or 38% below that for the uncompensated coil. The expansion of the field is

$$H_z(\rho, \theta) = H_0\{1 + [F_0E_6(\alpha, \beta) - F_cE_6(\alpha_c, \beta_c)](\rho/a_1)^6P_6(u) + \cdots\} \tag{141}$$

$$H_z(\rho, \theta) = H_0[1 - 0.43 \times 10^{-2}(\rho/a_1)^6P^6(u) + \cdots] \tag{142}$$

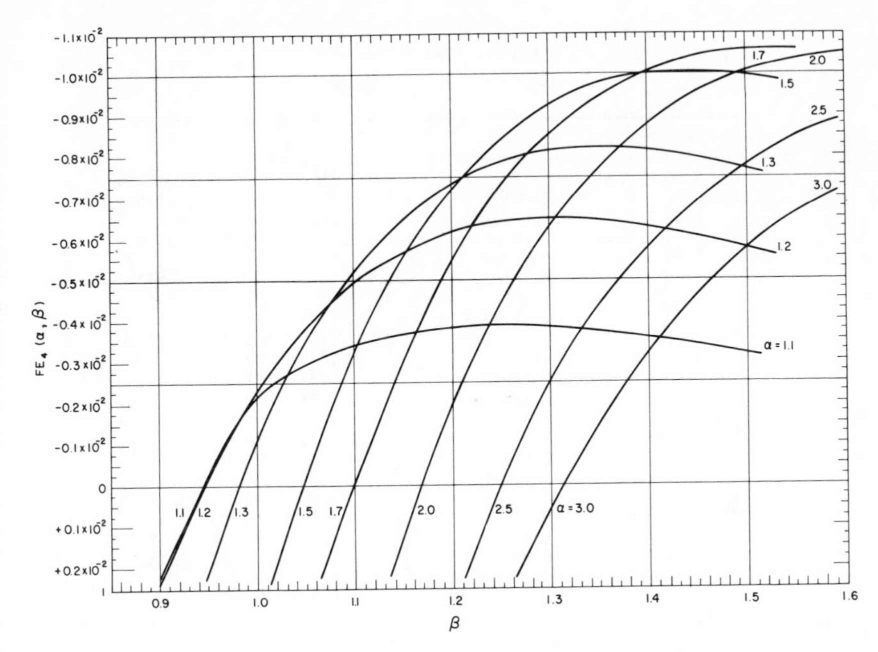

FIG. 48. An enlargement of Fig. 47 in the range $0.9 < \beta < 1.6$, which is the range in which inside compensation notches must usually be chosen. Lines of constant α are plotted versus β.

The severe penalty in field occurs because the original coil is especially short and therefore inhomogeneous.

We can see from Fig. 47 of $FE_4(\alpha, \beta)$ that any main coils of β greater than about 1.0 will have a negative $FE_4(\alpha, \beta)$, and, as the compensating coil must be of the same sign, the compensating coil must likewise have $\beta > 1.0$. It is for this reason that this particular region is enlarged in Fig. 48.

Cancellation of both of the first two expansion terms will again maximize the homogeneity in the center. Again, as with Helmholtz coils, a smaller mean-square deviation over a given axial interval will be achieved by slight adjustment of the compensating coil parameters.

The exact solution is quite sensitive to errors in choice or construction of α_c or β_c. For example, near our solution of $\alpha_c = 1.87$ and $\beta_c = 1.33$, the slope of FE_2 with respect to β (Fig. 45) is about 5% per percent change in β. Thus if 99% of the FE_2 component is to be canceled, α_c must be located within 0.2%. The slope of the FE_4 curve (Fig. 48) is slightly less, but still three times the change in β. Thus, residual terms, incompletely canceled because of constructional errors, can easily exceed the first legitimate term.

To design compensated coils with the notch on the outside diameter, it

TABLE X
UNIFORM-CURRENT-DENSITY
FOURTH-ORDER COILS

α	β	α_c	β_c	E_4	F	γ/a_1^3	G
2.00	2.00	2.0000	0.12004	−0.0789	0.9016	35.44	0.15145
2.00	3.00	2.0000	0.04131	−0.0233	1.0863	55.77	0.14546
2.00	4.00	2.0000	0.01671	−0.0088	1.1604	75.08	0.13391
2.00	5.00	2.0000	0.00776	−0.0039	1.1955	94.10	0.12324
2.00	6.00	2.0000	0.00402	−0.0020	1.2145	113.02	0.11424
3.00	2.00	3.0000	0.19187	−0.0686	1.5304	90.89	0.16053
3.00	3.00	3.0000	0.08122	−0.0253	1.9763	146.71	0.16316
3.00	4.00	3.0000	0.03790	−0.0107	2.1879	199.16	0.15504
3.00	5.00	3.0000	0.01931	−0.0052	2.2990	250.36	0.14530
3.00	6.00	3.0000	0.01063	−0.0028	2.3627	301.06	0.13617
4.00	2.00	4.0000	0.23976	−0.0597	2.0051	165.90	0.15568
4.00	3.00	4.0000	0.11590	−0.0253	2.7058	271.82	0.16412
4.00	4.00	4.0000	0.06061	−0.0120	3.0808	371.28	0.15989
4.00	5.00	4.0000	0.03365	−0.0062	3.2965	468.07	0.15237
4.00	6.00	4.0000	0.01971	−0.0035	3.4283	563.63	0.14440
5.00	2.00	5.0000	0.27058	−0.0530	2.3891	260.79	0.14794
5.00	3.00	5.0000	0.14229	−0.0244	3.3184	430.93	0.15986
5.00	4.00	5.0000	0.08087	−0.0125	3.8585	590.99	0.15872
5.00	5.00	5.0000	0.04820	−0.0069	4.1920	746.71	0.15341
5.00	6.00	5.0000	0.02989	−0.0041	4.4075	900.27	0.14690

is more convenient to write Eq. (139) in another form to normalize the two coils to their common dimension, namely, the outside radius, a_2 (Hart, 1969). The on-axis field would then be written:

$$H_z(z) = j\lambda a_2 \left\{ \left[\frac{F_0}{\alpha}(\alpha, \beta) - \frac{F_c}{\alpha_c}(\alpha_c, \beta_c) \right] + [F_0(\alpha, \beta)E_2(\alpha, \beta)\alpha - F_c(\alpha_c, \beta_c)E_2(\alpha_c, \beta_c)\alpha_c] \left(\frac{z}{a_2}\right)^2 + [F_0(\alpha, \beta)E_4(\alpha, \beta)\alpha^3 - F_c(\alpha_c, \beta_c)E_4(\alpha_c, \beta_c)\alpha_c^3] \left(\frac{z}{a_2}\right)^4 + \cdots \right\} \tag{143}$$

It is for this reason that the computer tables of Montgomery also give $F(\alpha, \beta)E_n(\alpha, \beta)\alpha^{n-1}$ (Montgomery, 1969). Using these terms rather than $F(\alpha, \beta)E_n(\alpha, \beta)$, we can proceed in exactly the same way to design compensated coils. In general, outside-notch coils are easier to construct, more homogeneous (i.e., the E_6 term is smaller), but less efficient in terms of field for a given conductor volume.

b. Computer solutions for fourth- and sixth-order coils. The graphical techniques of the previous section are useful to demonstrate the concept of compensated coils and to find approximate solutions for individual cases.

TABLE XI
UNIFORM-CURRENT-DENSITY SIXTH-ORDER COILS (INSIDE-NOTCH)

α	β	α_c	β_c	E_6	F	$\mathcal{V}/a_1^3$	G
2.00	2.00	1.3920	1.2639	−0.00613	0.6476	30.25	0.11774
2.00	3.00	1.0951	1.0414	−0.00315	1.0380	55.24	0.13965
2.00	4.00	1.0341	0.9674	−0.00143	1.1454	74.98	0.13228
2.00	5.00	1.0150	0.9323	−0.00070	1.1895	94.07	0.12264
2.00	6.00	0.0076	0.9128	−0.00037	1.2116	113.01	0.11398
3.00	2.00	1.8713	1.3327	−0.00434	1.0448	79.58	0.11712
3.00	3.00	1.2332	1.0547	−0.00341	1.8869	147.35	0.15545
3.00	4.00	1.0938	0.9708	−0.00189	2.1601	199.86	0.15279
3.00	5.00	1.0447	0.9332	−0.00103	2.2878	250.79	0.14446
3.00	6.00	1.0238	0.9130	−0.00058	2.3574	301.32	0.13580
4.00	2.00	2.3857	1.4177	−0.00282	1.2876	146.71	0.10630
4.00	3.00	1.3761	1.0813	−0.00309	2.5886	276.67	0.15563
4.00	4.00	1.1642	0.9830	−0.00204	3.0476	274.80	0.15742
4.00	5.00	1.0840	0.9394	−0.00124	3.2844	470.21	0.15147
4.00	6.00	1.0471	0.9164	−0.00076	3.4231	564.93	0.14402
5.00	2.00	2.9094	1.4958	−0.00189	1.4417	231.44	0.09477
5.00	3.00	1.5048	1.1089	−0.00267	3.1837	443.58	0.15116
5.00	4.00	1.2337	0.9986	−0.00200	3.8260	599.91	0.15621
5.00	5.00	1.1265	0.9488	−0.00134	4.1836	752.38	0.15252
5.00	6.00	1.0744	0.9222	−0.00088	4.4058	903.88	0.14655
1.50	2.00	1.1809	1.2584	−0.00624	0.3608	12.59	0.10169
1.50	2.50	1.0780	1.1203	−0.00414	0.4898	18.49	0.11389
1.50	3.00	1.0393	1.0459	−0.00263	0.5443	23.04	0.11341
1.50	3.50	1.0219	1.0003	−0.00168	0.5721	27.21	0.10967
1.50	4.00	1.0131	0.9702	−0.00111	0.5880	31.26	0.10518
2.00	2.00	1.3920	1.2639	−0.00613	0.6476	30.25	0.11774
2.00	2.50	1.1794	1.1167	−0.00461	0.9150	44.38	0.13735
2.00	3.00	1.0951	1.0414	−0.00315	1.0380	55.24	0.13965
2.00	3.50	1.0552	0.9966	−0.00211	1.1050	65.26	0.13678
2.00	4.00	1.0341	0.9674	−0.00143	1.1454	74.98	0.13228
2.50	2.00	1.6247	1.2932	−0.00529	0.8712	52.65	0.12006
2.50	2.50	1.2936	1.1269	−0.00457	1.2848	77.70	0.14576
2.50	3.00	1.1616	1.0452	−0.00339	1.4838	96.67	0.15091
2.50	3.50	1.0969	0.9979	−0.00241	1.5978	114.18	0.14953
2.50	4.00	1.0616	0.9676	−0.00170	1.6695	131.17	0.14577
3.00	2.00	1.8713	1.3327	−0.00434	1.0448	79.58	0.11712
3.00	2.50	1.4124	1.1447	−0.00424	1.6091	118.51	0.14782
3.00	3.00	1.2332	1.0547	−0.00341	1.8869	147.35	0.15545
3.00	3.50	1.1438	1.0033	−0.00257	2.0523	173.99	0.15559
3.00	4.00	1.0938	0.9708	−0.00189	2.1601	199.86	0.15279

Here we present selected computer solutions for fourth- and sixth-order coils (Tables X and XI) to complement graphical solutions (Figs. 44–47). Since the functions are quite linear for all but the most exacting applications, solutions can be located by interpolation.

One can solve for minimum-volume coils directly from these tables (Girard and Sauzade, 1964). Helmholtz coils, for example, can be designed with the field and its homogeneity coefficients $E_2 = 0$ and $E_4(\rho/a_1)^4 = \Delta H/H_0$ as three constraints, and the four variables of inner radius, outer radius, gap, and total length adjusted to satisfy these constraints with the minimum volume of material (Hitchock, 1965). It is often useful to

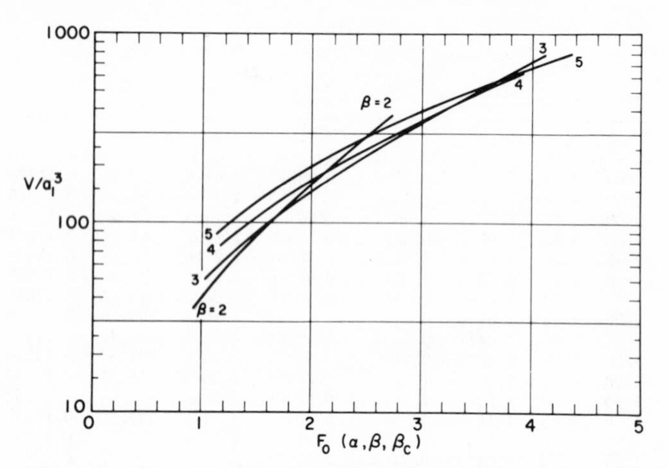

FIG. 49. The coil volume necessary to achieve desired $F(\alpha, \beta)$ in a Helmholtz coil, where $F_0(\alpha, \beta, \beta_c)$ represents the $F(\alpha, \beta)$ of the coil with gap. Lines of constant β are shown with α and β_c, chosen in each case to achieve the value of F_0 and cancel the second error coefficient.

plot the functions in Tables VIII and IX versus volume, as in Figs. 49 and 50, to examine how they vary with the shape of the coil.

Often β is chosen larger than the β for minimum volume, in order to decrease the magnitude of the expansion terms that are to be canceled so as to reduce the sensitivity of the homogeneity to possible errors in the

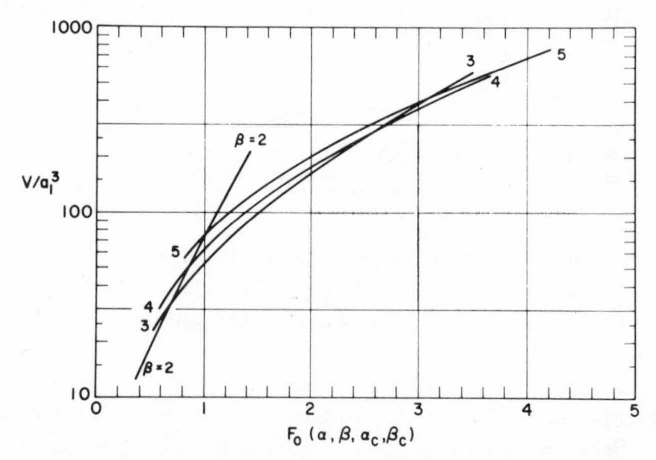

FIG. 50. Coil volume necessary to achieve a desired $F(\alpha, \beta)$ in an inside-notch sixth-order compensated coil, where $F_0(\alpha, \beta, \alpha_c, \beta_c)$ represents the $F(\alpha, \beta)$ of the coil with notch. Lines of constant β are shown with α, α_c, and β chosen in each case to achieve the value of F_0 and cancel the first two error coefficients.

TABLE XII
OUTSIDE-NOTCH SIXTH-ORDER
MINIMUM-VOLUME COILS

α	β	α_C	β_C	E_6	F_0*	Z†
1.5209	3.0	1.1254	1.5649	0.3280×10^{-2}	0.4275	235.0
1.6043	3.2	1.1286	1.5629	0.3045	0.515	177.2
1.6888	3.4	1.1308	1.5610	0.2921	0.601	140.0
1.7743	3.6	1.1325	1.5591	0.2781	0.690	114.7
1.8605	3.8	1.1336	1.5572	0.2665	0.785	96.6
1.9479	4.0	1.1347	1.5561	0.2560	0.875	83.19
2.0353	4.2	1.1352	1.5543	0.2465	0.970	72.91
2.1233	4.4	1.1356	1.5526	0.2382	1.065	64.89
2.2119	4.6	1.1359	1.5514	0.2315	1.161	58.49
2.3009	4.8	1.1361	1.5501	0.2245	1.260	53.25
2.3902	5.0	1.1362	1.5489	0.2185	1.360	48.98
2.4800	5.2	1.1364	1.5481	0.2135	1.465	45.36
2.5693	5.4	1.1362	1.5463	0.2080	1.555	42.29
2.6596	5.6	1.1362	1.5456	0.2040	1.650	39.66
2.7501	5.8	1.1362	1.5449	0.2005	1.750	37.39
2.8398	6.0	1.1359	1.5430	0.1970	1.850	35.40

*$H_z(0,0) = j\lambda a_1 F_0$; $H_z(\rho,\theta) = H_0\left[1 + E_6\left(\frac{\rho}{a_2}\right)^6 P_6(u) + \cdots\right]$.

Note use of outside radius in $(\rho/a_2)^n$ expansions.

†Volume = $Z(H_0/j\lambda)^3$ in cubic centimeters.

construction of the compensating notch or gap. A superconducting coil particularly may be made relatively long (β greater than about 4) to reduce the residual field and diamagnetic perturbations of the field from the superconducting material (Girard and Sauzade, 1964; Grivet and Sausade, 1965). Such coils may be sufficiently long for the outside-notch design to be attractive of minimum-volume coils of this type, as given in Table XII (Girard and Sausade, 1964).

c. Construction of a high-homogeneous-field coil for NMR spectroscopy. What is achieved in practice often is quite different from what is calculated or predicted from theory. The construction of a very homogeneous-field NMR magnet is a case in point; it requires a most painstaking effort on the part of those actually engaged in winding the magnet. The following description of a procedure is based on one currently used by Westinghouse Cryogenics Systems Department in winding their ultrahomogeneous-field NMR magnets—better than a part in 10^8 in a field of 60 kOe over a 0.5-cm-radius sphere (Lane, 1972).

The magnet is basically a sixth-order outside-notch type. Since its i.d. typically is 6.35 cm, we have

$$[\Delta H/H]_{\text{at } z=0.5 \text{ cm}} = E_6(0.5/3.17)^6$$

With $E_6 \sim 10^{-3}$ (Table XI), we estimate *theoretically*

$$[\Delta H/H]_{z=0.5 \text{ cm}} \sim 10^{-8}$$

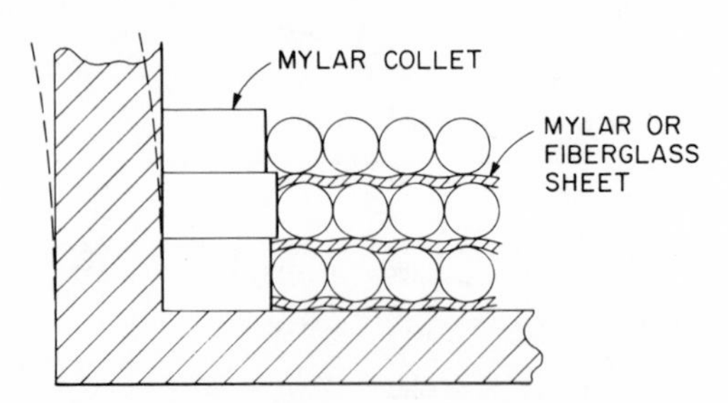

FIG. 51. Typical winding technique to achieve a high-homogeneity field distribution. Mylar collets prevent the coil from bowing out as shown by the dotted lines; Mylar sheets prevent the wires from shifting laterally.

In practice, their magnet without correction by a number of shim coils (a total of 14 sets) has a homogeneity on the order of only 10^{-5} over a 0.5-cm-radius sphere. Even to achieve the homogeneity of 10^{-5}, however, a very careful and stringent winding scheme must be followed, as described below:

(1) A superconducting wire is wound layer by layer, interleaved by two sheets of 25-μM-thick Mylar. The Mylar sheets not only provide good electrical insulation but, more important, keep the wire in place, preventing it from shifting laterally. The tension of the wire is kept at 1500 g. (Three different sizes of superconducting cores actually are used to make the most efficient use of the conductors—the largest one in the inner part where the field is highest, the smallest one outside where the field is lowest. The overall diameters of the wires are all 0.5 mm.)

(2) To prevent the ends of the coil form from bowing out as the layers build up, a Mylar collet is placed between the end of each layer and the coil form (see Fig. 51).

(3) Dimensions of the coil are taken at every layer.

(4) After winding is done, dimensions are fed into a computer to determine the necessary number of tuns for the end-correction winding.

(5) Dimensions are once more fed into the computer, and, if necessary, additional windings are introduced.

The magnet is now ready to be tested in liquid helium. Its persistent mode is also tested. All superconducting joints are mechanical—two wires, after copper is removed, are pressed flat and are clamped tightly into grooves of a metal block, about $\frac{1}{3}$-cm cube (see Fig. 52). As all of the wires used are single-core conductors, this type of joint poses no problem. Recently it has also been proved experimentally that this kind of pressure-joint technique can be used for making joints with two multifilamentary Nb–Ti wires (Leupold and Iwasa, 1974).

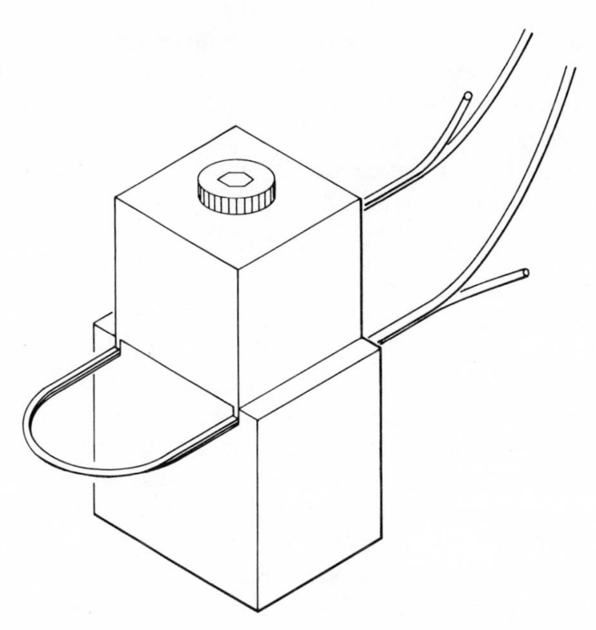

FIG. 52. Schematic sketch of a mechanical superconducting joint. Wires to be joined are passed through the grooves in a metal block about $\frac{1}{3}$-cm cube and pressed tightly.

The magnet, if satisfactory at all, is already so homogeneous that there is no way to measure its homogeneity except by NMR techniques. Therefore, it next is tested in the NMR spectroscope—the magnet has a 5-cm-diam room-temperature access. At this stage, as mentioned above, the magnet has a homogeneity of a part in 10^5, not 10^8 as theory says.

To improve its homogeneity further, now come four sets of superconducting shim coils between the diameters of 5 and 6.35 cm. These shim coils are to eliminate x, y, z, and z^2 field gradients, the z axis being the axis of the magnet. These shim coils achieve a homogeneity of a few parts in 10^7.

Finally, ten sets of copper shim coils are placed outside the superconducting magnet to correct the x, y, z, z^2, z^3, z^4, xy, xz, yz, and $(x^2 - y^2)$ components of gradients, bringing the homogeneity to the guaranteed minimum of a part in 10^8.

D. Thermal and Magnetic Stresses

1. *Thermal stresses*

One of the major design problems in large-volume superconducting magnets is that of minimizing thermal stresses. Thermal stresses arise within a magnet whenever the temperature distribution is nonuniform, as when the magnet is brought from room temperature to 4.2 K, the normal operating temperature. In addition, materials of different thermal expansions induce thermal stresses.

Thermal stress for a given change in length, $\Delta L/L$, induced by a temperature gradient, is expressed by

$$S_T = (\Delta L/L)E \tag{144}$$

where E is Young's modulus.

As an example of a design solution coping with thermal stresses, we give here that of Purcell, who was a principal designer of the Argonne National Laboratory's 18-kG, 4.8-m-i.d. bubble chamber magnet. The dimensions of the magnet are: 4.8-m i.d., 5.3-m o.d., and 3-m length. As the conductor is heavily loaded with copper to insure stabilized operation, for the purpose of this discussion we can think of it as a copper-wound magnet: $(\Delta L/L)_{Cu} \approx 1.5 \times 10^{-5}$/K near room temperature; $E_{Cu} \approx 20 \times 10^6$ lb/in.[2] Thus from Eq. (150) we observe that there will be a vertical stress on the order of a few hundred pounds per square inch for each degree of temperature gradient between the top and bottom of the magnet. Purcell solved this problem in the following manner—reproduced verbatim here (Purcell, 1969).

> In order to limit the ΔT to 10°K [between the ends of the magnet], the cold gas used for cooling is introduced into each coil compartment through a manifold that extends completely around the circumference of the coils. This manifold has small holes spaced about every 6 in. to spray the cold gas downward and mix it with the warm gas within the container. Initially, liquid nitrogen will be used to cool this incoming gas so that the inlet temperature will be about 80°K. The helium refrigerator compressor will be used to circulate this gas during cool-down. The gas is returned to the compressor through another 360° manifold at the bottom of each coil package. Introducing the cooling gas in this manner sets up convection currents within each coil compartment, resulting in a very large mass flow of gas. The average temperature of this large gas flow is only slightly below that of the coils, resulting in a uniform coil temperature. A second method of improving temperature uniformity is thermally shorting the ends of the aluminum spacer blocks together. This is done with a heavy copper braid and is shown in Fig. [53].

When a two-material composite undergoes a change in temperature, thermal stresses are introduced in each material because of the difference in thermal contraction. They are given by

$$S_1 = \left(\frac{\Delta L_1 - \Delta L_2}{L}\right)\left(\frac{E_1E_2}{E_1A_1 + E_2A_2}\right)A_2 \tag{145}$$

$$S_2 = \left(\frac{\Delta L_1 - \Delta L_2}{L}\right)\left(\frac{E_1E_2}{E_1A_1 + E_2A_2}\right)A_1 \tag{146}$$

FIG. 53. Photograph of a portion of a coil assembly (Argonne National Laboratory's 12-ft bubble-chamber magnet) showing the mode of clamping of a coil stack to the girder ring with stainless-steel tie bolts, the thermal clamps, the electrical interconnections between pancakes, and the mechanical clamping arrangement for the ends of the conductor.

where subscripts 1 and 2 refer, respectively, to materials 1 and 2; $\Delta L/L$, the thermal contraction; E, Young's modulus; and A, the cross section. In the case of Nb–Ti copper composite conductors, for example, this thermal stress is sometimes important because it can increase the electrical resistivity of copper, resulting in an increase in the copper required for stability. For Nb–Ti copper composite conductors, $\Delta L/L$ between room temperature and 4.2K and E are, respectively, 3.5×10^{-3} and 20×10^6 lb/in.^{2} for copper and 1.3×10^{-3} and 12×10^6 lb/in.^{2} for Nb–Ti, and in Fig. 54 S_{Cu} is plotted as a function of $A_{\mathrm{Cu}}/A_{\mathrm{Nb-Ti}}$ (Montgomery *et al.*, 1969b). For large copper-to-superconductor ratios, the stress is small, and little effect on resistivity from thermal contraction alone would be expected above a ratio of 2 to 1. The effect of stress and strain on the resistivity of copper wire is shown in Fig. 55 (Montgomery *et al.*, 1969b).

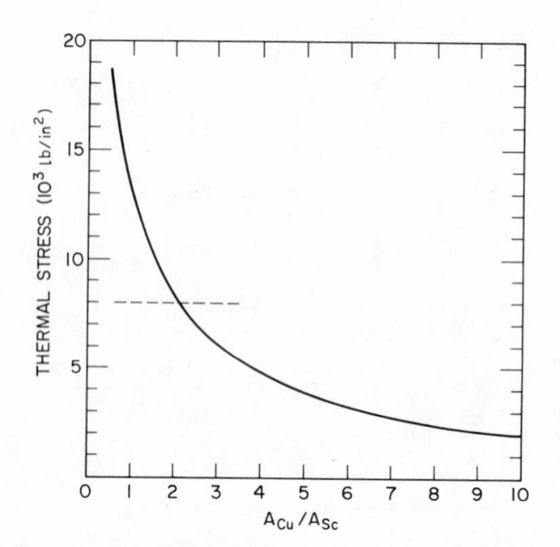

FIG. 54. Thermal stress induced in the copper portion of a Cu–Nb–Ti composite conductor as a function of the area ratio of copper to superconductor.

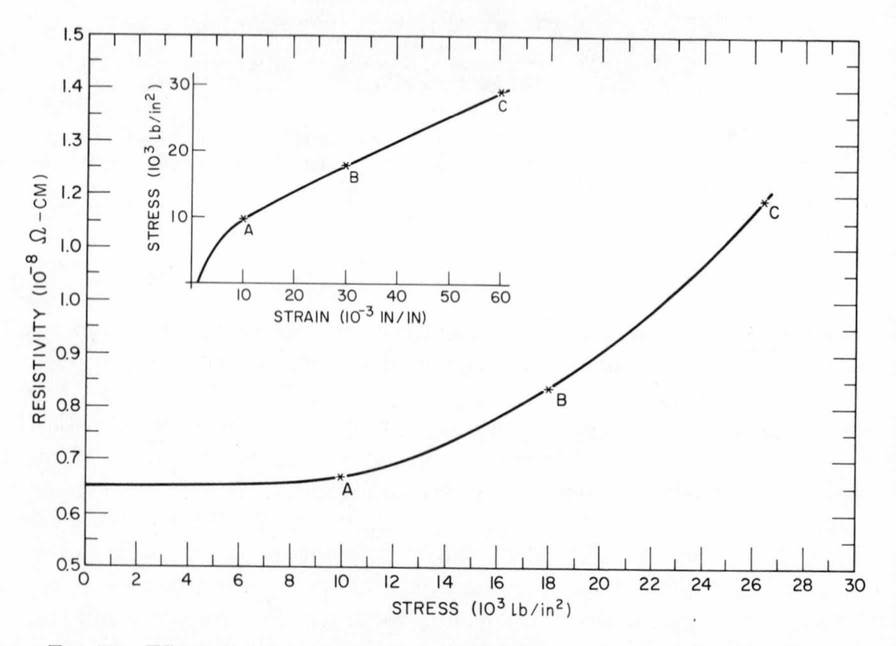

FIG. 55. Effect of stress and strain on the resistivity of copper magnet wire. A 1% strain is required before any effect on resistivity is noted.

2. *Magnetic stresses in solenoidal magnets*

In solenoid magnets, the current flows at right angles to the field, and the Lorentz-force interaction between the current and the field results in stresses within the coil which tend to burst the coil radially outward and to crush it axially (Martson, 1969; Westendorp and Kilb, 1969). We present here two approaches to compute the radial force: (1) a simple solution based on a magnetic pressure approximation, and (2) a more general solution with a set of computer-generated solutions for a limited range of α, β, and j. A simple approach similar to the magnetic pressure approximation will be used to compute the axial force.

a. Radial force. i. MAGNETIC PRESSURE APPROXIMATION—"THIN COILS." The force per unit length (N/m) on the element shown in Fig. 56a can be written:

$$\mathbf{F} = \mathbf{I} \times \mathbf{B} \tag{147}$$

where **I** (A) is the current through the element and **B** (T) is the magnetic induction present at the element.

We now imagine our element to be a loop in an external field carrying current in the direction shown in Fig. 56b. This results in a tensile stress in the material of

$$\sigma_{\mathrm{t}} = a_1 IB/A_{\mathrm{Cu}} \tag{148}$$

where A_{Cu} (m^2) is the loop cross section, a_1 (m) the inner radius, and σ_{t} ($\mathrm{N/m}^2$) is the tangential component of stress.

We have thus far assumed that the element of loop carrying the current was exposed to an external field from some other source. This other source could simply be the other turns in a coil or the self-field produced by the loop carrying the current.

We consider next not a loop but the thin solenoid ($\alpha \approx 1$) in Fig. 56 and assume that the field giving rise to the forces is that produced by the solenoid itself. In this self-field case, however, the magnitude of the field drops from B at the inside to some much lower value at the outside of the winding. We need not concern ourselves with the exact variation at the moment, but only note that if the coil has a length greater than its diameter the field at the outside of the winding is nearly zero. The average field at the winding is then very nearly $B/2$. As in Eq. (148), the tangential stress can now be written:

$$\sigma_{\mathrm{t}} \approx (a_1/d)(I/l)(B/2) \tag{149}$$

where d (m) is the thickness of the solenoid. For a solenoid whose length

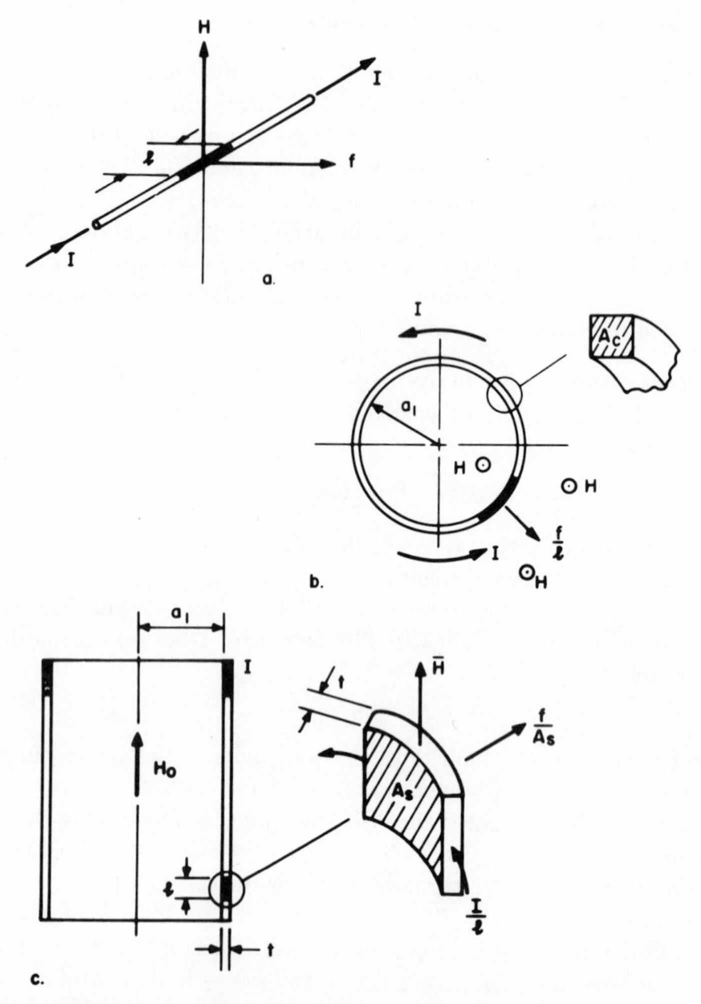

FIG. 56. The Lorentz force relations between field and current. (a) A wire carrying a current I. (b) A loop of cross-sectional area A_c and carrying a current I in a uniform transverse external field. (c) A thin shell generating a self-field H_0.

is much greater than its diameter, we have, from Eq. (9),

$$I/l \approx H = B/\mu_0 \tag{150}$$

Substituting Eq. (150) into Eq. (149), we obtain

$$\sigma_t \approx (a_1/d)\ (B^2/2\mu_0) \tag{151}$$

If a_1/d were 10, for example, the yield point of soft copper ($7 \times 10^7\ \text{N/m}^2$) would be exceeded by a field of approximately 4.5 T.

ii. STRESSES WITHIN A THICK UNIFORM-CURRENT-DENSITY COIL. Most often in practice we wish to examine not a current sheet but a coil whose build is comparable to its inner diameter. There are several levels of sophistication upon which this problem can be approached. The exact solution requires first knowledge of the local body forces resulting from the local value of field and current, and then knowledge of the modulus of the composite conductor–insulator material.

In a wire or tape-wound coil the forces on any given turn will be a sum of the forces, some resulting from the body of that turn, and others resulting from the other turns pushing on it. The exact solution of the multiturn coil is particularly complicated by the inexact knowledge of how the insulators will behave (Gersdorf *et al.*, 1965; Middleton and Trowbridge, 1967).

One simplified approach to this problem is to calculate the tangential stress in any given turn based on the assumption that the turns do not interact appreciably; we assume that the stress in one turn is not significantly affected by any neighboring turns. Following Eq. (149) and noting that $I/1d = \lambda j$ (A/m^2), the average current density, we can write

$$\sigma_t(r) = r\lambda j\bar{B} \tag{152}$$

In a uniform-current-density coil, in which λj is constant, the local stress will depend on the product of the radius and the field. The relationship between the current density and the central field in a coil can also be substituted into Eq. (152) to eliminate the current density, namely,

$$\lambda j = \frac{H_0}{a_1 F(\alpha, \beta)} \tag{153}$$

where H_0 is the central field. Substituting Eq. (153) into Eq. (152), we have

$$\sigma_t(r) = \left(\frac{r}{a_1}\right) \frac{H_0 \bar{B}}{F(\alpha, \beta)} \tag{154}$$

One interesting phenomenon is that in a region of increasing stress turns tend to separate from each other rather than "piling up," and our assumption of zero radial force is therefore valid over at least the initial portions of the region inside that of maximum hoop stress (Gersdorf *et al.*, 1965). The turn experiencing maximum tangential stress according to our simple assumption, however, may actually be backed up by the coil outside that turn, because the tangential stress is decreasing in that region of the coil. Whenever this is the case, we would expect the maximum stress to be smaller; the approximation we have just discussed then is an upper limit on the tangential stress.

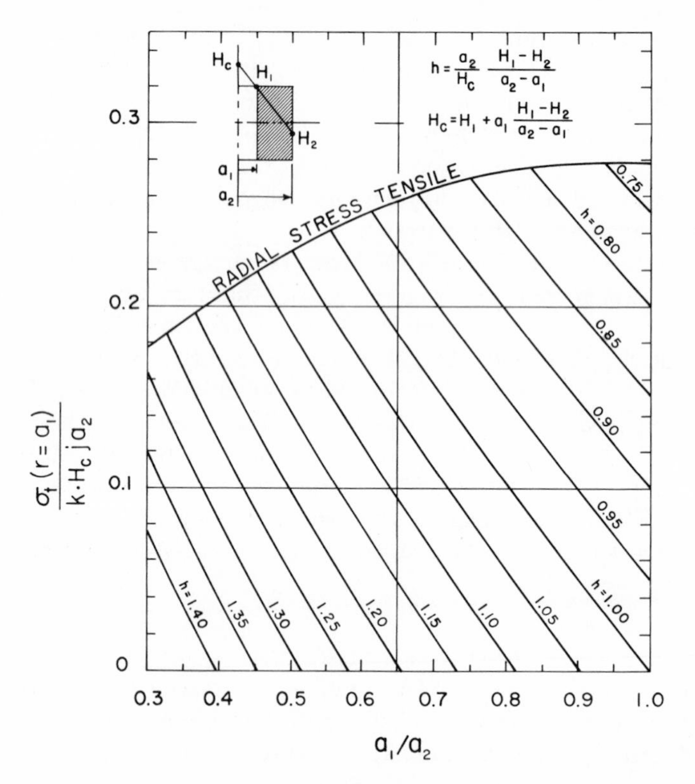

Fig. 57. Peak stress in uniform-current-density coils in which the outer elements support the inner. The intersection of a_1/a_2 with the calculated h contour gives the peak stress at the coil i.d. provided the intersection occurs below the cutoff tensile stress contour. If the intersection is outside the range, Eq. (160) should be used instead. Note that the field H_c is not the central field but is defined from the slope. The value of k should be $(1/9.8) \times 10^{-6}$ for H (Oe), j (A/cm²), a_2 (cm), and σ (kg/cm²).

An exact solution for the uniform-current-density case can be derived under the assumption that the field decreases linearly with the radius (Marston, 1969; Westendorp and Kilb, 1969); such a solution is shown in Fig. 57 for certain combinations of field gradient and dimensions and for which the outer elements of the coil support the inner elements or when the coil is bonded from outside. For cases lying outside this range, Eq. (154) should be used.

The solution given in Fig. 57 is also useful in dealing with composite magnets—magnets made up of one or more coils nested inside each other. In such a composite magnet, H_1, H_2, and H_0 of Fig. 57 are fields at the respective locations generated by the entire magnet: the outside field of the

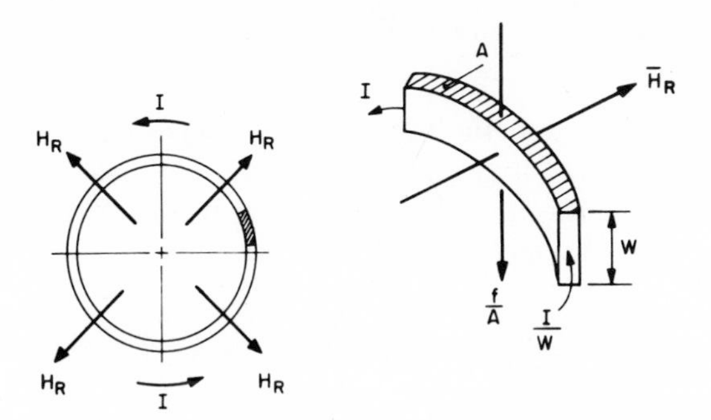

FIG. 58. The axial force on an element resulting from the radial field H_r.

innermost magnet, H_2, for example, could be quite large, rather than a small negative value as in the case of a single coil.

b. Axial forces. The radial component of the field gives rise to an axial compressive stress. Referring to Fig. 58, we have

$$F_a/A = P_a = w\lambda j\bar{B} \tag{155}$$

where P_a (N/m^2) is the pressure, A (m^2) the unit normal to the force, $\bar{B}$ (T) the average radial field, λj (A/m^2) the average current density, and w (m) the axial width of a turn.

The radial component of the field is strongest near the ends of the coil and decreases to zero at the midplane, and the self-force is therefore greatest on the end turns. However, unless the turns are actually separated from each other, the compressive stress is cumulative and is maximum at the midplane. The magnitude of the axial compressive stress is generally less than 25% that of the tangential stress and is, therefore, not critical in the calculation of maximum combined stresses in the conductors (Gersdorf *et al.*, 1965). However, in coils in which turns are separated radially by soft insulators or an actual gap, as in large superconducting magnets, the axial force should be compared with the product of the radial compression and the coefficient of friction between the turn and the interlayer insulation to see if the turns will remain stationary.

3. *Magnetic body forces in circular-aperture dipoles and quadrupoles*

We examine the magnetic body forces resulting from the Lorentz interaction in constant-winding-thickness dipoles and quadrupoles with circular-aperture iron shields (Iwasa and Weggel, 1972). Results for magnets of variable winding thickness should be similar. The body forces give a good indication, especially in "thin coils," of the level of magnetic stresses within the winding.

a. Dipoles. The magnetic field within the winding of a dipole of constant winding thickness is given by Eq. (52). To this we must add the field contribution of the iron shielding, given by Eq. (72). The resulting total field (considering only the dominant term) is

$$\frac{H_{r,\theta}}{H_0} = \left[\left(\frac{b_1 - r}{b_1 - a_1}\right) D_1 \pm \left(\frac{r^3 - a_1^3}{b_1^3 - a_1^3}\right)\left(\frac{b_1}{r}\right)^2 \tilde{D}_1 + \left(\frac{b_1}{b_s}\right)^2 \tilde{D}_1\right]\begin{Bmatrix}\sin\theta\\ \\ \cos\theta\end{Bmatrix} \tag{156}$$

The body forces in the dipole act as shown in Fig. 59a. We define two com-

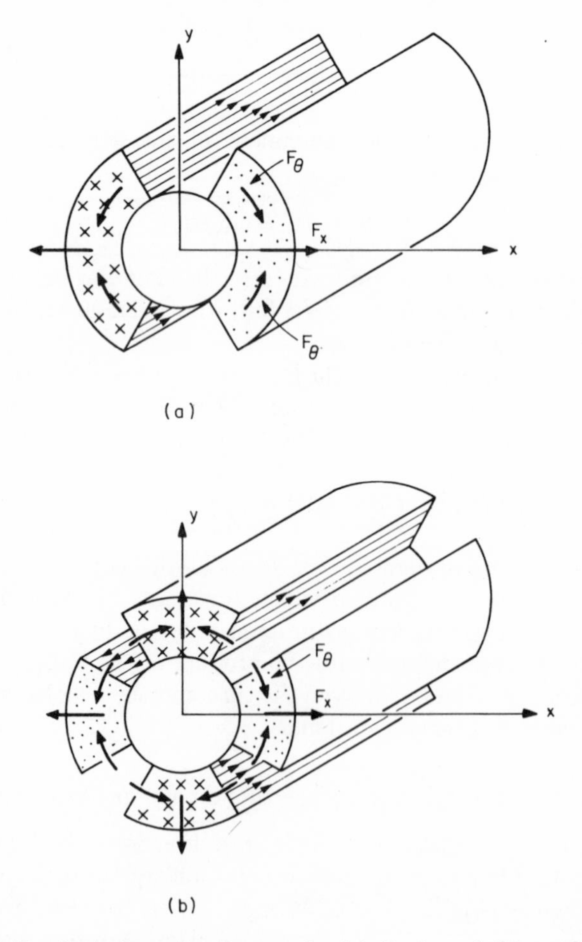

FIG. 59. Body-force distributions for (a) dipole, and (b) quadrupole.

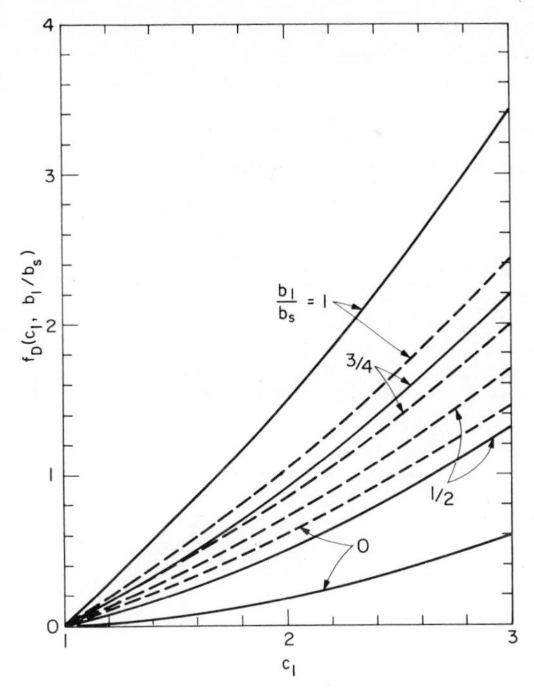

FIG. 60. Body-force coefficients as a function of c_1 for several values of b_1/b_s for circular-aperture, iron-shielded dipole of constant winding thickness. ——: f_{Dx}; – – –: f_{Dx}; $K = 1$.

ponents of the body force—one in the x direction, F_x, which tends to burst the magnet, and the other in the θ direction, F_θ, which tends to crush the winding circumferentially. These factors, per unit length of magnet, are

$$F_x = \int_{a_1}^{b_1} r\,dr \sum_{k=1}^{K} \int_{-\alpha_k}^{\alpha_k} d\theta\, j_k H_\theta \cos\theta \tag{157}$$

and

$$F_\theta = \int_{a_1}^{d_1} r\,dr \sum_{k=1}^{K} \int_{0}^{\alpha_k} d\theta\, j_k H_r \tag{158}$$

Combining Eqs. (156)–(158), we obtain

$$F_x = 10^{-4}\, j_1 H_0 D_1 a_1^2 f_{Dx}(c_1, K) \tag{159}$$

and

$$F_\theta = 10^{-4}\, j_1 H_0 D_1 a_1^2 f_{D\theta}(c_1, K) \tag{160}$$

where the forces are in N/m, H_0 (Oe) is given by Eq. (49), j_1 (A/cm^2) is the current density in the first block, and a_1 (cm) is i.d. f_{Dx} depends on K, although only weakly, whereas $f_{D\theta}$ does not. Both are plotted as a function of c_1 and for given values of b_1/b_s in Fig. 60. Note that iron shielding greatly increases F_x.

b. Quadrupoles. An identical approach computes body forces in a quadrupole of constant winding thickness. The total field within the winding is

$$\frac{H_{r,\theta}}{H_0} = \left[\pm \left(\frac{\ln b_1/r}{\ln c_1} \right) \left(\frac{r}{a_1} \right) Q_1 + \left(\frac{r^4 - a_1^4}{b_1^4 - a_1^4} \right) \left(\frac{b_1}{r} \right)^3 \tilde{Q}_1 \pm \left(\frac{a_1 b_1^3}{b_s^4} \right) \left(\frac{r}{a_1} \right) \tilde{Q}_1 \right] \begin{Bmatrix} \cos 2\theta \\ \sin 2\theta \end{Bmatrix} \quad (161)$$

The resulting body forces are shown in Fig. 59b. Again we decompose them

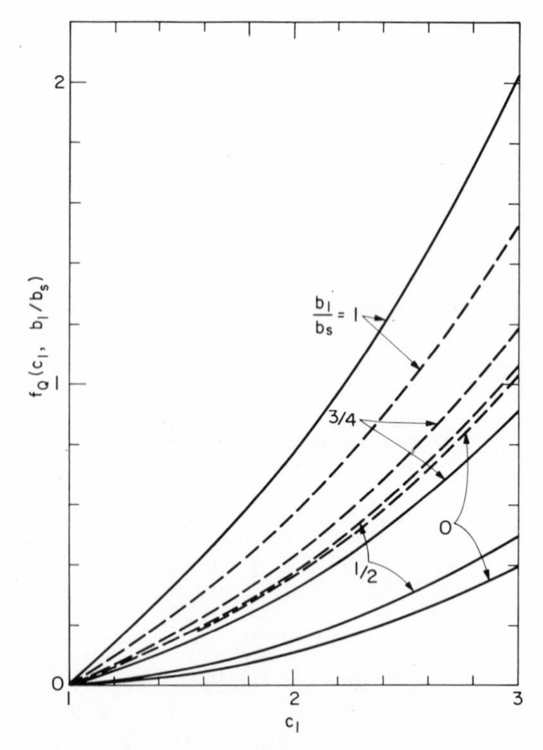

FIG. 61. Body-force coefficients as a function of c_1 for several values of b_1/b_s for circular-aperture, iron-shielded quadrupole of constant winding thickness. ——: f_{Qx}; - - -: $f_{Q\theta}$; $K = 1$.

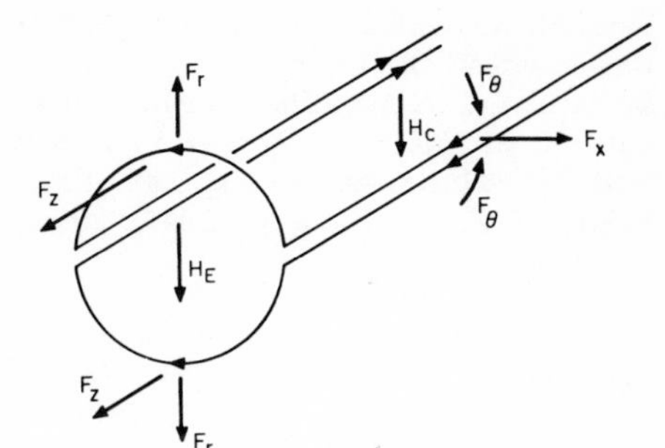

FIG. 62. Body-force distribution of a dipole with an end.

into two components:

$$F_x = 10^{-4} j_1 H_0 D_1 a_1^2 j_{Qx}(c_1, K) \tag{162}$$

and

$$F_\theta = 10^{-4} j_1 H_0 D_1 a_1^2 f_{Q\theta}(c_1) \tag{163}$$

where the forces again are in N/m. f_{Qx} and $f_{Q\theta}$ are plotted in Fig. 61.

c. Body forces at the ends of dipoles and quadrupoles. So far we have assumed each dipole and quadrupole to be infinitely long. In real magnets, the length is of course finite, and, therefore, all of the analyses given above, it should be remembered, are valid only for the region more than a bore-radius distance from each end of the magnet. The end effects (Meuser, 1971) are particularly troublesome in considering body forces. Figure 62 illustrates the body-force distribution in a typical dipole: the forces F_r and F_z arise solely because of end effects. We can estimate F_r by considering the Lorentz force between the two semicircular current loops which form each end; F_z is a result of the vertical field, H_E, at the end (see Fig. 62), H_E being about $\frac{1}{2}$ of the central field. A similar analysis leads to approximate values for F_r and F_z for quadrupoles.

References

Appleton, A. D. (1969). *Cryogenics* **9,** 147.
Aron, P. R., and Ahlgren, G. W. (1968). *Advan. Cryogen. Eng.* **13,** 21.
Asner, A. (1969). *Proc.* 1968 *Summer Study Superconducting Devices Accelerators* (A. G. Prodell, ed.), p. 866. Brookhaven Nat. Lab., Upton, New York.
Bean, C. P. (1962). *Phys. Rev. Lett.* **8,** 250.
Bean, C. P., and Schmitt, R. W. (1963). *Science* **140,** 26.
Benz, M. G. (1968). *J. Appl. Phys.* **39,** 2533.

Benz, M. G. (1969). *J. Appl. Phys.* **40**, 2003.
Berlincourt, T. G. (1963b). *Brit. J. Appl. Phys.* **14**, 749.
Berlincourt, T. G., and Hake, R. R. (1963a). *Phys. Rev.* **131**, 140.
Beth, R. A. *Proc.* 1968 *Summer Superconducting Devices Accelerators* (A. G. Prodell, ed.), p. 843. Brookhaven Nat. Lab., Upton, New York.
Blewett, J. P., and Hann, H. (1972). Brookhaven Nat. Lab. Rep. BNL 16716, Upton, New York.
Brechna, H. (1969). *Proc. 1968 Summer Study Superconducting Devices Accelerators* (A. G. Prodell, ed.), p. 478. Brookhaven Nat. Lab., Upton, New York.
Cheremnyk, P. A., Churakov, G. F., Keilin, V. E., Klimenko, E. U., Kuliamzin, A. N., Monoszon, N. A., Novikov, S. I., Ostroumov, U. N., Rozdestvenski, B. V., Sabanski, I. I., Samoilov, B. N., and Chernoplekov, N. A. (1973). The Kurchatov Institute of Atomic Energy IAE-2316.
Chester, P. F. (1967). *Progr. Phys.* **30**, Part II, 561.
Coffey, H. T., Hulm, J. K., Reynolds, W. T., Fox, D. K., and Span, R. E. (1965). *J. Appl. Phys.* **36**, 128.
Critchlow, P. R., Gregory, E., and Zeitlin, B. (1971). *Cryogenics* **11**, 3.
Crow, J. E., and Suenaga, M. (1972). IEEE Conf. Rec., IEEE Cat. No. 72 CHO 682-5 TABSC, p. 472.
Dahl, P. F., Morgan, G. H., and Sampson, W. B. (1969). *J. Appl. Phys.* **40**, 2083.
de Latour, C. (1973). *IEEE Trans. Magn.* **MAG-9**, p. 314.
DiSalvo, F. (1966). Avoc Everett Res. Lab. Rep. AMP-206.
Echarri, A., and Spadoni, M. (1971). *Cryogenics* **11**, 274.
Ferguson, R. C., and Phillip, W. D. (1967). *Science* **157**, 257.
Fernandez-Moran, H. (1965). *Proc. Nat. Acad. Sci.* **53**, 445.
Fernandez-Moran, H. (1966). *Proc. Nat. Acad. Sci.* **56**, 801.
Fietz, W. A., Beasley, M. R., Silcox, J., and Webb, W. W. (1964). *Phys. Rev.* **136**, A335.
Foner, S., McNiff, E. J., Jr., Matthias, B. T., Geballe, T. H., Willens, R. H., and Corenzwit, E. (1970). *Phys. Lett.* **31A**, 349.
Foner, S., McNiff, E. J., Jr., Vieland, L. J., Wicklund, A., Miller, R. E., and Webb, G. W. (1972). IEEE Conf. Rec., IEEE Cat. No. 72 CHO 682-5 TABSC, p. 404.
Foner, S., McNiff, E. J., Jr., Gavaler, J. R., and Janocko, M. A. (1974). *Phys. Lett.* **47A**, 485.
Frei, E. H. (1970). *CRC Crit. Rev. Solid State Sci.* **1**, 381.
Frei, E. H. (1973). Drs. R. Rand and J. A. Mosso of Univ. of California at Los Angelos, private communication.
Furuto, Y., Suzuki, T., Tachikawa K., and Iwasa, Y. (1974). *Appl. Phys. Lett.* **24**, 34.
Garrett, M. W. (1962). *In* "High Magnetic Fields" (H. Kolm, B. Lax, F. Bitter, and R. Mills, eds.), p. 14. MIT Press, Cambridge, Massachusetts.
Gavaler, J. R. (1973). *Appl. Phys. Lett.* **23**, 480.
Gersdorf, R., Muller, F. A., and Roeland, L. W. (1965). *Rev. Sci. Instrum.* **36**, 1100.
Gilbert, W., Voelker, F., Acker, R., and Kaugerts, J. (1972). IEEE Conf. Rec., IEEE Cat. No. 72 CHO 682-5 TABSC, p. 486.
Girard, B., and Sauzade, M. (1964). *Nucl. Instrum. Methods* **25**, 269.
Green, M. A. (1971). Lawrence Radiat. Lab. Eng. Note M 4373, UCID-3493.
Grivet, P., and Sauzade, M. (1965). *Proc. Int. Symp. Magn. Technol.*, p. 517.
Hale, J. R., and Williams, J. E. C. (1968). *J. Appl. Phys.* **39**, 2634.
Haller, T. R., and Belanger, B. C. (1971). *IEEE Trans.* **NS-18**, 671.
Hampshire, R. G., Sutton, J., and Taylor, M. T. (1969). Bull. Int. Inst. Refrig. Comm. I, London, Annex, p. 251.

Hancox, R. (1965). *Phys. Lett.* **16**, 208.
Hart, H. R., Jr. (1969). *Proc. 1968 Summer Study on Superconducting Devices Accelerators* (A. G. Prodell, ed.), p. 571. Brookhaven Nat. Lab., Upton, New York.
Hechler, K., Horn, G., Otto, G., and Saur, E. (1969). *J. Low Temp. Phys.* **1**, 29.
Heiniger, F., Bucher, E., and Muller, J. (1966). *Phys. Condens. Mater.* **5**, 243.
Hitchock, H. C. (1965). *Proc. Int. Symp. Magn. Technol.*, p. 111.
Inoue, K., and Tachikawa, K. (1972). IEEE Conf. Rec., IEEE Cat. No. 72 CHO 682-5 TABSC, p. 415.
Inoue, K., Tachikawa, K., and Iwasa, Y. (1971). *Appl. Phys. Lett.* **18**, 235.
Iwasa, Y. (1969). *Appl. Phys. Lett.* **14**, 200.
Iwasa, Y., and Montgomery, D. B. (1971). *J. Appl. Phys.* **42**, 1040.
Iwasa, Y., and Weggel, R. J. (1972). *Proc. Int. Conf. Magn. Technol., 4th.*, p. 607.
Iwasa, Y., Montgomery, D. B., and Polzer, H. W. (1970). *Appl. Phys.* **14**, 2476.
Kim, Y. B., Hempstead, C. F., and Strnad, A. R. (1963). *Phys. Rev.* **129**, 528.
Kohr, J. G., Eager, T. W., and Rose, R. M. (1972). *Met. Trans.* **3**, 1177.
Kolm, H. H., and Thornton, R. D. (1972). IEEE Conf. Rec., IEEE Cat. No. 72 CHO 682-5 TABSC, p. 76.
Lane, S. (1972). Private communication.
Leupold, M. J., Iwasa, Y., and Montgomery, D. B. (1972). IEEE Conf. Rec., IEEE Cat No. 72 CHO 682-5 TABSC, p. 308.
Leupold, M. J. and Iwasa, Y. (unpublished report, 1974).
Lohberg, R., Eagar, T. W., Puffer, I. M., and Rose, R. M. (1973). *Appl. Phys. Lett.* **22**, 69.
Lubell, M. S. (1972). *Cryogenics* **12**, 340.
Lubell, M. S., Chandrasekhar, B. S., and Mallick, G. T. (1963). *Appl. Phys. Lett.* **3**, 79.
Lubell, M. S., Long, H. M., Luton, J. N., Jr., and Stoddart, W. C. T. (1972). IEEE Conf. Rec., IEEE Cat. No. 72 CHO 682-5 TABSC, p. 341.
Marston, P. G. (1969). *Proc. 1968 Summer Study Superconducting Devices Accelerators* (A. G. Prodell, ed.), p. 709. Brookhaven Nat. Lab., Upton, New York.
McInturff, A. D., Jr. (1969). *J. Appl. Phys.* **40**, 2080.
McInturff, A. D. (1972). IEEE Conf. Rec., IEEE Cat. No. 72 CHO 682-5 TABSC, p. 395.
McInturff, A. D., Dahl, P. F., and Sampson, W. B. (1972). *J. Appl. Phys.* **43**, 3546.
Meuser, R. B. (1971). Lawrence Radiat. Lab. (Berkeley) Engineering Note M4370, UCID-3509.
Middleton, A. J., and Trowbridge, C. W. (1967). *Int. Conf. Magn. Technol., 2nd* (H. Hadley, ed.), p. 140. Rutherford Lab.
Montgomery, D. B. (1969). "Solenoid Magnet Design." Wiley, New York.
Montgomery, D. B., Weggel, R. J., Leupold, M. J., Yody, S. B., and Wright, R. L. (1969a). *J. Appl. Phys.* **40**, 2129.
Montgomery, D. B., Williams, J. E. C., Pierce, N. T., Weggel, R. J., and Leupold, M. J. (1969b). *Advan. Cryogen. Eng.* **14**, 88.
Otto, G., Saur, E., and Wizgall, H. (1969). *J. Low Temp. Phys.* **1**, 29.
Polgreen, G. R. (1970). *In* "New Applications of Modern Magnets," p. 268. Boston Tech. Publ., Boston, Massachusetts.
Popley, R. A., Sambrook, D. J., Walters, C. R., and Wilson, M. N. (1972). IEEE Conf. Rec., IEEE Cat. No. 72 CHO 682-5 TABCS, p. 516.
Powell, J. R. (1974). In "Superconducting Machines and Devices" (S. Foner and B. B. Schwartz, eds.), p. 1. Plenum Press, New York.
Powell, J. R., and Danby, G. R. (1966). Amer. Soc. Mech. Eng. Paper 66-WA/RR-5.

Purcell, J. R. (1969). *Proc. 1968 Summer Study Superconducting Devices Accelerators* (A. G. Prodell, ed.), p. 765. Brookhaven Nat. Lab., Upton, New York.
Purcell, J. R. (1972), IEEE Conf. Record, IEEE Cat. No. 72 CHO 682-5 TABCS, p. 246.
Sampson, W. B., Britton, R. B., Dahl, P. F., McInjurff, A. D., Morgan, G. H., and Robins, K. E. (1970). *Particle Accelerators* **1**, 173.
Schrader, E. R. (1969). *J. Appl. Phys.* **40**, 2076.
Schrader, E. R., Freedman, N. S., and Fakan, J. C. (1964). *Appl. Phys. Lett.* **4**, 105.
Smith, J. L., Jr., Kirtley, J. L., Jr., Thullen, P., and Woodson, H. H. (1972). IEEE Conf. Rec., IEEE Cat. No. 72 CHO 682-5 TABSC, p. 145.
Smith, R. V. (1969). *Proc. 1968 Summer Study Superconducting Devices Accelerators* (A. G. Prodell, ed.), p. 249. Brookhaven Nat. Lab., Upton, New York.
Stekly, Z. J. J., and Zar, J. L. (1965). *IEEE Trans.* **NS–12**, 367.
Stekly, Z. J. J., Lucas, E. J., De Winter, T., Strauss, B., and DiSalvo, F. (1968). *J. Appl. Phys.* **39**, 2641.
Suenaga, M., and Sampson, W. B. (1971). *Appl. Phys. Lett.* **18**, 584.
Suenaga, M., and Sampson, W. B. (1972). IEEE Conf. Rec., IEEE Cat. No. 72 CHO 682-5 TABSC, p. 481.
Swartz, P. S., and Bean, C. P. (1968). *J. Appl. Phys.* **39**, 4991.
Tachikawa, K., and Iwasa, Y. (1970). *Appl. Phys. Lett.* **16**, 230.
Tachikawa, K., and Tanaka, Y. (1966). *Jpn. J. Appl. Phys.* **5**, 834.
Thullen, P., and Smith, J. L., Jr. (1970). *Advan. Cryogen. Eng.* **15**, 132.
Tuck, J. L. (1971). *Nature (London)* **233**, 593.
Weggel, R. J., and Montgomery, D. B. (1972). *Proc. Int. Conf. Magn. Tech., 4th* p. 18.
Westendorp, W. F., and Kilb, R. W. (1969). *Proc. 1968 Summer Study Superconducting Devices Accelerators* (A. G. Prodell, ed.), p. 714. Brookhaven Nat. Lab., Upton, New York.
Williams, J. E. C. (1967). *Conf. Product. Appl. Intense Magn. Fields* p. 281.
Wilson, M. N., Walters, C. R., Lewin, J. D., Smith, P. F., and Spurway, A. H. (1970). *Brit. J. Phys. D. Appl. Phys.* **3**, 1515.
Wipf, S. L. (1967). *Phys. Rev.* **161**, 404.
Wipf, S. L. (1969). *Proc. 1968 Summer Study Superconducting Devices Accelerators* (A. G. Prodell, ed.), p. 511. Brookhaven Nat. Lab., Upton, New York.
Wong, J., Fairbanks, D. F., Randall, R. N., and Larson, W. L. (1968). *J. Appl. Phys.* **39**, 2518.
Yasukochi, K., Ogasawara, T., Usui, N., and Ushio, S., Jr. (1964). *Phys. Soc. Japan* **19**, 1964.
Yodh, S. B., Pierce, N. T., Weggel, R. J., and Montgomery, D. B. (1968). *Med. Biol. Eng.* **6**, 143.

General References

I. Books

J. E. C. Williams, "Superconductivity and its Applications." Pion, London, 1970.
D. B. Montgomery, "Solenoid Magnet Design." Wiley, New York, 1969.
R. B. Scott, "Cryogenic Engineering." Van Nostrand-Reinhold, New Jersey, 1959.
W. R. Smythe, "Static and Dynamic Electricity." McGraw-Hill, New York, 1950.
J. P. Den Hartog, "Strength of Materials." Dover, New York, 1961.

II. Review Articles

T. G. Berlincourt, *Brit. J. Appl. Phys.* **14,** 749 (1963).
Y. B. Kim, *Phys. Today* **September,** 21. (1964).
D. B. Montgomery, *IEEE Spectrum* **February,** 103. (1964).
C. Laverick, *Cryogenics* **5,** 152 (1965).
P. F. Chester, *Rep. Progr. Phys.* **30,** Part II, 561 (A. C. Stickland, ed.). Inst. Phys. Phys. Soc., London, 1967.
W. B. Sampson, *IEEE Trans.* **—MAG-4,** 99 (1968).
G. Bogner, *Proc. Int. Cryogen. Eng. Conf., 3rd,* p. 35. LLIEFE Science Technol. Publ., Guildford, 1970).
J. Hulm, D. J. Kasun, and E. Mullan, *Phys. Today* **August,** 48 (1971).
Z. J. J. Stekly, *J. Appl. Phys.* **42,** 65 (1971).
M. S. Lubell, *Cryogenics* **12,** 340 (1972).

III. Conferences

Appl. Superconductivity (Years Held: 1966, '67, '68, '70, '72; '74).
Int. Conf. Magn. Technol. (1965, '67, '70, '72; '75).
Particle Accelerator Conf. (1965, 67, '69, '71, '73).
1968 Summer Study Superconducting Devices Accelerators.

Chapter 7

Superconductive Machinery

THEODOR A. BUCHHOLD†

GE Research & Development Center
Schenectady, New York

I. Superconductive Bearings

A. Bearings of Type I

Normal magnetic bearings, which are stable in all directions, cannot be built unless a control system is added. However, the phenomenon of super-

† Present address: 62 Wiesbaden, Uranusweg 6, West Germany.

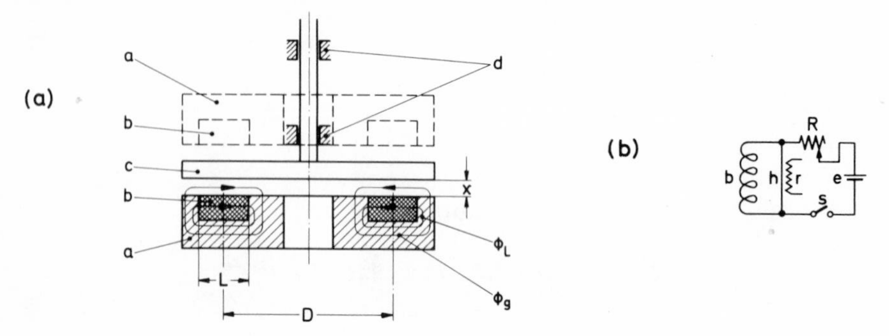

FIG. 1. (a) Model of a superconductive bearing, Type I. (b) Scheme to short-circuit coil (b).

conductivity makes entirely stable bearings possible (Buchhold, 1960, 1961). These must be made of a Type-I superconductor such as lead or niobium, which magnetic flux does not penetrate. A slight flux penetration, for distances of the order of 10^{-4} to 10^{-5} mm, is neglected. Nb is preferred as a bearing material, since it has the high critical field of about 1500 Oe. Hard superconductors such as Nb_3Sn with much higher critical fields are not used, since the flux can penetrate deeply into the material.

Two types of bearing, Type I and Type II, are possible and will be discussed (Buchhold, 1964a). Figure 1a shows a cross section of a flat cylinder of soft iron (a) containing a superconductive coil of Nb wire (b). Above the coil is a flat disk (c) guided by a mechanical bearing (d). If the current in the coil (b) is increased, countercurrents are produced in the disk which prevent the flux penetration. The flux ϕ_g is therefore compelled to flow in the gap x along the disk surface. If now the coil (b) is short-circuited with a superconductive wire, the entire linked flux ϕ stays constant:

$$\phi = \phi_g + \phi_L = \text{const} \tag{1}$$

where ϕ_L is leakage flux passing through the coil. If the gap x decreases, the flux density B increases, since the flux ϕ is constant. The compressed flux lines produce a repelling upward force on the disk. The specific magnetic force is, to a good approximation,

$$f = (B/5000)^2 \text{ kg cm}^{-2} \tag{2}$$

B is the flux density in gauss. Thus for $B = 1200$ G, for instance, $f = 0.057$ kg/cm^{-2}.

If the coil surface is $A = \pi DL$ (D is average coil diameter and L coil width), the entire upward force is

$$F = A(B/5000)^2 \text{ kg} \tag{3}$$

The flux going through the gap x can be written as

$$\phi_g = kBx$$

where k is a constant. If the reluctance of the iron is neglected, the leakage flux ϕ_L within the coil is proportional to the current I and the flux density B and can be expressed as

$$\phi_L = kBx_i$$

where x_i can be thought of as an imaginary fixed gap. The entire linked flux is

$$\phi = \phi_g + \phi_L = kB(x + x_i) \qquad \text{or} \qquad B = \phi/k(x + x_i) \tag{4}$$

The coil current I is proportional to B and thus changes with the variable gap x even though the flux ϕ remains constant. With decreasing gap, the flux density increases up to a critical value where the bearing material becomes resistive. With practical Nb bearings, fields of 1000–1300 G should not be exceeded.

Figure 1b shows a scheme for the excitation and short-circuiting of the coil (b). A superconductive wire (h), which can be made resistive with a heater (r), is in parallel with the coil (b). If (h) is made resistive, the resistor R can be adjusted to produce any desired coil current i. If the current through the heater (r) is switched off, (h) becomes superconductive, and the coil (b) is short-circuited. The switch (s) can now be opened.

The force F which is proportional to B^2 can, with Eq. (4), be written as

$$F = K/(x + x_i)^2 \tag{5}$$

where K is another constant. Figure 2a shows the force as a function of the gap x for zero and for finite flux leakage. The flux leakage produces a smaller slope in the F versus x curve and thus a smaller bearing stiffness.

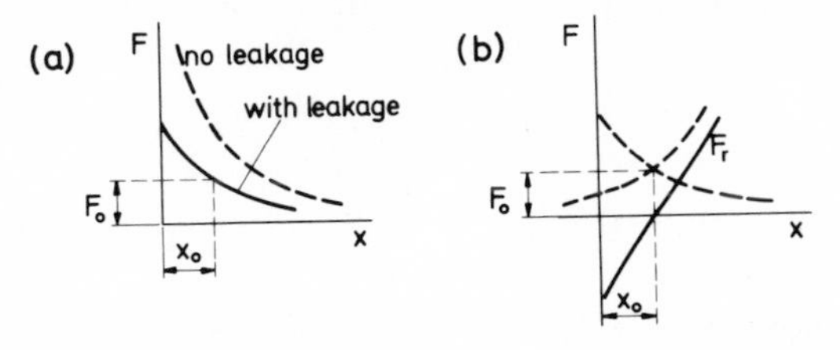

FIG. 2. (a) Force as a function of displacement. (b) Resultant force of two opposed coils.

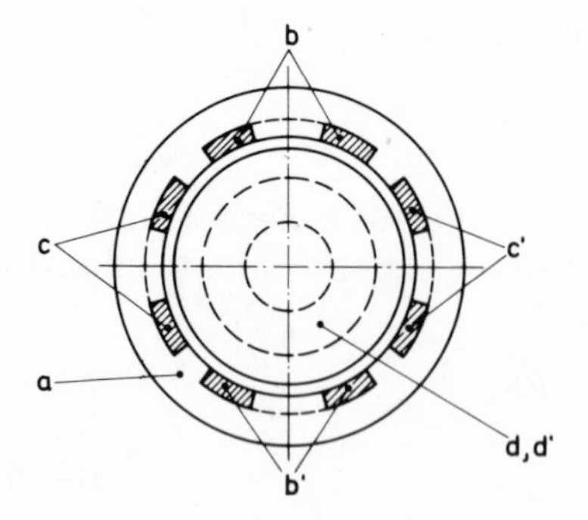

FIG. 3. Bearing for supporting a sphere.

The stiffness or spring constant c for x in the vicinity of x_0 is given by

$$c = \frac{dF}{dx} = -\frac{2K}{(x + x_i)^3} \tag{6}$$

Using Eqs. (5) and (6) setting $x = x_0$ and neglecting the minus sign gives

$$c = 2F_0/(x_0 + x_i) \tag{7}$$

Let a second coil containing cylinder be placed above the disk (shown dashed in Fig. 1a). The resulting force F_r, which is the difference force of both systems, is shown in Fig. 2b. The spring constant for x near x_0 is now found to be twice the value given by Eq. (7), i.e.,

$$c = 4F_0/(x_0 + x_i) \tag{8}$$

This bearing is seen to be stable against upward as well as downward vertical displacements.

Figure 3 shows a bearing carrying a sphere which is stable in all directions. The housing (a) which is made of iron carries three pairs of coils. The coils b, b′ and c, c′ provide stability against vertical and side forces, and the coils d, d′ give stability perpendicular to the plane of paper.

In order to have great stiffness, the gap x_0 and the leakage, which is given by x_i, should be as small as possible. x_0 can be made very small, but x_i cannot, since the leakage flux passing through the coil is practically constant and is not negligible. It therefore looks attractive to try to build a bearing which does not depend on the leakage.

B. BEARINGS OF TYPE II

Figure 4 shows a usable form of such a bearing (Buchhold, 1964a). On each side of a shaft (a) is a flange (b) of Nb. Each coil (c) is surrounded by an iron yoke (d) and by a Nb cover (e) which must contain a slit (not

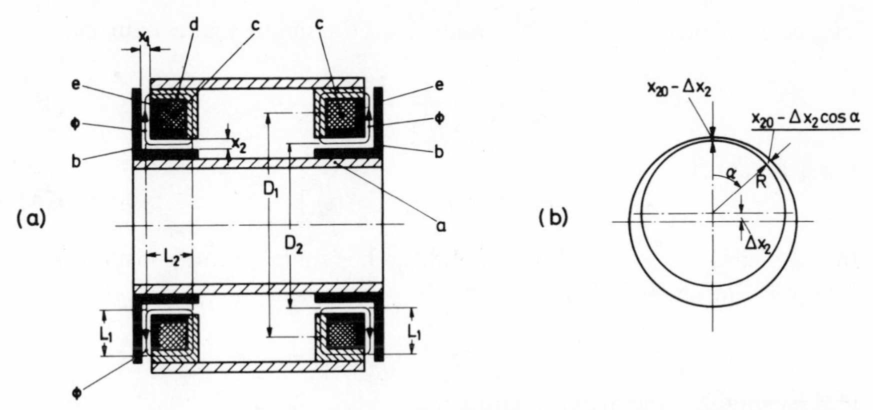

FIG. 4. (a) Superconductive bearing, Type II. (b) Displacement of bearing shaft.

shown) so as to avoid a superconductive short. If the bearing is in the center position, let the gaps x_1 and x_2 be x_{10} and x_{20}, respectively. The flux in the variable gap x_1 of length L_1 has the reluctance

$$r_1 = L_1/\mu_0\pi D_1 x_1 \tag{9}$$

with $\mu_0 = 4\pi/10$ and $\pi D_1 x_1$ being the average cross section for the flux. The reluctance of the gap x_{20} of length L_2 is

$$r_2 = L_2/\mu_0\pi D_2 x_{20} \tag{10}$$

The horizontal force for the left bearing will be written as

$$F = kB^2 \tag{11}$$

where k is a constant. The spring constant for the left bearing is

$$c = \frac{dF}{dx_1} = 2kB\frac{dB}{dx_1} \qquad \text{or} \qquad c = \frac{2F}{B}\frac{dB}{dx_1} \tag{12}$$

The flux density in the gap x_1 is

$$B = (\mu_0 IN/L_1)[r_1/(r_1 + r_2)] \tag{13}$$

where IN is in ampere turns. Therefore, if r_1 is the variable,

$$dB = \frac{\mu_0 IN}{L_1}\frac{r_2}{(r_1 + r_2)^2}dr_1 \tag{14}$$

With

$$\frac{dr_1}{r_1} = -\frac{dx_1}{x_{10}} \tag{15}$$

and, combining Eqs. (13), (14), and (15) (the minus sign is omitted),

$$\frac{dB}{dx_1} = \frac{B}{x_{10}} \frac{1}{1 + (r_1/r_2)} \tag{16}$$

From Eq. (12),

$$c = 2F/x_{10}[1 + (r_1/r_2)] \tag{17a}$$

In the vicinity of x_{10}, x_{20}, $F = F_0$, and the horizontal spring constant c_h is, for both bearings together,

$$c_h = 4F_0/x_{10}[1 + (r_1/r_2)] \tag{17b}$$

If, for example, r_1 is made equal to r_2,

$$c_h = 2F_0/x_{10} \tag{18}$$

Note that c_h is not influenced by any leakage. To obtain high stiffness, x_{10} can be made small.

Assume now that the bearing shaft is moved upward a distance Δx_2 and that x_{20} becomes x_2. Around the circumference (Fig. 4b), the gap changes to

$$\Delta x = \Delta x_2 \cos \alpha \tag{19}$$

A small surface area $RL_2\, d\alpha$ located at the angle α will be considered. The increase in force perpendicular to the surface is

$$d(\Delta F) = c'(\Delta x_2 \cos \alpha)$$

when ΔF is force for $x_2 = x_{20}$. c' can be obtained from Eq. (17a) if r_1 and r_2 are replaced by r_2 and r_1, F_0 by $f_0RL_2\, d\alpha$, and x_{10} by x_{20}. One obtains

$$d(\Delta F) = \frac{2f_0RL_2\, d\alpha\, \Delta x_2 \cos \alpha}{x_{20}[1 + (r_2/r_1)]} \qquad \text{where} \quad f_0 = \left(\frac{B}{5000}\right)^2 \tag{20}$$

To obtain the vertical resultant ΔF_v of all of the surface elements, $d(\Delta F)$ has to be multiplied with $\cos \alpha$ and integrated from $\alpha = 0$ to $\alpha = 2\pi$, i.e.,

$$\Delta F_v = \int_0^{2\pi} \frac{2f_0RL_2\, d\alpha\, \Delta x_2 \cos^2 \alpha}{x_{20}[1 + (r_2/r_1)]}$$

Thus,

$$\Delta F_v = 2\pi RL_2f_0/x_{20}[1 + (r_2/r_1)] \tag{21}$$

Defining F_0 as $2\pi RL_2f_0$, the total spring constant in the vertical direction

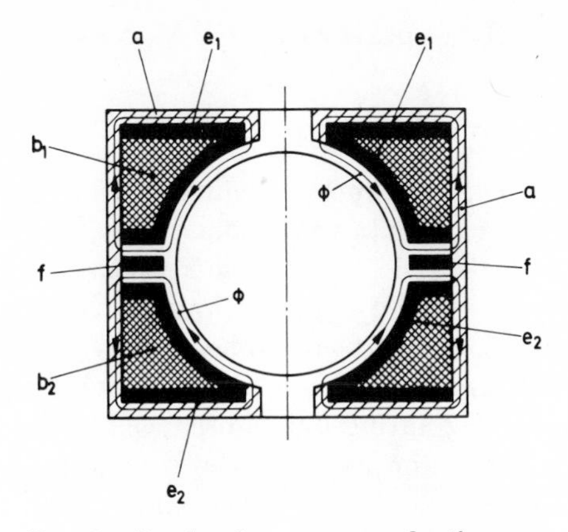

FIG. 5. Bearing for a superconductive gyro.

for both bearings is

$$c_v = 2F_0/x_{20}[1 + (r_2/r_1)] \tag{22}$$

and, for $r_1 = r_2$,

$$c_v = F_0/x_{20} \tag{23}$$

Superconductive bearings need some damping, e.g., by eddy currents. The flux of the bearing of Fig. 4 contains an ac component if the bearing vibrates. Therefore, eddy currents are produced in the iron yoke which provide damping.

Figure 5 shows the application of the bearing principle described above to the support of a superconductive sphere (Buchhold, 1963a). The parts (a) of the housing are made of soft iron. The two coils (b_1) and (b_2) are surrounded by the two niobium walls (e_1) apd (e_2), which contain a vertical slit (not shown) to avoid a superconductive short. The flux ϕ due to the upper coil passes through the gap formed by the ball and the wall (e_1) and through the constant gap formed by (e_1) and the superconductive ring (f). This bearing stably supports the sphere in all directions. It was used in a cryogenic gyro (Buchhold, 1962) with the ball spinning in a vacuum at about 12,000 rpm. Tests showed that the freely spinning ball had so little friction that it would take more than one year to come to a stop.

With available materials, superconductive bearings can be built only for moderate forces because the flux density is limited to 1000–1300 G. Static bearings with Nb, which had a cold-worked surface, have worked with flux densities up to 3000 G. However, no experiments were made under rotating conditions. It is possible that the flux would then gradually enter into the material and that the bearing would fail.

II. Superconductive Motors

A. Reluctance Motor (Buchhold, 1960)

Figure 6a shows a superconductive cylinder, e.g., of Nb rotating in a superconductive bearing (not drawn). Around the cylinders are conductors (d), which carry currents and produce a flux which cannot penetrate into the material. At the surface, the magnetic lines are compressed and produce forces F perpendicular to the surface and hence provide no torque (Schoch, 1961). If the conductors are part of a polyphase winding, similar to that of an induction motor, a rotating flux is produced. This causes a pressure force on the surface but no torque. However, if the rotor is noncircular, for instance, as shown in Fig. 6b, a torque is produced, and the rotor follows the rotating field as a kind of synchronous motor. Motors based on this principle are practically lossless and have been built and used to drive a cryogenic gyro.

In the following, a simplified theory of such a motor is given which shows what parameters are of importance. If the rotor of Fig. 6b is turned clockwise, the inductance of the conductors changes. The inductance is smallest at an angle $\alpha = 0°$ but is at a maximum for $\alpha = 30°$. Figure 6c shows the inductance as a function of the angle α. As an approximation, the inductance is assumed to be sinusoidal and to be expressible as

$$L = L_0 - \Delta L \cos p\alpha \tag{24}$$

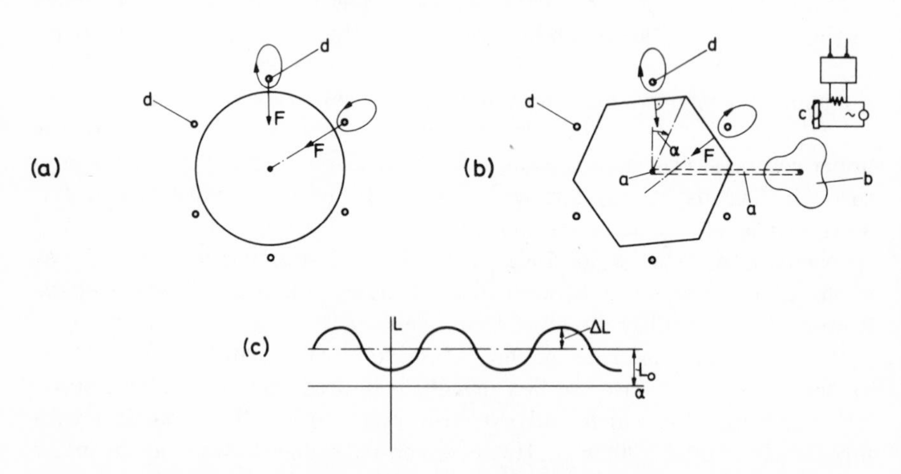

FIG. 6. Principle of a superconductive reluctance motor. (a) Magnetic pressure producing no torque; (b) magnetic pressure producing a torque; (c) inductance as a function of angle of rotation.

where L_0 is the average inductance, ΔL its maximum variation, and p the number of faces of the polygon. A known formula for the torque T yields

$$T = \frac{1}{2}\frac{dL}{d\alpha} i^2 \tag{25}$$

which is independent of the direction of the current (i). Let a current

$$i = I_m \sin(\omega t + \varphi) \tag{26}$$

flow through the conductors. φ is a phase angle whose optimum values will be determined. Since the torque is proportional to i^2, the angle α of the rotating motor will change by $2\pi/p$ twice per current cycle. Thus,

$$\alpha = 2\omega t/p \tag{27}$$

Equations (24)–(26) give

$$T = \tfrac{1}{2}p \sin p\alpha[I_m^2 \sin^2(\omega t + \varphi)]\,\Delta L$$

Substituting for $p\alpha$ from Eq. (27) yields, after some algebra,

$$T = \tfrac{1}{8}p\,\Delta L\, I_m^2[2 \sin 2\omega t - \sin(4\omega t + 2\varphi) + \sin 2\varphi] \tag{28}$$

The average torque is

$$T_0 = \tfrac{1}{8}p\,\Delta L\, I_m^2 \sin 2\varphi \tag{29}$$

which has a maximum value at $\varphi = 45°$ given by

$$T_{0max} = \tfrac{1}{8}p\,\Delta L\, I_m^2 \tag{30}$$

to attain maximum average torque, the current has to be

$$i = I_m \sin(\omega t + 45°) \tag{31}$$

Equation (30) shows that ΔL should be made as big as possible, e.g., by proper shaping of the surfaces. To satisfy Eq. (30), the current must have the right frequency and phase angle. This can be achieved (see Fig. 6b) by connecting a properly shaped disk (b) with the rotor axis (a). A coil (c) is excited with high frequency and is modulated by the rotating disk (b). Electronic means produce the proper frequency and phase angle φ. φ can be adjusted by turning the disk on its axis.

The rotor, according to Eq. (28), has a variable torque with a positive average value. At certain angles α, the torque can be zero or even negative. Therefore, the rotor does not always start. If the load is small with an additional displaced winding, an initial push can be given which starts the rotor. Better starting conditions are obtained with a two-phase system, which also gives a larger torque. The two currents of such a system are separated by 90° and must be a function of the rotor position; this doubles the size of the electronically controlled power supply. It looks tempting to eliminate the electronics and use a normal power supply whose frequency can be regulated starting from zero. Tests have shown that such a motor tends to oscillate and will fall out of synchronism.

The described motor can be built only for small powers, since ΔL is limited as is the current I, which must not produce flux density above a critical value on the surface of the motor. If this is exceeded, the material becomes resistive, and the motor ceases to function. A motor of a few watts was built to bring a gyro rotor, spinning in vacuum, up to 12,000 rpm.

B. Motor with Induced Rotor Currents

Figure 7 shows a two-pole version of another type of superconductive motor. The stator consists of the two concentric parts (a) and (b) of iron lamination. Part (a) contains a two-phase winding which produces an approximately sinusoidal current distribution and hence a sinusoidal flux ϕ_1 of which a few lines (d) are shown. Within the gap is a short-circuited superconductive winding (e), which can turn around the center axis. Before starting the motor, no current flows in the winding (e); therefore, the flux linked with it is zero. According to Eq. (1), the flux in a superconducting short-circuited winding stays constant and therefore has to remain zero in (e). This is achieved by countercurrents in (e) which produce a flux ϕ_2 that cancels the part of the primary flux ϕ_1 which links the coil (e). The induced current, together with the flux density B, produces a torque in a clockwise direction. Since the flux ϕ_1 rotates, the coil (e) follows.

Figure 8a shows as a function of the circumference the instantaneous current density distribution J expressed in ampere conductors per centimeter of circumference, and the flux density distribution B:

$$J = J_{\mathrm{m}} \cos(\pi x/l) \tag{32}$$

where l is the circumferential length of a pole

$$B = B_{\mathrm{m}} \sin(\pi x/l) \tag{33}$$

If the gap is δ, it follows that

$$B_{\mathrm{m}} = \frac{4\pi}{10} J_{\mathrm{m}} \frac{2}{\pi} \frac{l}{2\delta} \qquad \text{or} \qquad B_{\mathrm{m}} = 0.4 J_{\mathrm{m}} \frac{l}{2\delta} \tag{34}$$

If the rotor conductors are located at c_1, c_2 (see Fig. 8b), the flux linkage with them is zero, and no current flows in the rotor. However, if the con-

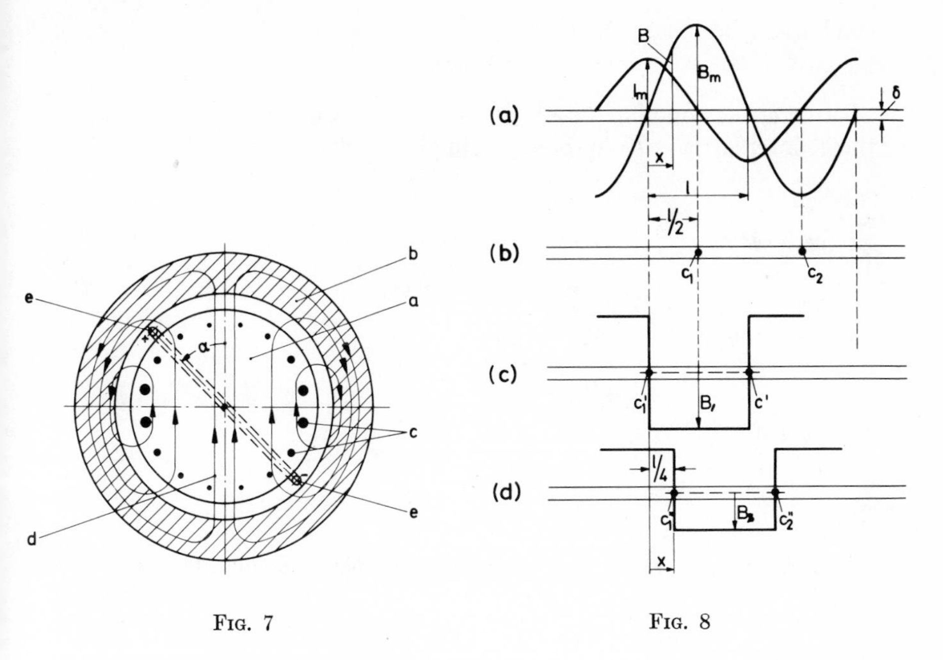

FIG. 7 FIG. 8

FIG. 7. Motor showing induced rotor current in (e). Different thicknesses of dots (c) indicate sinusoidal current distribution.

FIG. 8. (a) Flux density b and current density J in the motor of Fig. 7; (b) rotor is at c_1, c_2, flux linkage zero; (c) rotor is at c_1', c_2', flux linkage maximum; (d) rotor is at c_1'', c_2'', flux linkage intermediate.

ductors are at c_1', c_2' as shown in Fig. 8c, the flux ϕ_1 is linked:

$$\phi_1 = \frac{2}{\pi} B_m lh \tag{35}$$

where h is the length of the conductor. Countercurrents I_m flow which produce a rectangular flux of the same amount but in the opposite direction with the density B_1

$$\phi_1 = \frac{2}{\pi} B_m lh = I_m \frac{4\pi}{10} \frac{lh}{2\delta} \qquad \text{or} \qquad I_m = \frac{10}{\pi^2} B_m \delta = \sim B_m \delta \tag{36}$$

Since at c_1', c_2' the conductors are exposed to the flux density $B = 0$, no force is developed. When the conductors are shifted a distance x to c_1'', c_2'', where x corresponds to an angle $\alpha = (x/l)\pi$, the flux ϕ_1, linked with the conductors, becomes

$$\phi_1' = \phi_1 \cos \alpha \tag{37}$$

Therefore, the opposing flux shown with the flux density B_2 in Fig. 8d and the countercurrents I are correspondingly smaller. Therefore,

$$I = I_m \cos \alpha \tag{38}$$

At the location c_1'', c_2'' the flux density is

$$B = B_m \sin \alpha \tag{39}$$

and both conductors produce a force

$$F = 2B(I/10)h = (I_m/10)B_m h \sin 2\alpha$$

or, using Eq. (36),

$$F = B_m^2 \frac{\delta h}{10} \sin 2\alpha \text{ dyne} \tag{40}$$

If the motor has p pole pairs instead of two poles and the rotor has a lever arm R, the torque is

$$T = (p/10)RB_m^2 h\delta \sin 2\alpha \tag{41}$$

The maximum torque is obtained at $\alpha = 45°$ corresponding to $x = l/4$:

$$T_{max} = (p/10)RB_m^2 h\delta \text{ dyne cm} \tag{42}$$

Instead of a one-turn winding, a distributed winding with many turns can be used, with the beginning and end short-circuited. The stator consists of iron laminations. If the flux density is not too high, the iron losses can be kept small and can be tolerated for many applications. A motor with iron produces much higher torques than one using superconductors only. To avoid quenching of the rotor winding, the maximum flux density is limited. The motor operates synchronously and the proper frequency is best produced in a manner similar to that described previously.

C. Low-Temperature Induction Motor (Buchhold, 1965)

The motors described above were of the synchronous type and need a complicated power supply to produce the drive frequency. A simpler motor of the induction type using superconducting bearings will now be described.

Because of material limitations, superconducting bearings cannot carry as much load and are not as stiff as normal bearings. For this reason, an iron rotor should be avoided. If an iron rotor is at the center of the stator, magnetic forces acting on the surface cancel. However if the rotor is dis-

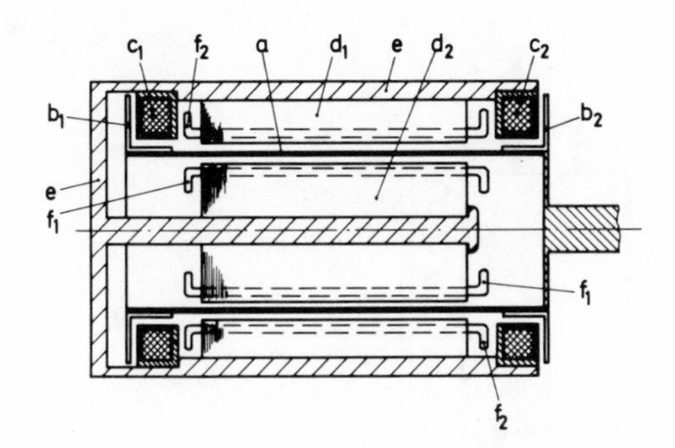

FIG. 9. Induction motor with superconductive bearing.

placed from the center, a force difference results with increases with displacement. This condition, together with limited stiffness of the bearing, could lead to rotor suspension instability. Therefore, an iron-free rotor was used in this motor consisting of a lightweight thin aluminum shell. Figure 9 shows a cross section of the motor. The rotor consists of a thin aluminum cylinder (a) with niobium flanges (b_1) and (b_2) at its ends. The flanges together with the coils (c_1), (c_2) form a superconductive bearing as described in Fig. 4. The stator consists of the two parts (d_1) and (d_2) of iron lamination which are attached to a housing (e). The motor has two stator windings (f_1), (f_2) which are wound in a conventional three-phase manner using superconductive Nb wires. The three phases of the windings (f_1), (f_2) are connected in series. When energized from a conventional three-phase supply, a rotating flux is produced, and currents are induced in the shell (a), and the rotor follows with a small slip. The rotor losses can be kept small since, at helium temperatures, the aluminum has a conductivity which is at least 100 times higher than at normal temperatures. To keep the iron losses in the stator small, a low-flux density has to be applied. The currents in the stator winding have to be kept below the quenching current of the Nb wire.

Such a motor was built and operated according to expectations. If an impeller is connected to the rotor axis, such a motor works as a pump and can circulate liquid helium.

D. OSCILLATING MOTOR (Buchhold and Darrel, 1965)

In certain experiments with superconductive devices, e.g., with a superconductive gyro, a continuous helium transfer from a helium storage container is preferred. Figure 10a shows an oscillating motor built as a pump which can be inserted through the neck of a helium storage container into

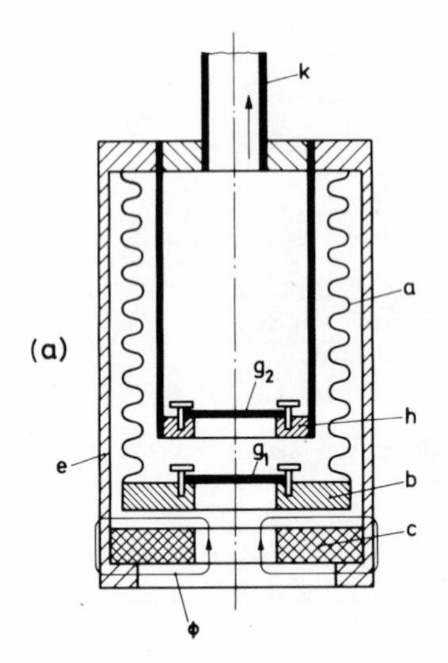

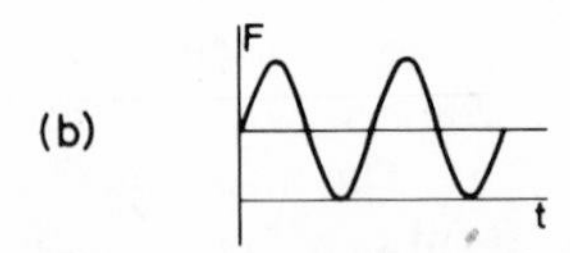

FIG. 10. (a) Oscillating motor. (b) Motor force as a function of time.

the liquid and which can pump helium to a cryostat in a regulated manner. The transfer line should be nitrogen cooled. In Fig. 10a, the bellow (a) of nickel carries an armature ring (b) of Nb. Below (b), a coil (c) of Nb wire is attached to the housing (e). If the coil (c) is excited with an ac current of 60 Hz, an ac flux ϕ is produced which passes along the lower surface of the ring (b). If the area of the ring surface is A and the maximum average flux density is B_m on the armature (b) in an upward direction, a force

$$F = \tfrac{1}{2}(B_m/5000)^2(1 + \sin 2\omega t)A \text{ kg} \tag{43}$$

is produced (see Fig. 10b). The force contains a constant term and a term which changes sinusoidally with double the drive frequency. If the spring constant of the bellow and the mass of the ring (b) are dimensioned so that the system is in resonance with this frequency, strong oscillations of the armature take place. The armature (b) has a valve plate (g_1) of nylon. A similar valve plate (g_2) is placed on the ring (h).

As the armature ring (b) moves upward, it closes the valve (g_1). The liquid helium is pushed through the upper valve (g_2) into the discharge tube (k), which is connected to the transfer line. When the armature (b) moves downward, the liquid opens the lower valve (g_1), and some liquid from the storage container flows into the pump chamber. The weight of the columns of the fluid in the discharge tube (k) closes the upper valve (g_2) and prevents a backflow.

A pump as described was built for 60 Hz with an outside diameter of 0.9 in. and worked satisfactorily. A pumping speed of 20 liter/h was achieved against the pressure of a liquid column of height 1 m. For a 2-m column, the pumping speed was about 10 liter/h. By changing the exciting current, the helium flow can be regulated.

III. AC Losses

A. Losses in Type-I Superconductors (Buchhold and Molenda, 1962)

In many machinery applications, the surface of a superconductor is exposed to a variable flux density. If, for example, a superconductive bearing carries a load, the flux density of a rotating surface element varies with each revolution. At the surface of the armature of Fig. 10 and also at the windings of a superconductive transformer, the flux density alternates sinusoidally. A varying flux density causes ac losses. Theory predicts that an ideal superconductor should have immeasurably small losses in the 60–400-Hz region of frequencies used in ac machinery. Measurements with soft superconductors (Type I) such as lead and niobium show small but noticeable losses. This section and the following treat these types of losses in both soft and hard superconductors.

A magnetic flux at the surface of niobium, which is the material of choice for low-loss superconductive machinery, penetrates a distance of about 6×10^{-6} cm into the material. If, close to the surface, the material has imperfections (Buchhold, 1963), such as impurities, dislocations of the lattice, or very tiny voids, fluxoids are pinned, and a certain decrease in field strength is required before the flux lines move out of the faults. This pinning causes hysteresis losses. The losses depend on the surface conditions and can differ by a factor of several hundred. In order to have low losses, the surface field should be below H_{c1}, which for Nb is between 1000 and 1400 Oe, depending on the impurities. Wipf (1968) has made a study of ac losses in Nb and Nb alloys and has compiled the loss data published by different authors.

For the operation of a superconducting gyro, carried by a superconducting bearing, and spinning in a vacuum, the amount of surface losses is very important. At low temperature the heat transfer is very small, and if the

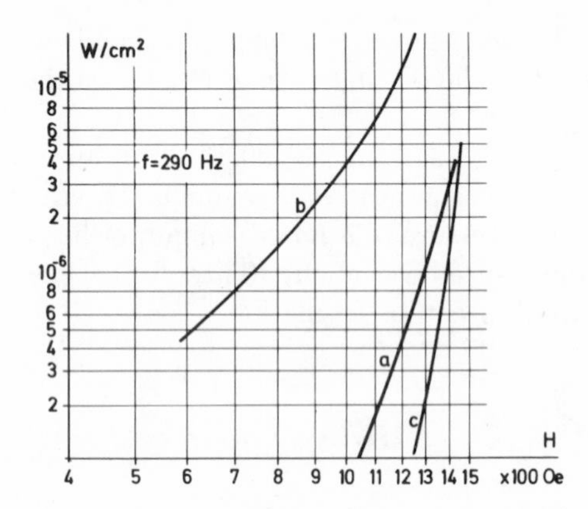

FIG. 11. Losses at 290 Hz for high-quality Nb. Curve a: Nb annealed after manufacturing; curve b: Nb lapped after machining; curve c: Nb annealed and then slightly polished.

losses exceed a given limit the sphere warms up, the material becomes resistive, and the gyro ceases to function. The author, J. Molenda, and R. L. Rhodenizer (Buchhold and Rhodenizer, 1969) made many tests with differently treated Nb. Figure 11 shows a few measurements for cylindrical samples, 5–6 mm diam, of electron beam molten niobium. Because of the gyro application, the applied frequency was 290 Hz. Since the losses are hysteresis losses and proportional to the frequency, losses at 60 Hz would be smaller than the data of Fig. 11 by a factor 60/290 = 0.206. In Fig. 11, curve (a) belongs to a sample which was annealed after manufacturing; curve (b) refers to a surface lapped after machining; sample (c) was machined, then annealed, and finally lapped with great care. The highest losses occur with sample (b), because, with the method of lapping used, a considerable amount of impurities and strain was probably introduced into the surface. The lowest losses belong to sample (c), which after annealing was slightly polished with great care. For most practical applications, the surfaces are much worse than those of Fig. 11, and the losses are considerably higher. In the region of 800–900 Oe and 60 Hz, a value of 10^{-5} W/cm² may be assumed as a conservative value which for most practical applications will not be exceeded.

Figure 12 shows the temperature dependence of the losses for the sample (b) of Fig. 11. The losses tend to infinity if the critical field belonging to a given temperature is approached.

An interesting effect was observed for a machined Nb sample, whose surface was apparently slightly cold-worked. If the sample was cooled down quickly, the losses were small. However, if the cool-down was slow, e.g.,

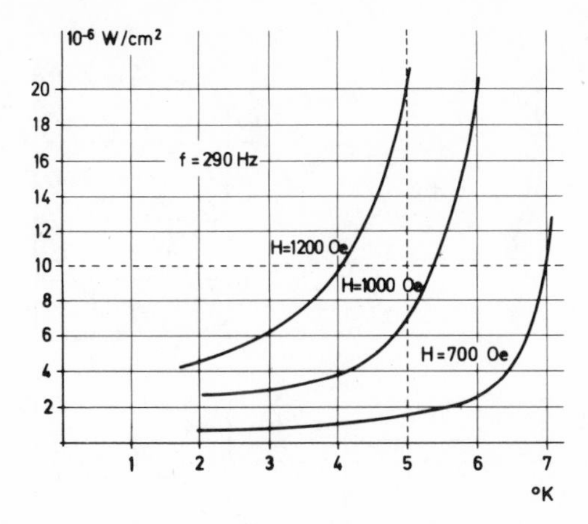

FIG. 12. Losses for Nb as a function of temperature.

overnight cooling, the losses could be more than 100 times higher. After such a sample was warmed up and then cooled down quickly, the losses were small again. After the sample was annealed, it did not show a rate effect. This effect may have practical consequences, e.g., for a superconductive ac cable when the cool-down is slow.

As mentioned before, the losses are hysteresis losses and are proportional to the frequency up to several thousand Hertz, provided the temperature is constant. However, if the inner part of a lossy coil warms up, the losses increase more rapidly than linearly with frequency.

B. LOSSES IN TYPE-II SUPERCONDUCTORS (Bean, 1962, 1964)

Contrary to a Type-I superconductor where the magnetic field and the surface currents penetrate only slightly into the material, flux applied to a hard superconductor (Type II), such as niobium–titanium or niobium–tin, penetrates deeply into the material. Figure 13a shows part of the cross section of a Type-II superconductor with a magnetic induction B at the surface. For a small field, the flux penetration and also the penetration of the surface currents are only appreciable within the penetration depth. Figure 13a shows the current density J, which is perpendicular to B and to the plane of the paper. For a small field, the current density J follows the curve (a) and for a larger field the curve (b). The maximum current density of curve (b) will be called the critical current density J_c of the material. If the flux density on the surface increases further, the widely used phenomenological model originated by Bean assumes that the current

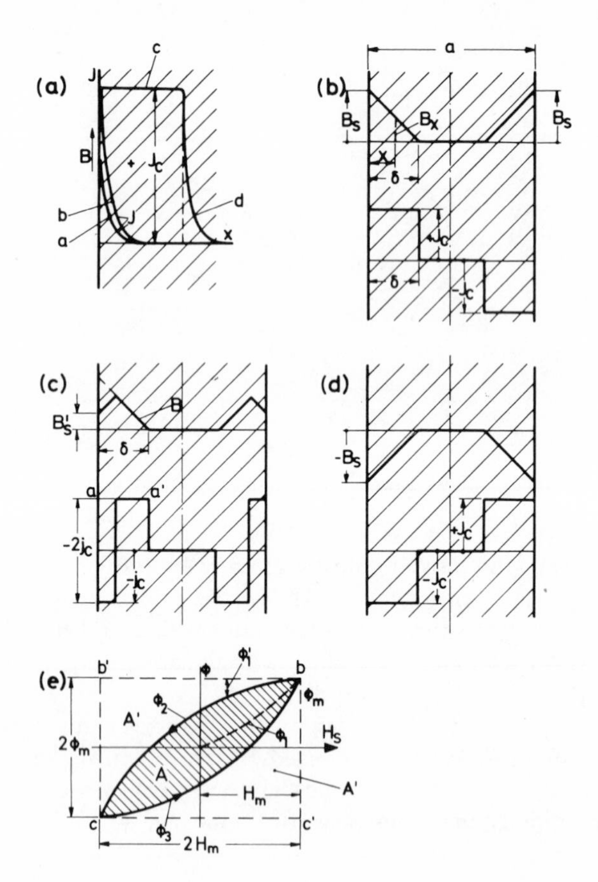

FIG. 13. Loss behavior of Type-II superconductors according to the Bean model. (a) Current density as a function of distance; (b) flux density B and current density J_c; (c) flux density B and current density for decreasing surface field; (d) the same as (b) but with B and J_c negative; (e) hysteresis loop of trapped flux as a function of applied field.

density J_c stays approximately constant and moves into the material (see curve c). A tendency of the current density to exceed J_c is prevented because then the material temporarily becomes slightly resistive. The magnetic field (gauss) on the surface is

$$B_s = \mu_0 \int_0^\infty J \, dx \tag{44}$$

where J is in A/cm and $\mu_0 = 4\pi/10$. Figure 13b shows a slab of material of thickness (a) which is exposed on both sides to the surface field B_s. This causes the current density J_c to move from both sides into the material by the distance δ. The tail (d) of Fig. 13a, which is small in practice, is neglected in the following calculation. If the surface induction is

$$B_s = \mu_0 J_c \delta \tag{45a}$$

the induction B at a distance x from the surface will be

$$B = \mu_0 J_c(\delta - x) \tag{45b}$$

If B_s is reduced to B_s', the new distribution of the flux and current density is as shown in Fig. 13c. Adjacent to the left surface the current density becomes negative, and B decreases. Finally, if $(-B_s)$ is reached (Fig. 13d), the current and flux distribution is similar but of opposite polarity to that shown in Fig. 13b.

For a region of 1 cm² surface area, the penetrated flux is (see Fig. 13b)

$$\phi = B_s\delta/2$$

and, with (45),

$$\phi = B_s^2/2\mu_0 J_c \tag{46a}$$

If we define

$$H_s = J_c\delta = B_s/\mu_0 \tag{46b}$$

then

$$\phi = \mu_0 H_s^2/2J_c \tag{47}$$

where H_s is the magnetic surface field (A/cm).

The losses will be calculated assuming that δ is small compared with the width of the slab. Figure 13e shows the flux ϕ as a function of H_s. Initially, ϕ follows the curve ϕ_1. After the maximum value H_m is reached, the field is decreased and causes the flux to follow the curve ϕ_2. This can be proved by superimposing the current density $-2J_c$ on the current density J_c which extends from $x = 0$ to $x = \delta$ [see line (aa′) in Fig. 13c]. At $(-H_m)$, the flux becomes $(-\phi_m)$. Now the field strength increases from $(-H_m)$ to $(+H_m)$, and the flux follows the quadratic curve ϕ_3. The hatched area A between ϕ_2 and ϕ_3 multiplied by 10^{-8} gives the hysteresis loss in Joules per cycle per cm² surface. The area (A) is obtained from the rectangle (bb′), (cc′) by subtracting twice the area (A′) which is located between (bb′) and ϕ_2. Since ϕ_2 is a quadratic curve,

$$A' = \tfrac{1}{3}(2H_m)(2\phi_m) = \tfrac{4}{3}H_m\phi_m$$

Therefore,

$$A = 4H_m\phi_m - 2(\tfrac{4}{3}H_m\phi_m) = \tfrac{4}{3}H_m\phi_m \tag{48}$$

The hysteresis loss is thus

$$W_h = \tfrac{4}{3}H_m\phi_m 10^{-8}$$

Using Eqs. (46), we find that

$$W_h = (5/12\pi^2 J_c) B_m^3 10^7 = 4.22 (B_m^3/J_c) 10^{-9} \tag{49}$$

again in joules per cycle per square centimeter surface. The above result is only an approximation, since J_c varies with B. The current densities also vary according to the material and are between 10^{+5} and 10^{+6} A/cm^2.

Curves (a) and (b) of Fig. 13a show that for small flux densities the flux and the screening current stay within the London penetration depth. This flux and current move deeper into the material only after a critical surface field and screening current density are reached. Therefore, a better approximation is obtained if in Eq. (49) B_m is replaced by $B_m - \Delta B$, where ΔB is not constant. For Nb–25% Zr and small fields, ΔB is about 500; for $B = 1000$, it rises to about 200 but decreases for larger fields to about 100. Therefore, for larger fields ΔB can be neglected. If $B = 1000$ G and $J_c = 3 \times 10^{+5}$ A/cm^2 a loss $W = 0.72 \times 10^{-3}$ J/cm^2 is obtained which, for $f = 60$ Hz, gives the loss $W_{60} = 0.43 \times 10^{-3}$ W/cm^2. This loss is large compared with the conservative value of 10^{-5} W/cm^2 given in Section III.A for Nb. For higher fields, the losses increase rapidly. Thus the hope of using hard superconductors for ac fields of many thousand gauss can not be realized at this time, since the losses become unacceptably large (Chant *et al.*, 1970).

We shall now analyze the case (Hart and Swartz, 1966) where the thickness (a) of the slab is relatively small and the flux density B' is increased until flux just penetrates to the center as shown in Fig. 14a. If the surface flux density B_s is increased further, the density B within the slab increases also. It is assumed that the current density J_c stays constant. The flux density B will be assumed large enough so that B within the slab can be assumed approximately constant. (B has to stay below the critical value which drives the material resistive.) Figure 14d shows B and the voltage e to produce the current density J_c as a function of time t. It is assumed that even a small change in e near $e = 0$ is sufficient to bring J_c from a negative to a positive value. Figure 14b shows the slab seen in the direction of the magnetic lines, which is perpendicular to the plane of the paper. At a distance x from the center in an element of width dx and length 1 cm a voltage e is induced:

$$e = -\frac{dB}{dx} \times 10^{-8} \tag{50}$$

which produces a loss $eJ_c'\,dx$. Figure 14c shows that the average voltage is $e_m/2$, where e_m is the voltage close to the surface. Therefore, the loss in the

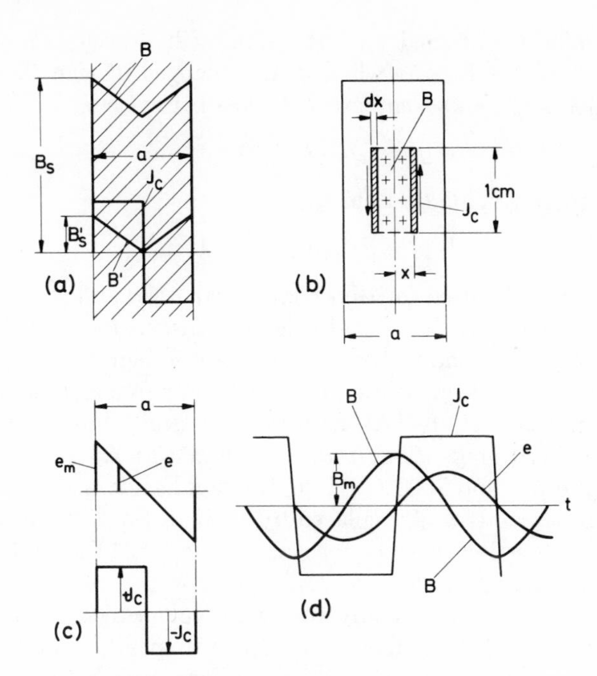

FIG. 14. Flux and current density for a material of thickness a. (a) Flux density B for increasing surface density B_s; (b) induced current density J_c; (c) voltage e and current density; (d) e, J_c, and B as a function of time t.

slab at time t is

$$W_t = \frac{e_m}{2} J_c a \tag{51}$$

Since from Eq. (50)

$$e_m = -\frac{a}{2}\frac{dB}{dx} \times 10^{-8}$$

the loss for one cycle is

$$W_h = \frac{J_c a}{2} \oint e_m \, dt = \frac{J_c a^2}{4} \times 10^{-8} \oint - dB$$

Thus for the first half-cycle where B varies between $+B_m$ and $-B_m$,

$$\frac{W_h}{2} = \frac{J_c a^2}{4} \times 10^{-8} \int_{+B_m}^{-B_m} - dB = \frac{1}{2} J_c a^2 B_m 10^{-8}$$

The same value is obtained for the following half-cycle, where B varies between $-B_m$ and $+B_m$ and where J_c has changed its sign. Therefore, for a slab of thickness a and 1-cm^2 area, the loss per cycle is

$$W_h = J_c B_m a^2 \times 10^{-8} \text{ J/cm}^2 \tag{52}$$

or, referred to unit volume (1 cm^3),

$$W_{h0} = J_c B_m a \times 10^{-8} \text{ J/cm}^3 \tag{53}$$

This equation can be used for thin rectangular wires and is approximately true for round wires. It shows that small dimensions reduce ac losses.

For the hard superconductors, the losses are so high that even at 60 Hz material heats up and quenches. However, lower frequencies are possible. Coils of 5-mil wires of Nb–25% Zr were successfully built to drive power cryotrons in flux pumps. The frequency used was 7.5 Hz, and the flux changed between 0 and 7000 G. Here, the flux has dc and ac components both equal to $B_m/2$. If B_m is replaced by $B_m/2$ in Eq. (53), it can only be used for a rough estimate, since for small values of B the assumptions for its derivation are not fulfilled.

For larger cross sections, many thin wires connected in parallel can be used. However, precautions have to be taken to avoid circulating currents which would lend to extra losses. Within power cryotrons described later, each of the wires connected in parallel is arranged in a bifilar fashion.

To avoid instability caused by flux jumps, it is proposed to built the wires of conductors made of many parallel twisted filaments embedded in a matrix of copper. These conductors may be needed for large coils (e.g., special atom smashers) which have to be charged and discharged at a low frequency.

The theory presented here is over-idealized but has the merit of simplicity and brings out the physical parameters which determine losses. A good review of the subject of ac losses is given by Wipf (1968).

IV. Superconductive DC Generation (Flux Pumps)

If large dc currents need to be supplied to a device immersed in liquid helium, lead-ins with large cross sections are required to reduce Joule heating. This, however, increases the heat flux from room temperature so that again more helium tends to evaporate. It therefore looks attractive to produce larger dc currents at low temperature so that large lead-ins can be avoided. Many such devices, called flux pumps, have been proposed. Only those few which have shown practical results or which are of a special interest will be discussed here.

A. Superconductive Dynamo (Admiraal and Volger, 1962)

A novel idea for producing dc currents with superconductors was proposed by Prof. J. Volger. Figure 15 shows the principle. A superconductive disk (a) of lead or niobium, fixed in space, is connected with the coil L through wires g and g_0. A permanent magnet (b) emitting flux ϕ can rotate around the axis (c). When the device is cooled down to 4.2°K, the disk becomes superconductive except for the spot (f), where the flux penetrates the disk, and the material stays resistive. When the magnet (b) moves, the spot (f) follows, emf's (e) are produced in the disk, and currents flow (dashed lines) which are short-circuited by the superconductor. Outside the disk, the flux lines ϕ cut the wire g and induce a voltage in it. The wire g stays always superconductive. Since the magnetic lines only cut the wire during a part of a revolution, the induced voltage is not constant. Its average value is (T being the time for one revolution)

$$\bar{V} = \phi 10^{-8}/T$$

In terms of the revolution frequency $f = 1/T$,

$$\bar{V} = f\phi 10^{-8} \text{ V} \tag{54}$$

This voltage produces a current I_0 which flows through the coil L and the superconductive part of the disk. The current I_0 increases in the inductance according to

$$\bar{V} = L\frac{dI_0}{dt} \qquad \text{or} \qquad I_0 = \frac{\bar{V}t}{L} \tag{55}$$

Since with increasing current I_0 the coil is charged with the flux LI_0, such a device may be called a flux pump. The voltage $\bar{V}$ in the wire (g) and the

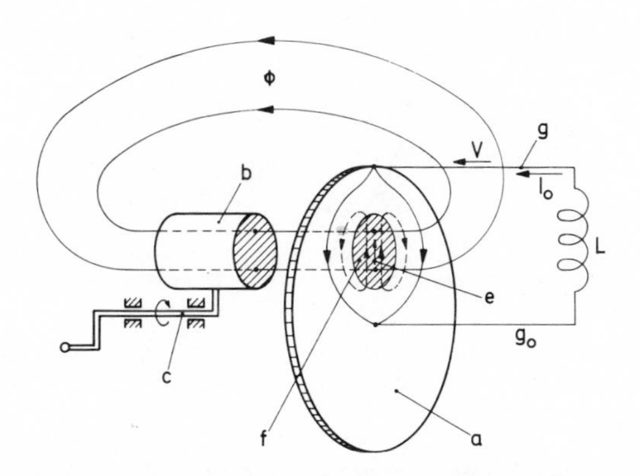

FIG. 15. Superconductive dynamo (Volger's principle).

voltages e in the spot are on the average opposite but do not cancel, since the voltages e are short-circuited.

The voltages e produce losses W. The voltage is given by

$$e = Bvl10^{-8} \text{ V} \tag{56}$$

where B is the flux density, v is the velocity of the spot at its center, and l is the length of an element. With dx being the width of an element, ρ being the resistivity, and D being the thickness of the disk, the loss per element is

$$dW = \frac{e^2D}{\rho l}\,dx = \frac{B^2v^2Dl}{\rho}\,dx\,10^{-16} \qquad \text{or} \qquad W = \frac{B^2v^2D10^{-16}}{\rho}\int l\,dx$$

The area of the spot is $A = \int l\,dx$; therefore,

$$W = \frac{B^2v^2D}{\rho}\,A10^{-16} \text{ W} \tag{57}$$

Figure 16a shows a modification of the principle of Volger's device. Above a segment (a) moves a magnet (b). When the flux of (b) cuts the wire (g), a voltage is again induced with the average value given by Eq. (54). The currents, produced in the resistive spot, are short-circuited. The width of

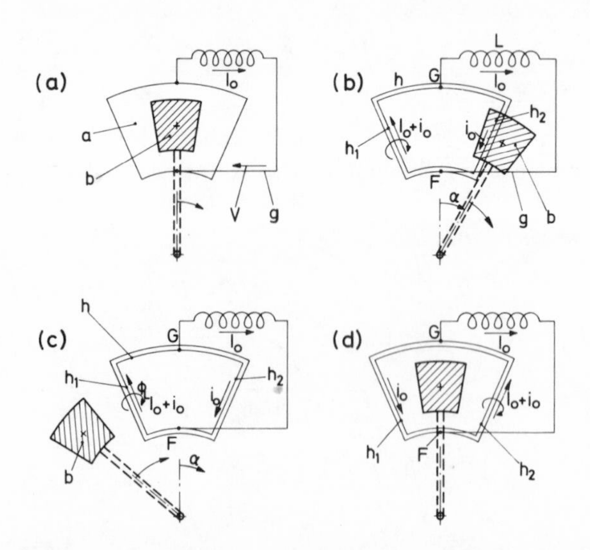

FIG. 16. Modifications of the dynamo principle. (a) Disk of Fig. 15 replaced by segment a; (b) segment replaced by wires h, h_1, and h_2, magnet b above h_2; (c) magnet b is outside of loop; (d) magnet b is at center.

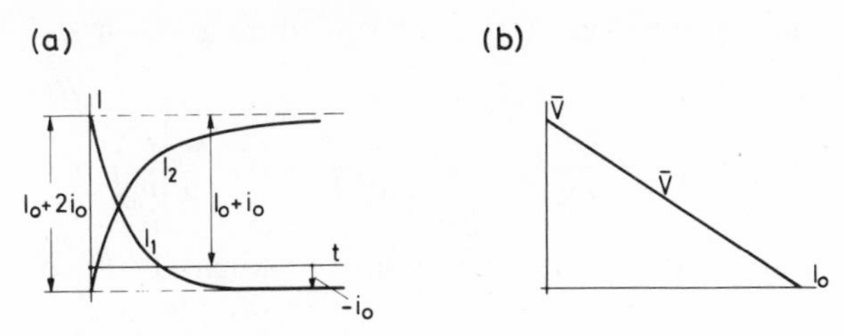

FIG. 17. (a) Current changes in the wire loop of Fig. 16. (b) Mean voltage $\bar{V}$ in the flux pump of Fig. 16 as a function of current I_0.

the segment limits the maximum current I_m, or else the superconductor quenches.

Equations (54) and (57) can be made more exact. For easier understanding, the segment (a) of Fig. 16a is replaced by a superconducting loop made of the wires (h), (h_1), (h_2) (see Figs. 16b–d). In Fig. 16b, the moving magnet (b) has made the wire (h_2) resistive. In it a current

$$i_0 = (Bvl/r)\,10^{-8} \tag{58}$$

is produced, where r is the resistance of (h_2). The sum of i_0 and the coil current I_0 flows in the left wire (h_1), which is superconducting. When the angle α of Fig. 16b increases, flux lines cut the wire (g) and produce the voltage $\bar{V}$. When the magnet (b) arrives in the position of Fig. 16c, both wires (h_1) and (h_2) are superconducting and still have the same current distribution as in Fig. 16b. The current $I_0 + i_0$ is linked with a flux ϕ_L. The flux around (h_2) is small and not shown. It is assumed that the inductance L_h of (h_1) and (h_2) is constant. The effect of the magnet (b) on the inductance of these elements is neglected. When the magnet (b) arrives at (h_1), this wire becomes resistive, and its current I_1 changes exponentially (see Fig. 17a) from $I_0 + i_0$ into $(-i_0)$. At the same time, in (h_2) the current I_2 changes from $(-i_0)$ into $I_0 + i_0$. The decrease of I_1 is equal to the increase of I_2, since during one cycle I_0 is approximately constant, assuming $L_h \ll L$. The currents I_1 and I_2 are

$$I_1 = (I_0 + 2i_0)\exp\left(-\frac{t}{T_0}\right) - i_0, \qquad I_2 = (I_0 + 2i_0)\left[1 - \exp\left(-\frac{t}{T_0}\right)\right] - i_0 \tag{59}$$

where $T_0 = 2L_h/r$. The current changes of I_1 and I_2 require a voltage ΔV at the terminals (FG) which is

$$\Delta V = L_h \frac{dI_2}{dt} \qquad \text{or} \qquad \Delta V = (I_0 + 2i_0)\,\frac{1}{T_0}\exp\left(-\frac{t}{T_0}\right)$$

If T is the time for one cycle and t_0 the time during which (h_1) is resistive, the average voltage is

$$\langle \Delta V \rangle_{av} = \frac{1}{T}(I_0 + 2i_0)\frac{1}{T_0}\int_0^{t_0} \exp\left(-\frac{t}{T_0}\right) dt$$

Since $T_0 \ll t_0$, the value of the integral is approximately T_0. With $1/T = f$, it follows that

$$\langle \Delta V \rangle_{av} = f(I_0 + 2i_0) L_h \tag{60}$$

The voltage ΔV is zero when the magnet (b) moves within the loop (Fig. 16d). However, as the magnet (b) passes (h_2), the same voltage $\langle \Delta V \rangle_{av}$ is produced. Therefore, at the terminals (FG) the entire average voltage is

$$\langle \Delta V \rangle_{av} = 2fL_h(I_0 + 2i_0) \tag{61}$$

$\langle \Delta V \rangle_{av}$ is opposite the voltage $\bar{V} = f\phi 10^{-8}$ induced in (g). Therefore, the net voltage at the terminals of the coil L is only

$$\bar{V}_L = f\phi 10^{-8} - 2fL_h(I_0 + 2i_0), \qquad \bar{V}_L = f[\phi 10^{-8} - 4L_h i_0 - 2L_h I_0] \tag{62}$$

which shows that the voltage $\bar{V}_L$ decreases with increasing current I_0 due to the inductance L_h (Fig. 17b). Equation (62) gives the maximum current I_m if $\bar{V}_L$ is assumed to be zero.

From Eq. (59), the loss W in wire (h_1) can be calculated as $I_1^2 r$. Since (h_1) is only resistive during the time t_0 when the magnet (b) passes over it, the average loss per cycle becomes

$$\begin{aligned} W_1 &= \frac{r}{T}\int_0^{t_0}\left[(I_0 + 2i_0)\exp\left(-\frac{t}{T_0}\right) - i_0\right]^2 dt \\ &= \frac{r}{L}\left[(I_0 + 2i_0)^2\int_0^{t_0}\exp\left(-\frac{2t}{T_0}\right) dt \right. \\ &\qquad \left. - 2i_0(I_0 + 2i_0)\int_0^{t_0}\exp\left(-\frac{t}{T_0}\right) dt + \int_0^{t_0} i_0^2\, dt\right] \end{aligned}$$

Since $T_0 \ll t_0$, the first exponential integral is $T_0/2$ and the second T_0. Thus,

$$W_1 = \frac{r}{T}\left[(I_0 + 2i_0)^2\frac{T_0}{2} - 2i_0(I_0 + 2i_0)T_0\right] + i_0^2 r\frac{t_0}{T}$$

With $T_0 = 2L_h/r$ and some algebra, it follows that

$$W_1 = fI_0^2 L_h[1 - (2i_0/I_0)^2] + i_0^2 r(t_0/T)$$

If $i_0 \ll I_0$,

$$W_1 = f[I_0^2 L_h + i_0^2 r t_0] \tag{63}$$

The total loss for (h_1) and (h_2) is

$$W = 2f[I_0^2L_h + i_0^2rt_0] \tag{64}$$

This shows the great influence of the inductance on the losses. The loss problem for different flux pumps was also treated in a simplified way by Newhouse (1968) in his paper "On Minimizing Flux Pump Heat Dissipation."

The working principles of the pumps of Figs. 15 and 16a are the same as those of Figs. 16b–d. Therefore, Eqs. (62) and (64) can be used to describe the operation of the first two pumps, provided that the various parameters are chosen so as to provide the best possible agreement with test results.

Figure 18 shows an experimental device (Van Houwelingen and Volger, 1968) based on the principle illustrated in Fig. 15. A cylindrical foil (a) of lead or niobium is attached to the iron housing (b). A rotor (c) of iron is shaped so that it has four poles (d). Two coils (f) produce magnetic lines (shown dashed) which penetrate the foil at four resistive spots adjacent to the four poles. If the rotor (c) rotates, the four resistive spots follow, and in the resistive areas voltages are produced which are short-circuited by superconductive material. The moving flux lines cut the leads (h_1) and (h_2) and induce voltages here. The four leads (h_1), belonging to the four poles, are connected in parallel, as are the leads (h_2).

Figure 19 shows the principle of an arrangement proposed by Wipf (1963) which gives higher voltages because it consists of segments connected

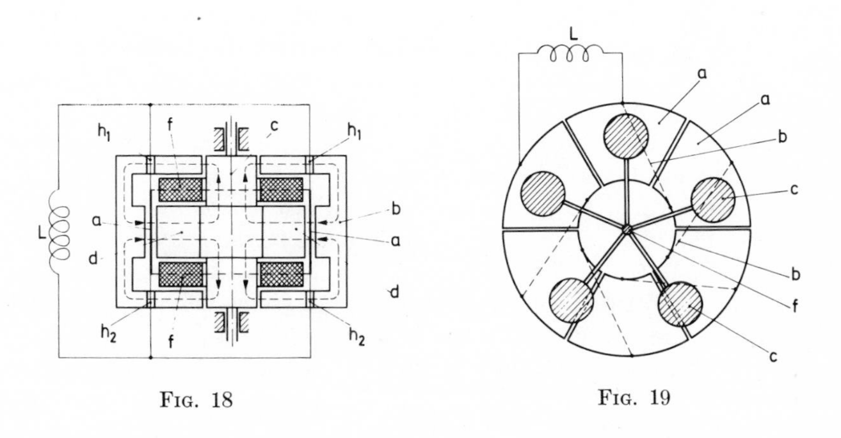

FIG. 18

FIG. 19

FIG. 18. Practical version of a dynamo.
FIG. 19. Modification of dynamo principle for higher voltage.

in series. Five poles (c) are shown which can rotate around the axis (f). The flux from the poles cuts the superconducting wires(b) and induces voltages which add in series to energize the coil L. The number of the poles is smaller than the number of segments; this produces a more even voltage than would be produced if the number of poles and segments were equal.

If the rotation of such a device stops, the current continues as a persistent current. The rotor has to oppose a torque proportional to the flux and the current I_0. If the rotor is turned in the opposite direction, the coil becomes discharged. The flux pumps described so far have relatively high losses, and their efficiencies usually can not exceed 50%. The loss problem has been treated by several authors such as Newhouse (1968), van Houwelingen and Volger (1968), and Wipf (1965).

Flux pumps may be used as test devices to produce large dc currents. Here the efficiency does not matter. One such pump was built by Volger and co-workers which produced about 12,000 A when short-circuited. Such a pump can also be used for very large coils to compensate for any small ohmic losses caused by nonsuperconducting connections. These coils are charged with a conventional power supply and then maintained with a flux pump so that the input leads used for charging can be interrupted. (Such a circuit, used for another type of flux pump, is shown in Fig. 29.)

B. DC Transformer (Giaever, 1965)

A new principle for producing dc without moving magnets or alternating currents was proposed by I. Giaever. Figure 20 shows the principle. The transformer consists of two closely placed thin films of Sn, (a_1) and (a_2). If a steady current I_1 flows through (a_1), fluxoids are formed of which two, (b_1) and (b_2), are shown in Fig. 20. The magnetic field lines (d) of the current I_1 move the fluxoids toward the center of the film, where they cancel each other and disappear. The moving fluxoids produce small currents within their resistive areas which are short-circuited by superconductive regions. The conditions are very similar to the moving flux spots of the devices described in the preceding section. Since fluxoids disappear continuously, flux lines (d) move from the outside to the centers of the loop shown in Figs. 20a and 20b and cut the wire (f_1). This induces a voltage V which can be measured with the voltmeter. The same voltage is produced in the circuit consisting of the wire (f_2) and the resistor R. This produces a current I_2 which also flows through the film (a_2) but in the opposite direction to I_1.

Figure 20c shows the relationship of the voltage V and the magnetizing current I_M. I_M is equal to I_1 if $I_2 = 0$. A minimum current I_M' is needed to produce a voltage. If a secondary current I_2 flows, the primary current

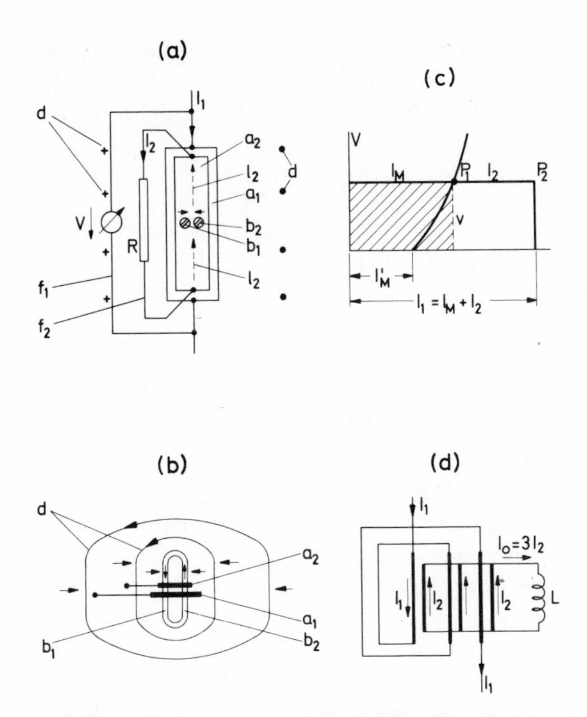

FIG. 20. DC transformer of I. Giaever. (a) Arrangement of transformer; (b) front view; (c) voltage characteristic; (d) arrangement to multiply current I_2 by a factor of three.

increases by the same amount. Thus $I_1 = I_M + I_2$, which is a relation similar to that found for a conventional transformer. However, I_M is a resistive current and causes losses. The magnetizing current of a normal transformer is basically lossless.

Figure 20d shows schematically three units whose primaries are connected in series and whose secondaries are in parallel to obtain a higher secondary current.

The efficiency of a dc transformer is very low. To produce a voltage V, the primary current I_M is needed [point (P_1) in Fig. 20c]. This causes the loss

$$W = VI_M \tag{65}$$

equal to the hatched area in Fig. 20c. If a current I_2 flows in the secondary circuit, an additional power is needed. Therefore, the input power is $V(I_M + I_2)$, and the efficiency is

$$\eta = I_2/(I_M + I_2) \tag{66}$$

η is low, since I_M cannot be made sufficiently small at present. So far, dc transformers have only been built for very small currents and voltages. It appears unlikely that this very elegant principle can be developed for larger powers.

C. Superconductive Rectification Using Self-Controlled Cryotrons (Olsen, 1958, 1959)

By use of a transformer and rectifiers, small ac currents can be changed into large dc currents. Olsen has pointed out that it appears attractive to build such devices for low temperatures. In that case, only small ac currents have to be conducted to the low-temperature transformer (Fig. 21a). This allows the use of thin wires which keeps the heat influx to the low-temperature part of the system small.

To keep Joule heating losses small, all windings and conductors are superconducting, and a specially connected type of cryotron, (b), is used which acts as a rectifier. In this device, the cryotron gate (b) is connected in series with the control coil (c) and is also subjected to a constant bias field H_0 by a second coil (d).

Figure 21b shows the resistance characteristic of such a cryotron rectifier. Gate resistance starts when the net applied field exceeds H_c. The bias field H_0 is adjusted to a value just below H_c so that a small additional field ΔH_1, produced by the current i_0 of the control coil (c_1), makes the cryotron element (b_1) resistive. During the half-cycle shown in Fig. 21a, a large current I_2 flows in the cryotron (b_2) which is superconductive, since H_0 and ΔH_2 are opposite. A voltage $2V$ exists across the resistive cryotron

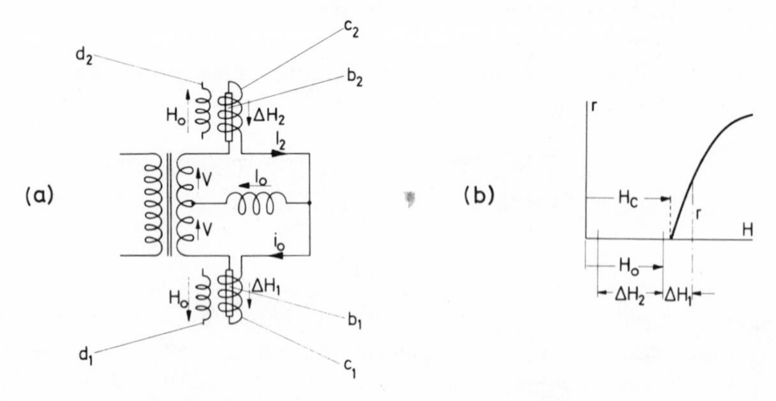

Fig. 21. Rectifier of J. L. Olsen. (a) Connection Scheme; (b) resistance characteristic of cryotron.

(b_1) and produces i_0. During the following half-cycle, (b_1) is superconductive and (b_2) resistive. The loss is

$$W = 2Vi_0 \tag{67}$$

To keep W low, i_0 should be as small as possible. H_0 must be close to H_c, and the cryotron gate resistance characteristic must be steep. A steep characteristic can be obtained if the cryotron (b) consists of a wire or a tape of lead which is aligned in the direction of the field. Niobium does not have a steep enough slope and also requires a much higher field to make it resistive.

It seems that this system can be built only for small powers and moderate currents, since it is restricted to cryotrons of lead. Furthermore, if all losses are considered, it is doubtful that high efficiency can be obtained. This will become more clear in the next section, where the various loss mechanisms are discussed in detail.

It is desirable that coil L should not only be charged but also discharged with good efficiency. This cannot be done with the cryotron elements of Fig. 21a, even if the primary voltage is reversed. To discharge a coil with good efficiency, controlled cryotron elements are needed which in their operation resemble grid-controlled rectifiers. Such a system, which operates with high efficiency, is discussed in the following section.

D. Superconductive Rectification Using Externally Controlled Cryotrons and Saturable Reactors (Buchhold, 1964b)

We now describe a system, developed by the author, whose efficiency approaches the theoretical maximum and which allows a coil to be charged and discharged efficiently. At normal temperature, a rectification system using grid-controlled rectifiers fulfills these requirements. For the present system, it was necessary to develop a superconductive controlled rectifier, a so-called power cryotron, which can be made resistive with a magnetic field when the current tends to reverse. Since the properties of a cryotron do not depend on current polarity, a polarity sensing device, in the form of a saturable reactor, was used. Such a reactor keeps the cryotron current before switching very small, so that the loss caused by switching is practically zero.

To begin with, for easier understanding, an ideal lossless controlled rectifier will be assumed. Figure 22a shows the scheme. The rectifiers (a_1), (a_2) can be fired with pulses coming from the phase shifter p. The transformer T is supplied with a rectangular voltage waveform and gen-

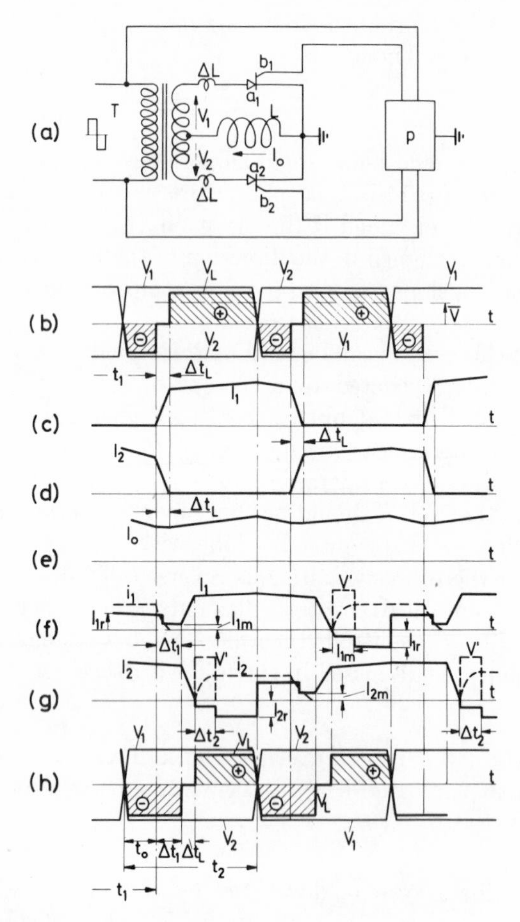

FIG. 22. Rectification with controlled rectifiers. (a) Schematic assuming ideal controlled rectifiers; (b)–(e) time dependence of voltages and currents assuming ideal controlled rectifiers; (f)–(h) current and voltage dependence using cryotrons and saturable reactors.

erates the secondary voltages V_1, V_2. The arrows are drawn in the positive direction. In Figs. 22c and 22d, up to the time t_1 the current I_1 is zero, I_2 is nonzero, and the rectifier (a_2) is conductive, whereas (a_1) is not. Even when the voltage V_2 is negative (see Fig. 22b), a positive current I_2 is maintained by the large inductance L. At time t_1, the phase shifter p causes (a_1) to become conductive, and, since the voltage V_1 is positive, the current shifts from the lower to the upper circuit. During the time Δt_L, the current I_1 increases to its full value, and I_2 decreases to zero. If the leakage reactance per branch of the circuit is ΔL, this time interval is given by the expression

$$\Delta t_{\mathrm{L}} = I_0 \frac{\Delta L}{V} \tag{68}$$

During this time interval, the voltage V_{L} across the coil is zero. I_0, which is the sum of I_1 and I_2, is constant during Δt_{L}. The voltage V_{L} across the coil is shown by the heavy lines in Fig. 22b which surround the hatched areas. The average dc voltage $\bar{V}$ is seen to be positive (in Fig. 22b). The influence of the positive and negative parts of V_{L} shows up in the different slopes of I_1, I_2, and I_0. The dc current I_0 increase with time is given by

$$I_0 = \frac{\bar{V}t}{L} \cdot \qquad \text{or} \qquad \bar{V}I_0 = \frac{I_0^2 L}{2} \frac{2}{t} = \frac{2E}{t}$$

where E is the coil energy. The maximum pumping power is

$$P = VI_0 = 2E/t$$

Thus,

$$t = 2E/P \tag{69}$$

which shows that the time t required to charge a coil is determined by the power P. It is therefore of little value when, as in so many flux pump publications, only the current but not the voltage and especially the power are given.

If the phase-shifting pulse occurs earlier, the positive area of Fig. 22b becomes larger, the negative area becomes smaller, and $\bar{V}$ increases. If the phase shift is delayed, $\bar{V}$ can be made negative, the coil discharges, and the coil energy is transferred to the primary side of the transformer.

Although superconductive controlled rectifiers are not available, they can to a certain degree be replaced by power cryotrons. Such a cryotron may consist of a long thin superconductive tape (a), e.g., niobium, which is folded in a bifilar fashion (Fig. 23b), so as to minimize the inductance and volume. Figure 23a shows the cryotron seen in the direction of the arrow (A) in Fig. 23b. A coil (b) surrounds the tape. An iron lamination (c) is used to keep the reluctance of the magnetic path small. The coil and the lamination are not shown in Fig. 23b. When the coil (b) is excited, a magnetic flux penetrates the tape and makes it resistive. The length of the tape determines its resistance, and the width controls the current capacity. It was found that the best results were obtained when the tape was formed of many parallel insulated 5-mil wires. Commercially obtained Nb wires gave undependable performance which depended on small impurities that could not be controlled. Therefore, a Nb wire was selected which was

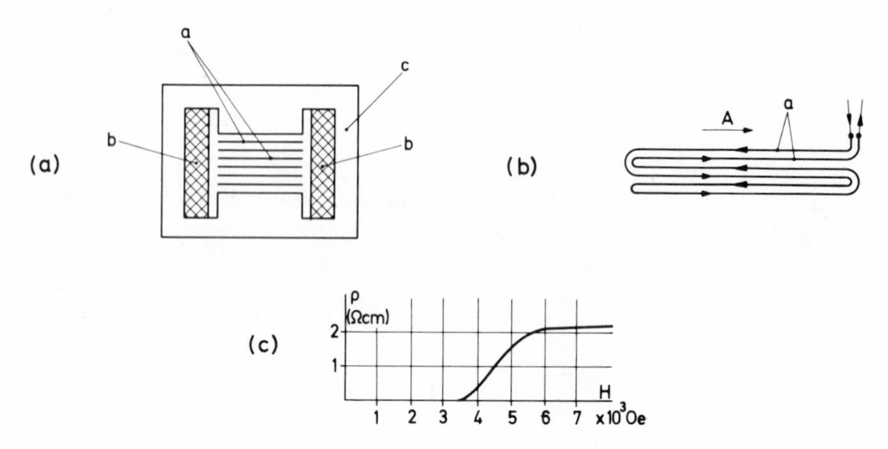

FIG. 23. A power cryotron. (a) Front view of cryotron; (b) side view of cryotron; (c) resistance characteristic of a cryotron.

slightly alloyed so that small impurities did not matter. Figure 23c shows the resistance characteristic of a Nb wire slightly alloyed with tungsten. Since fields of the order of $H = 6000$ Oe are necessary to drive this wire normal, coil (b) has to be made of Nb–Zr or Nb–Ti.

Figure 24a shows the actual flux pump circuit in which the hypothetical controlled rectifiers of Fig. 22a are replaced by cryotrons (a_1), (a_2) and their control coils (b_1), (b_2). At the time t_1, the current I_1 increases and I_2 decreases (Figs. 22c and 22d). At t_1, cryotron (a_1) should become superconducting; however, cryotron (a_2) should stay superconducting during Δt_L but then be made resistive. These changes are obtained by changing the currents through the cryotron coils (b_1) and (b_2). Unfortunately, these coils have inductance so that any change in the current through them requires time. During this time, Joule heating occurs in the cryotrons which tends to keep (a_1) resistive. Therefore, controlled cryotrons alone are not sufficient to provide an operating flux pump.

To avoid these difficulties, saturable reactors (c_1), (c_2) are used in series with the cryotrons (a_1), (a_2) as shown in Fig. 24a. The magnetizing curves $\phi = f(I)$ for both reactors are shown in Fig. 24b. Each reactor has a primary winding (e) and a so-called retarding winding (f). (f) is in series with the cryotron control winding (b).

Before the time t_1, the cryotron (a_1) is resistive because its coil (b_1) is excited with the current i_1 which is produced by the control unit (u) and which also flows through the retarding winding (f_1). Figure 24a shows the currents and voltages for this condition. Because V_1 is positive, a small current I_{1r} flows through the resistive cryotron (a_1) and the winding (d_1).

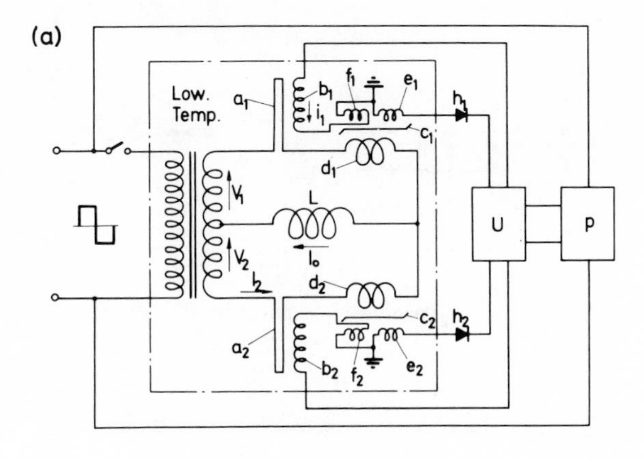

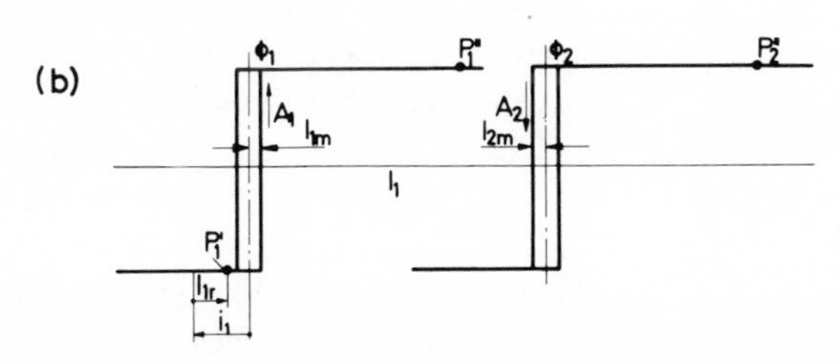

FIG. 24. The author's flux pump using power cryotrons and saturable reactors. (a) Schematic; (b) flux characteristic of saturable reactors.

This current is small compared with I_1 and for matters of clarity is very much exaggerated in Fig. 22f. The ampere turns produced by i_1 in the winding (f_1) are somewhat larger than the opposing ampere turns due to I_{1r} in the winding (d_1) (Fig. 22f). The reactor (c_1) is therefore subjected to a net field just sufficient to keep it in a state of slight negative saturation [see point (P_1') in Fig. 24b]. The large current I_2 keeps its reactor in positive saturation [see point (P_2'')]. At time t_1, the phase shifter (p) together with the control unit (u) (which mainly contains amplifiers) causes the coil (b_1) to be de-energized, and the current i_1 in (b_1) and (f_1) decreases (see Fig. 22f). After a slight decrease of i_1, the reactor comes out of negative saturation, and a voltage equal to $2V_1$ appears across winding (d_1). The flux ϕ_1 increases and moves along the characteristic [arrow (A_1)] up to positive saturation and then to point (P_1''). During the time of the flux increase (see Fig. 24b), the current I_1 is limited to the magnetizing current I_{1m} which is small enough for the Joule heating in cryotron (a_1) to be insignificant. The time Δt_1 has to be larger than the time needed for the

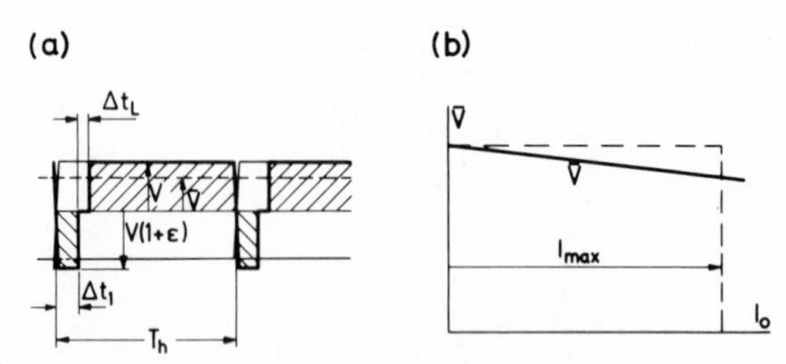

FIG. 25. (a) Diagram to determine maximum dc voltage $\bar{V}$. (b) Voltage $\bar{V}$ as a function of I_0.

current i_1 to decrease to zero (see Fig. 22f). A rectifier (h_1) prevents the voltage, produced in the winding (e_1), from affecting the control unit (u).

After the period Δt_1, the current I_1 increases to its full value [point (P_1'')], and I_2 decreases from (P_2') (Fig. 24b) to the left until the small magnetizing current ($-I_{2m}$) is reached. The core leaves positive saturation and moves on the left side of the magnetizing curve [arrow (A_2)] until negative saturation is reached. During the time Δt_2 of this change, a voltage V' is produced in the winding (e_1) which causes the control unit (u) to energize control winding (b_2). The time Δt_2 is sufficient for the current i_2 to reach its end value (see Fig. 22g). After Δt_2, the voltage $2V$ occurs across the resistive cryotron (a_2) which carries current I_{2r}. At time t_2, the voltages V_1 and V_2 reverse and I_{2r} also reverses and becomes positive. The period Δt_1 causes the negative part of the V_L curve (Fig. 22h) to become larger and the positive part to become smaller.

The combination of power cryotron and saturable reactor works well. Using a phase shifter, any voltage V between $+V_{max}$ and $-\bar{V}_{max}$ can be obtained. A persistent current continues to flow when the primary winding of the transformer is switched off.

The highest average dc voltage $\bar{V}$ is obtained if the phase shifter (p) operates as early as possible, that is, when the time t_0 of Fig. 22h is zero. Because of the delay time Δt_1, the voltage V_L still has a negative portion (see Fig. 25a). If the coil voltage V_L is positive and equal to V, it is smaller than the no-load voltage V_0 of the power supply due to its inner voltage drop. However, if the coil voltage is negative, power is delivered into the power supply, and the coil voltage becomes larger by ϵV and thus can be expressed as $-V(1 + \epsilon)$. (This increase is not shown in Fig. 22.) If the time of one half-cycle is T_h, the average dc voltage $\bar{V}$ is

$$\bar{V} = \frac{V[T_h - (\Delta t_1 + \Delta t_L) - (1 + \epsilon)\Delta t_1]}{T_h}$$

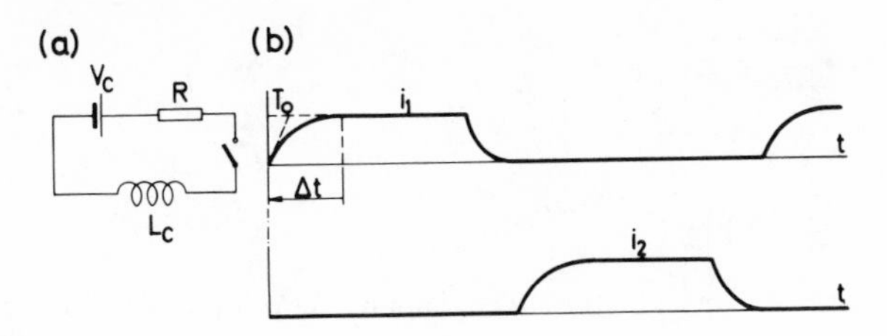

FIG. 26. (a) Cryotron control coil charging circuit. (b) Time dependence of coil currents.

with $\gamma_s = \Delta t_1/T_h$, $\gamma_L = \Delta t_L/T_h$,

$$\bar{V} = V[1 - \gamma_s(2 + \epsilon) - \gamma_L] \tag{70}$$

This equations hows that γ_L and expecially γ_s should be made as small as possible. Since Δt_1 cannot be reduced below certain limits, $T_h = 1/2f$ has to be made as large as possible Therefore, a frequency in the region of 7.5 to 10 Hz was chosen. With this low frequency, the iron loss of the transformer and the hysteresis losses of the cryotrons are kept low. Since Δt_L and γ_L are proportional to the current I_0, the voltage $\bar{V}$ suffers a slight decrease ($\sim$10%) with increasing current (Fig. 25b). The maximum current I_{max} is reached when the cryotrons quench.

We shall now study the room temperature losses associated with the cryotron control coils. It is assumed that each coil has inductance L_c in a circuit with resistance R (Fig. 26a). The time constant of such a circuit is

$$T_0 = L_c/R$$

It is assumed that the coil currents i_1, i_2 reach their final value in about $\Delta t = \sim 3T_0$. Thus,

$$T_0 = \frac{\Delta t}{3} = \frac{L_c}{R} = \frac{\frac{1}{2}L_c i^2}{\frac{1}{2}R i^2}$$

Since the coil energy is $E_c = \frac{1}{2}L_c i^2$ and the resistive loss per cryotron control which carries current for half of every cycle is $W_c = \frac{1}{2}R i^2$, we may write

$$W_c = \frac{3E_c}{\Delta t} \tag{71}$$

This shows that the cryotron loss is inversely proportional to Δt. For a typical flux pump, $f = 7.5$ Hz, $\Delta t = 9 \times 10^{-3}$ sec, $E_c = 1.5$ J, which gives

$$W_c = (3 \times 1.5)/(9 \times 10^{-3}) = 500 \text{ W}$$

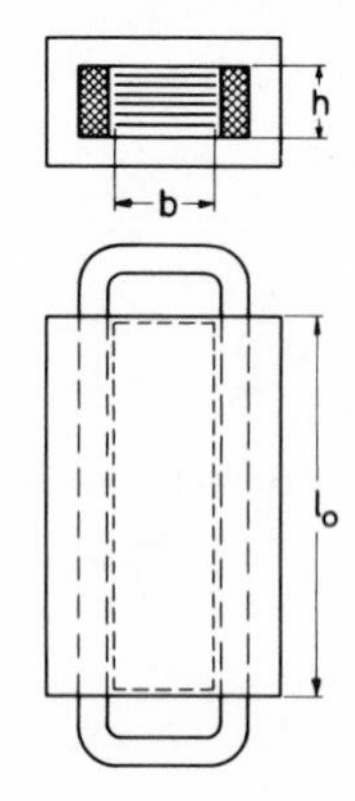

FIG. 27. Simplified cryotron used to calculate losses.

Both cryotrons together have a loss of 1000 W, which is high even though it is produced at normal temperature. A scheme was therefore used in which the energy of the interrupted coil is stored in a capacitor which a short time later discharges itself into the coil of the other cryotron. With such a circuit, the outside losses were reduced considerably. The low-temperature losses will now be studied. A cryotron has the following losses:

W_1, the hysteresis loss of the cryotron coil made of Nb–Zr or Nb–Ti;
W_2, the hysteresis loss of the cryotron tape made of slightly alloyed Nb wires; and
W_3, the ohmic loss of the cryotron tape.

For a given flux density in the cryotron (e.g., $B = 6000$ G), the loss W_1 is proportional to a constant K_1, to the frequency f, to the height h, and to the length l_0 of the lamination (see Fig. 27), i.e.,

$$W_1 = K_1 f l_0 h \tag{72}$$

Because the width d of the coil depends on the flux density, which is assumed to be given, it does not enter Eq. (72). The volume of the folded cryotron tape is

$$Q_0 = hbl_0 \tag{73}$$

which includes cooling channels and insulation. The active volume Q of the cryotron is

$$Q = Al \tag{74}$$

where A is the area of the parallel connected wires and l the entire length of the folded cryotron. Defining γ as the space factor,

$$\gamma = Q/Q_0 \tag{75}$$

we find that Eqs. (72), (74), and (75) give

$$W_1 = K_1 flA/\gamma b \tag{76}$$

The hysteresis loss W_2 in the cryotron is proportional to a constant K_2, the volume $Q = Al$, and the frequency f; thus,

$$W_2 = K_2 flA \tag{77}$$

Thc ohmic loss W_3 of each cryotron gate is caused by the voltage $2V$ which exists across the cryotron for half of each period. With gate resistivity ρ,

$$W_3 = (2V)^2 A/2l\rho = 2V^2 A/l\rho \tag{78}$$

The entire loss is

$$W = W_1 + W_2 + W_3 = [(K_1/f\gamma) + K_2]flA + (2V^2/l\rho)A \tag{79}$$

The first term is the entire hysteresis loss and is proportional to the tape length l. The second term is the ohmic loss and is proportional to $1/l$. If l is the variable, Eq. (79) has a minimum if the hysteresis loss $W_1 + W_2$ is equal to the ohmic loss W_3. Thus,

$$[(K_1/b\gamma) + K_2]fl = 2V^2/l\rho \tag{80a}$$

or

$$l_{\min} = 2^{1/2}V/\{[(K_1/b\gamma) + K_2]\rho f\}^{1/2} \tag{80b}$$

which is the tape length for minimum loss. Equation (80a) inserted in Eq. (79) yields

$$W_{\min} = 2 \times 2^{1/2} VA\{[(K_1/\gamma b) + K_2](f/\rho)\}^{1/2} \tag{81}$$

The full cryotron current is I_0 and has density J in the cryotron gate wires. Therefore,

$$A = I_0/J \tag{82a}$$

The maximum average dc voltage $\bar{V}$ of practical cryotrons is approximately

$$\bar{V} = \tfrac{2}{3}V \tag{82b}$$

and the output power is

$$P = \bar{V}I_0 \tag{82c}$$

Usiug Eq. (82) and remembering that a flux pump has two cryotrons, Eq. (81) becomes

$$\frac{W}{P} = \frac{8.4}{J\rho^{1/2}}\left[\left(\frac{K_1}{b\gamma} + K_2\right)f\right]^{1/2} \tag{83}$$

From Eqs. (82b) and (80b), the optimum tape length is

$$l_{\min} = 2.1\bar{V}/\{[(K_1/\gamma b) + K_2](f(\rho)\}^{1/2} \tag{84}$$

To obtain low losses, $J\rho^{1/2}$ and the space factor γ should be large, whereas

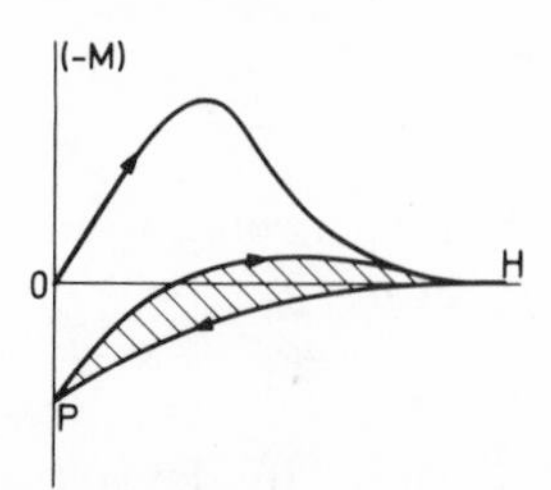

FIG. 28. Magnetization of cryotron wires.

the hysteresis factors K_1 and K_2 and the frequency f should be small. A cryotron with a large width b is advantageous. However, b is often determined by the maximum current and other design considerations. K_1 depends not only on the material but also on the geometry. If the cryotron is built as in Fig. 23a, the coil is exposed to a smaller flux density than the tape and has smaller losses. The hysteresis factor K_2 depends on the hysteresis loop of the cryotron wire. Figure 28 shows the negative magnetic moment $(-M)$ of the wire as a function of H. The shaded area determines the loss K_2, which should be as small as possible. After H has returned to zero [point (P)], the cryotron current I_0 starts to flow and should reduce the trapped flux as little as possible, since otherwise the hysteresis loss increases. Should the current cause the trapped flux to disappear suddenly, losses and a heat pulse would be produced and the current capacity reduced- Since a low frequency reduces the loss, $f = 7.5$ Hz was selected. A further decrease of f increases the dimensions of the transformer too much.

The transformer and the saturable reactors also have losses; these, however, are much smaller than the cryotron losses.

A number of flux pumps of the type described have been manufactured and delivered to customers where they performed satisfactorily. The power outputs were of the order of 50–60 W, and the rated currents ranged from several hundred up to 1800 A. The best overall efficiency obtained at maximum power output was 97.8%. If needed, flux pumps of 10,000 A and above can be built. One manufactured flux pump had its losses distributed as shown in Table I.

TABLE I

W_1, hysteresis loss in two cryotron coils	25%
W_2, hysteresis loss in two cryotron tapes	13%
W_3, ohmic loss in two cryotrons	50%
Iron loss in transformer	8%
Current loss in transformer	2%
Iron loss in two saturable reactors	2%
	100%

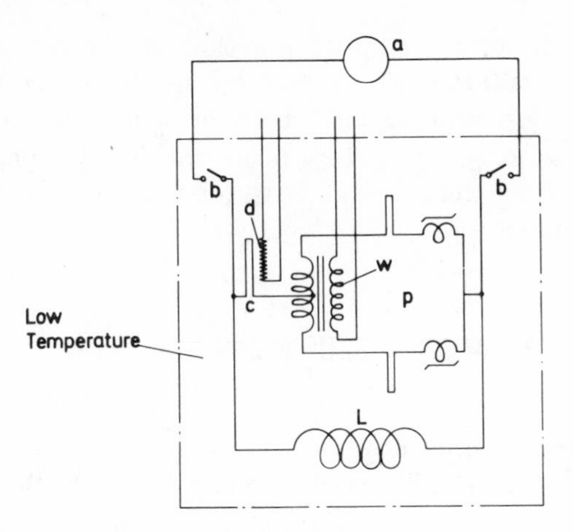

FIG. 29. Scheme of a flux pump which compensates for the current decay caused by nonsuperconducting joints in a large magnet.

According to the theory, Eq. (80a), the hysteresis losses $W_1 + W_2$ of the cryotron should be equal to the ohmic loss W_3 which, according to Table I, is not quite fulfilled. However, the loss minimum is flat, so that moderate deviations change the overall cryotron loss only a little. The cryotrons provide the main contribution (88%) to the entire loss. The author believes that efficiencies up to 99% are possible if more research is done in the field of power cryotrons and its materials.

The question arises whether Nb–Zr or Nb–Ti could be used as gate materials, since these have resistivities an order of magnitude larger than that of Nb used at present. However, to make Nb–Zr or Nb–Ti resistive requires a larger magnetic field which causes correspondingly large hysteresis losses. These materials can be made resistive with a heater. (Their transition temperature is about 10°K.) However, this produces additional losses which reduce the efficiency. Also, the switching frequency has to be kept much lower than 7.5Hz, which requires the use of larger transformers.

To control very large coils, pumps in the kilowatt region are needed which are not yet available. However, the following hybrid arrangement is possible with presently available components (Fig. 29). A large coil L is energized by a dc generator (a). The contacts (b), being in the cryostat, are closed. Parallel to the coil L is a flux pump (p) in series with a cryotron (c) which can be controlled by a heater. Because this cryotron does not have to operate quickly, it can be designed to have small losses. During charging of the coil L, the cryotron (c) is resistive and the pump (p) not operating. After completion of the charging, the cryotron (c) is made superconducting, and the current coming from the generator (a) is reduced to zero. The coil current I_0 must flow through the superconductive flux pump (p). The contacts (b) are now opened to reduce the heat influx. Finally, the primary winding (w) of the flux pump is energized, and the dc voltage produced is adjusted by phase shifting, to compensate for the voltage drop of the nonsuperconductive joints of the coil L. The coil current can be regulated slowly.

V. Magnetically Suspended Trains

For safety and maintenance reasons, the utmost speed limit for trains supported by wheels on rails may be around 200 mph. Beyond this speed, other concepts are needed which avoid wheels and rails and suspend the train above the track with air cushions or magnetic forces. Such trains would be in demand between major urban areas 300–500 miles apart, where the high speed of airplanes cannot be fully utilized because of the loss of time involved in travel to and from the airport.

One proposed suspension method is the air-cushion system in which a flow of slightly compressed air underneath the vehicle causes lift. Trains of this type may be able to travel with speeds between 200–350 mph. The needed ventilator for the airflow produces much noise. The air-cushion principle is already well established, and in the near future test vehicles will be available showing the possibillty of such high speeds.

Other proposals use magnetic forces which do not produce noise for suspension. These ideas are still in the study phase and are not yet developed as far as the air-cushion vehicle.

One type of suspension uses the attractive force of conventional electromagnets. At a certain distance above the ground, a beam of laminated iron is mounted along the track. The electromagnets of the vehicle are underneath this beam separated by a gap of about 1 in. The attractive forces of the magnets suspend the vehicle. Such a system is not stable, so that a control system is needed. Similar means are required for lateral guidance.

Another proposal utilizes the repulsive force of permanent magnets of opposite polarity. Permanent ferrite magnets are mounted along the track, and the vehicle carries permanent magnets of opposite polarity. For the side guidance, a similar arrangement is required. Since such a system is unstable with respect to horizontal displacements, additional means for lateral guidance are required.

Another system uses superconductive coils on the train which produce such strong magnetic fields that a much larger clearance of several inches—compared with about 1 in. for the previous systems—is possible between

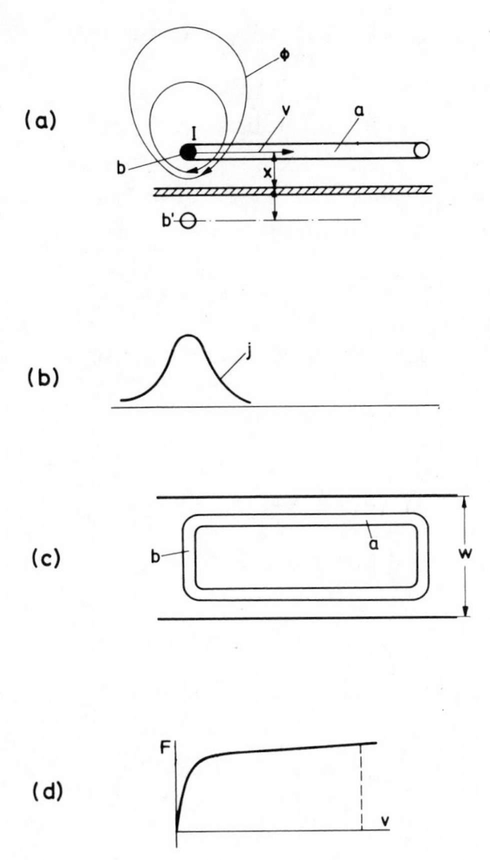

FIG. 30. Principle of magnetic train support system. (a) Flux around current carrying coil moving over conducting sheet; (b) current density; (c) top view of coil and sheet; (d) lift as a function of speed.

track and vehicle. Only superconductive systems will be discussed in the following using an approximate treatment.

In Fig. 30a, a superconductive coil (a), carrying a current I ampere turns, is moving with velocity v parallel to the surface of a conductive sheet (about 1 cm thick) in which currents are produced (Coffey *et al.*, 1969). With large speeds, the flux penetrates only into a thin layer near the surface, and the induced currents are surface currents. Figure 30a shows that, for a very small penetration depth, the flux ϕ of the conductor (b) is compressed and produces a magnetic pressure against the surface and a corresponding upward force on the conductor (b). This force is equal to that between the current I in (b) and its image current (b′). If the conductor is a distance x from the surface, the force F on the conductor per centimeter length is

$$F = 2.04 \frac{I^2}{2x} 10^{-8} \quad \text{kg/cm}^{-1} \tag{85}$$

If j is the current per centimeter width on the sheet surface (Fig. 30b), the ohmic loss per square centimeter of surface is

$$L_0 = j^2/\kappa\delta \tag{86}$$

where κ is the conductivity and δ the penetration depth. Should the surface be exposed to a sinusoidal flux of frequency f, the penetration depth would be

$$\delta = (\pi\kappa\mu f)^{-1/2} \tag{87}$$

where μ is the permeability. However, the surface flux of Fig. 30a is not sinusoidal. To obtain the penetration depth, an effective frequency f has to be introduced which depends on the coil geometry and on the fact that the sheet has a finite width (w) (see Fig. 30c). Certainly f will depend on the velocity v. Therefore, δ can be written as

$$\delta = k_1/(\kappa v)^{1/2} \tag{88}$$

where k_1 is a constant which can be determined by experiment. Using Eq. (88), Eq. (86) becomes

$$L_0 = (j^2/k_1)(v/\kappa)^{1/2} \tag{89}$$

The loss in the track is proportional to $v^{1/2}$. The surface flux density B (G) is connected with j by the formula

$$B = \frac{4\pi}{10} j = 1.256j \tag{90}$$

With good approximation, the pressure force on the surface is

$$p_0 = (B/5000)^2 \quad \text{kg/cm}^{-2} \tag{91}$$

Using Eq. (90),

$$p_0 = (1.256/5000)^2 j^2 = \sim(j/4000)^2$$

Substituting for j from Eq. (89), we obtain

$$p_0 = (1/4000)^2 L_0 k_1 (\kappa/v)^{1/2} = 6.25 \times 10^{-8} L_0 k_1 (\kappa/v)^{1/2} \tag{92}$$

The complete lift force F is obtained if L_0 is replaced by the entire surface loss L in Eq. (92); thus,

$$F = 6.25 L k_1 (\kappa/v)^{1/2} \times 10^{-8} \tag{93}$$

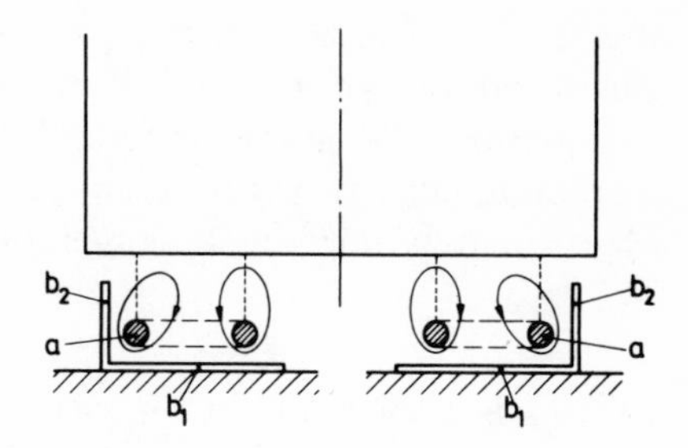

FIG. 31. Continuous sheet support system.

This shows that the lift is proportional to the loss L. This loss causes a drag D on the vehicle given by

$$Dv = L/k_2 \tag{94}$$

where k_2 depends on the units used in the analysis. With Eq. (93),

$$Dv = \frac{F}{6.25 \times 10^{-8}} \left(\frac{v}{\kappa}\right)^{1/2} \frac{1}{k_1 k_2}$$

Thus,

$$\frac{D}{F} = \frac{1.67 \times 10^7}{k_1 k_2} \frac{1}{(\kappa v)^{1/2}} = \frac{C}{(\kappa v)^{1/2}} \tag{95}$$

where C is another constant. Equation (95) shows that for a given lift F the drag D is inversely proportional to the square root of v. Equation (95) is valid only if v is so large that the penetration depth is small compared to the sheet thickness. For v becoming smaller, the penetration depth increases and F decreases until the flux penetrates the sheet completely and F vanishes as v becomes zero. Figure 30d shows the lift as a function of speed assuming x constant. Because of the weight W of the vehicle, a minimum speed is necessary to produce sufficient lift to compensate for this weight. Therefore, just like an airplane, the vehicle needs wheels to start. Figure 31 shows schematically the lower part of a vehicle to which two superconductive coils (a) are attached. The coils are contained in cryostats which are not shown. The cross section of the conductive sheets (b_1), (b_2) forms two right angles. Since the magnetic lines are not only compressed against the lower part (b_1) but also against the vertical part (b_2), both sides of the vehicle are subjected to horizontal forces which cancel each other. However, if the vehicle is moved sideways, the side forces are different, and the resultant force tends to center the vehicle. Therefore, the system is stable.

Even with the vehicle centered, the two equal but opposing side forces produce losses in the vertical parts of the sheets. For a speed of 250 mph, the overall ratio of drag to lift may be of the order of 10%.

Equation (85) gives the force F for a straight conductor moving above a ground plane. It will be assumed that a similar expression holds, at least approximately, for a rectangular coil; thus,

$$F = kI^2/2x \tag{96}$$

where k is a constant. In the vicinity of the equilibrium suspension height x (for which F is equal to the weight W), the restoring force per unit displacement is

$$c = -\frac{\Delta F}{\Delta x} = \frac{kI^2}{2x^2} = \frac{F}{x} \tag{97}$$

Since the vehicle has the mass $m = W/g = F/g$, where g is the acceleration of gravity, its vertical oscillations have a frequency of approximately

$$f = \frac{1}{2\pi}\left(\frac{c}{m}\right)^{1/2} = \frac{1}{2\pi}\left(\frac{g}{x}\right)^{1/2} \tag{98}$$

For instance, with $g = 981$ cm sec^{-2} and an assumed gap of 15 cm, Eq. (98) yields

$$f = \frac{1}{2\pi}\left(\frac{981}{15}\right)^{1/2} = 1.3 \text{ Hz}$$

Since the coil is in the persistent mode, not I but ϕ is constant, and f becomes somewhat increased. Since the inherent damping of the system is probably not adequate, it may be necessary to use an auxiliary damping system for the passenger compartment.

For safety reasons, the train will have several parallel coils working in the persistent mode. Should one coil fail, the remaining coils provide sufficient lift. For a vehicle of 100 passengers and a weight of 50,000 lb, Coffey *et al.* (1969) propose six quadratic coils (0.5 × 0.5 m), three on each side. These carry 300,000 ampere turns each. Since the coils are at 4.2°K, they must be well insulated. Figure 32 shows one kind of heat insulation. The conductor (a) consists of many Nb–Ti wires which are fully stabilized by Cu. The center cavity (b) of Fig. 32 contains liquid helium. The conductor (a) is within a tube (c) of Al which screens the conductor against varying fields caused by oscillation of the vehicle. To be able to transmit forces from the conductor to the outside container (g), which is at room temperature, thin disks (d_1) and (d_2) of epoxy material are placed at distances of about 1 ft. Between (d_1) and (d_2) is a thin thermal screen (e) of Al which is at the temperature of liquid nitrogen, supplied by two

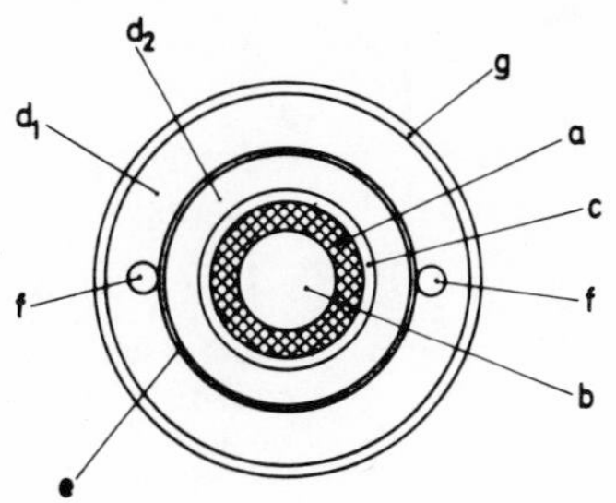

FIG. 32. Cross section of superconductive conductor.

pipes (f). The space between (c) and (g) is evacuated and filled with thermal superinsulation. With the additional nitrogen cooling, the evaporation of He is minimized. To replace the evaporated gas, the vehicle carries tanks of liquid He and N_2. These must be replenished periodically, perhaps once or twice a day. Such a system is more economical than one using a refrigerator on the train.

At high speeds, large surface currents of the order of the coil current I are induced in the sheet. These currents flow within the small penetration depth δ. Therefore, most of the sheet material is not utilized, and also the losses are high.

This can be improved (Powell and Danby, 1966) if, as shown in Fig. 33, the sheet on the track is replaced by a series of coils (b) in which the moving flux of the coil (a) of the vehicle induces currents i which, together with the current I of the vehicle, produce the lift. Figure 33a shows the top and Fig. 33b the front view of the coils. The coils (b) are connected in series with inductances (c) to keep the current i and therefore the track losses small without using resistive means. The lifting force is proportional to Ii. With i being only a fraction of I, to produce the same lift the vehicle needs more ampere turns than are required with the continuous sheet system. Since the coils are superconducting, their increased size does not lead to larger electric losses. However, the heat influx is increased so that more cryogenic liquids are required.

Thus the track loop system has smaller track losses and requires less

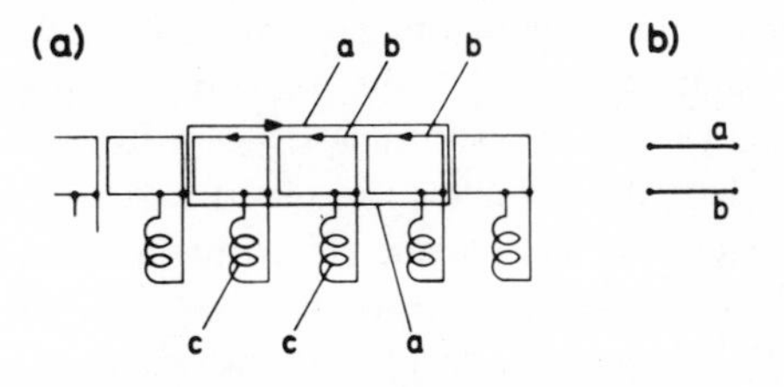

FIG. 33. Track loop system.

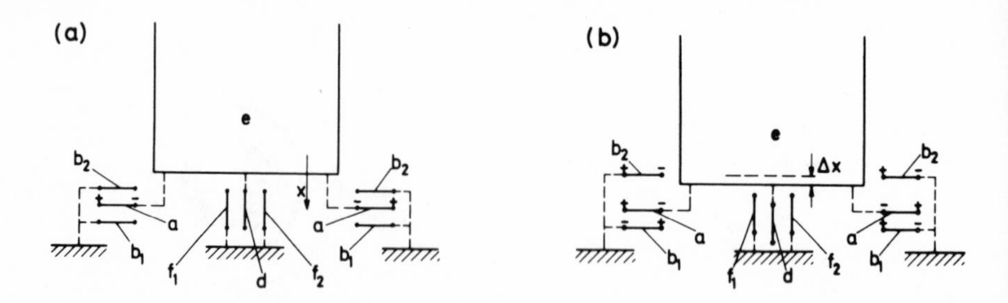

FIG. 34. Zero flux train support system.

material on the track but requires more superconductive coil material and cryogenic equipment on the vehicle.

Figure 34a shows an improved system—called the zero flux system (Powell and Danby, 1969)—which needs no track inductances and has increased stiffness and better damping of oscillations than the track loop system. Superconductive coils (a) carrying the current I are attached to the vehicle (e). Coils (b_1) and (b_2) are mounted to the ground. The flux of the coil (a) penetrates the coils (b_1) and (b_2). These coils are electrically connected in opposition so that the moving flux produces two coil voltages which cancel each other, provided that coil (a) is centered between (b_1) and (b_2). Since the sum of the opposing fluxes in (b_1) and (b_2) is zero, this arrangement is called the "zero flux system."

In Fig. 34b, the vehicle is shown displaced downward from the zero flux position. The flux through (b_1) is now larger than that through (b_2), and the induced voltages do not cancel. Thus a current i flows through both coils which, together with the current I of the train coil (a), produces lift. The current i is approximately proportional to the displacement Δx and increases until an equilibrium displacement is established for which the lift equals the weight of the vehicle. The system is very stiff, since i increases rapidly with x.

Figure 35a shows a side view of the vehicle coil (a) and the track coils (b_1) and (b_2). If no coil resistance is assumed, the induced current in the track coils adjacent to (a) is constant and equal to i. When the vehicle has moved to the position of Fig. 35b, the current in the end coils (b_1') and (b_2') is only $i/2$, since the flux linking these coils now has only half its maximum value. After the vehicle has moved to the position of Fig. 35c, the current in the track coils is i again. That means that the lift is not constant but exhibits pulsations. In practice, the coils (b_1) and (b_2) have resistance, so that the current i decays, as indicated by the dashed line in Fig. 35a. This decrease of i which is accentuated at small speeds reduces

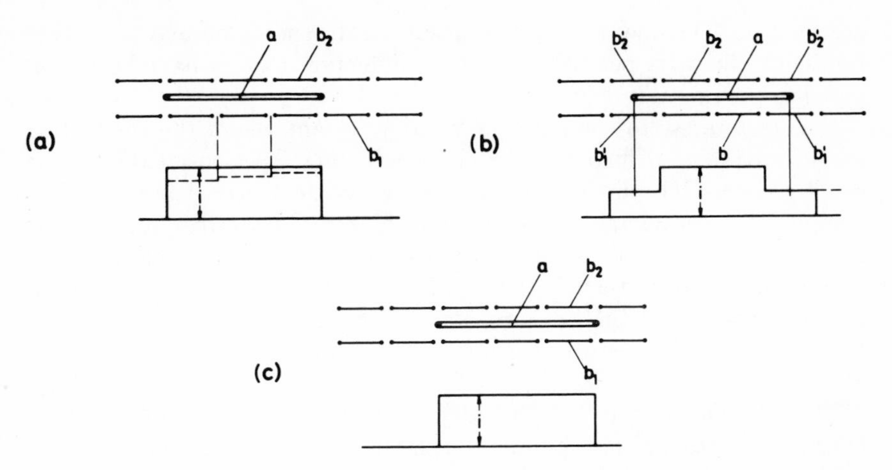

FIG. 35. Side view of vehicle and track coils in three positions with corresponding currents.

the lift. The decay can be reduced if, instead of one longer coil (a), several shorter coils of opposite polarity are used.

To provide lateral stability, the vehicle shown in Fig. 34 carries a superconducting coil (d) which fits between a series of oppositely connected coils (f_1) and (f_2) fastened to the track. No current flows in (f_1) and (f_2), and no losses are produced when the vehicle is centered. However, sideways movements produce currents in (f_1) and (f_2) which exert a restoring force.

A somewhat different arrangement is shown in Fig. 36, in which the vehicle carries two superconducting coils (a_1) and (a_2) and only one series of track coils (b). The coils (a_1) and (a_2) carry the same current but in opposite directions. The track coil (b) intercepts zero flux if it is midway between (a_1) and (a_2). However, if the vehicle moves downward, flux links with (b), and a current flows which interacts with the current through the coils (a_1) and (a_2) to produce lift.

Each of the magnetic suspension systems discussed above requires driving power to compensate for drag, for aerodynamic resistance, and to provide

FIG. 36. Another version of a zero flux system.

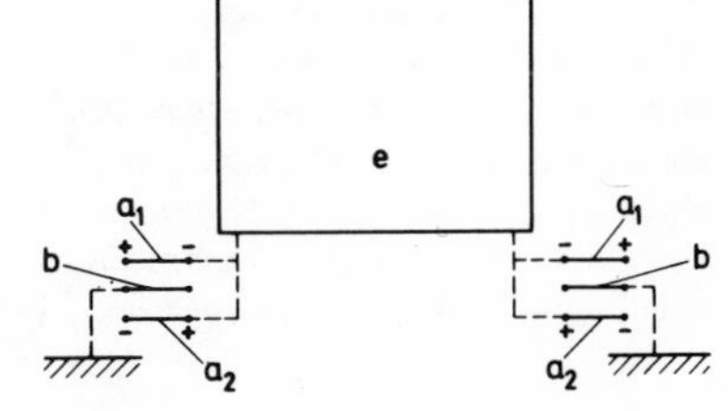

acceleration. The power required depends on the speed and may be several thousand kilowatts per vehicle. Linear induction motors have been proposed to provide this drive.

It is of interest to compare the relative advantages of the conducting sheet and the zero flux system. The conducting sheet suspension is the simpler system. The amount of superconductive coil material and cryogenic equipment is considerably less than for the zero flux system, and the lift force does not exhibit pulsations. With decreasing speed, the lift force declines more slowly than in the zero flux system, because the penetration depth increases and the sheet resistance decreases. This is in contrast to the situation in the zero flux system where the resistance is almost constant with velocity.

The zero flux system has the advantage of considerably smaller track losses. The restoring force per unit displacement and the inherent damping are much larger.

VI. Superconductive Machinery for Large Power

The devices discussed in this chapter are for small powers. At present, the application of superconductivity to large-power machinery, e.g., generators and motors, is limited to replacing the conventional dc excitation windings by superconductive coils of Type-II material. Higher flux densities can thus be obtained, and a reduction in size is possible. The problems to be solved are not so much connected with superconductivity as with the mechanical design of the coils and of the specially built cryostats required to keep them at low temperature. The least difficulties are offered by homopolar (acylic) motors and generators, so in the near future the application of superconductivity to these is promising. A reduction in size and weight of up to 50% seems possible.

References

Admiraal, P. S., and Volger, J. (1962). *Phys. Lett.* **2**, 257.
Bean, C. P. (1962). *Phys. Rev. Lett.* **9**, 93.
Bean, C. P. (1964). *Mod. Phys.* **36**, 31.
Buchhold, T. A. (1960). *Sci. Amer.* March.
Buchhold, T. A. (1961). *Cryogenics* **1**, No. 4.
Buchhold, T. A. (1962). *Symp. Celerina, Int. Un. Theoret. Appl. Mech.* August.
Buchhold, T. A. (1963). *Cryogenics,* 141. Sept.

Buchhold, T. A. (1964a). ASME Publ. 64-WA/PID-9.
Buchhold, T. A. (1964b). *Cryogenics*, 212. Aug.
Buchhold, T. A. (1965). *Cryogenics*, 216. Aug.
Buchhold, T. A., and Darrel, B. (1965). *Cryogenics*, 109. April.
Buchhold, T. A., and Molenda, P. J. (1962). *Cryogenics*. Dec.
Buchhold, T. A., and Rhodenizer, R. L. (1969). *IEEE Trans. Magn.* **5**, 429.
Chant, M. J., Halse, M. R., Lorch, H. O. (1970). *Proc. IEE* **117**, No. 7.
Coffey, H. T., Troy, W., Barbee, Jr., and Chilton, Fr. (1969). *Conf. Low Temp. Elec. Power, London.*
Giaever, I. (1965). *Phys. Rev. Lett.* **15**, 825.
Hart, H. R., Jr., and Swartz, P. S. (1966). Tech. Rep. AFML-TR-65-431, Sect. II.
Newhouse, V. L. (1968). *IEEE Trans. Magn.* **4**, No. 3.
Newhouse, V. L. (1972). *In* "Superconductivity" (R. D. Parks, ed.). Dekker, New York.
Newhouse, V. L., and Fruin, R. E. (1962). *Electronics* **35**, No. 18, pp. 31–36, May 4, 1962.
Olsen, J. L. (1958). *Rev. Sci. Instrum.* **29**, 539.
Olsen, J. L. (1969). *Conf. Low Temp. Elec. Power, London.*
Powell, J. R., and Danby, G. R. (1966). ASME Publ. 66-WA/R R-5.
Powell, J. R. and Danby, G. R. (1970). *Appl. Supercond. Conf., Boulder, Colorado.*
Schoch, K. F. (1961). *Advan. Cryogen. Eng.* **6**, 65.
van Houwelingen, D., and Volger, I. (1968). *Phillips Res. Rep.* **23**, 249.
Wipf, S. L. (1963). *Cryogen. Conf., Boulder, Colorado.*
Wipf, S. L. (1965). *Proc. Int. Symp. Magn. Technol., Stanford* p. 615; *AEC Conf., 1966* 650922.
Wipf, S. L. (1968). *Proc. 1968 Summer Study Superducting Devices* Brookhaven, Part II, pp. 511–543.

Chapter 8

RF Superconducting Devices

WILLIAM H. HARTWIG

Department of Electrical Engineering
The University of Texas at Austin
Austin, Texas

CORD PASSOW

Center for Nuclear Studies
Karlsruhe, Germany

I. Introduction

The essential elements of a successful technology are available to exploit superconductivity at high frequencies. Superconducting components have been designed and fabricated which perform a wide variety of electric circuit functions. Some of the most successful to be discussed in this chapter include high Q filters and tank circuits, resonators to control frequency for oscillators and particle accelerators, resonators for storing energy, transmission and storage lines, antennas, frequency converters and mixers, detectors, experiment chambers for studying properties of materials, and even amplifiers.

The performance of these devices and components is superior to their room-temperature counterparts because of their low losses and high energy density. The design of superconducting systems for communication, control, instrumentation, and research can now be carried out in a rational way supported by adequate scientific understanding. This is not to imply there are no problems, because there are many improvements in understanding and technique still needed.

This chapter summarizes the underlying theory as it applies to high-frequency effects and applications. Emphasis is given to those developments which appear, to the authors, to have made significant contributions to exploiting superconductivity.

The treatment is not exhaustive, and the reader is not expected to have a complete understanding of the theory of superconductivity. An attempt is made, however, to describe the phenomena to a point of appreciation by applied physicists and engineers.

The spectacular drop in surface resistance below T_c is the principle motivation to build high-frequency devices. At 1 GHz, the improvement by a superconductor can be 5 or 6 orders of magnitude over room-temperature copper. The magnetic and electric field intensities can be very high in superconducting resonators, and current densities can exceed 10^6 A/cm^2.

There is a variety of temperature dependencies and nonlinear effects which superconductors display and which normal conductors do not. These features can add value to the technology, and several investigators have utilized them in interesting ways. They can also exert constraints on some applications. One of the most useful is the Josephson effect, but it is of such profound and fundamental importance that it is treated in other chapters. With the exception of this quantum mechanical effect, this chapter describes several examples of behavior not available at ordinary temperatures.

High-frequency devices are limited in size by the wavelength of the rf fields. For this reason, most applications described in this chapter are at frequencies above 10 MHz where the device size is compatible with convenient Dewar dimensions and refrigeration loads. The existence of reliable helium refrigerators, described elsewhere in this book, makes it possible to achieve the improved performance of superconducting devices in a practical way.

II. Theory of Superconductors in High-Frequency Fields

A. Introduction

To calculate the properties of a device to generate, process, radiate, or store alternating electromagnetic waves, Maxwell's equations must be solved with respect to the particular boundary conditions given by the device geometry. To achieve an exact solution of the problem, it is necessary to solve the field equations inside the various volumes of space which are filled with different materials, and then to fit the resulting solutions to the area boundaries. This procedure is generally very difficult and usually leads to very long calculations in most of the known geometries. To avoid laborious calculations and to get solutions even for difficult device geometries, the concept of the surface impedance is useful. The basic idea of this concept is to first calculate the field outside of the metal; that means in vacuum or dielectric. The boundary conditions are rather simple, and the fields in the device can be calculated independent of the particular properties of the metal. To calculate the losses in the metal, the fields calculated for the ideal conductor are used. This can be done, since the fields on an ideal metal surface can be regarded (in a small area) as a plane wave traveling normal to the surface. The ratio of the power flux in the penetrating wave to the power flux in the incident wave then gives the surface impedance, in MKS units,

$$Z = (\mu_0/\epsilon_0)^{1/2} S(z = 0+)/S(z = 0-) \tag{1}$$

Here $S(z = 0-)$ is the Poynting vector just above the metal surface; $S(z = 0+)$ is the Poynting vector in the metal very close to the surface.

Equation (1) leads almost directly to an alternative expression

$$Z_s = R + iX = E(z = 0+)/H(z = 0+) \tag{2}$$

Here $E(z = 0+)$ is the electric field and $H(z = 0+)$ is the magnetic field, both taken on the inside of the metal surface. The approximation made in this concept of the surface resistance consists in neglecting a finite electric component at $z = 0-$ parallel to the metal surface. The magnetic field, however, has to be fitted to the boundary.

To calculate the reactance of a device, if the stored field energy in the area outside of the conductor is of the same order of magnitude as that inside the metal, the concept of the surface resistance has to be slightly modified. Then the field calculations are performed by fitting both field components. Usually the field inside the metal is assumed to be a decaying wave, with a decay length calculated from the surface impedance. Thus independent calculations of the surface impedance and of the fields are possible.

The concept of the surface resistance makes it possible to discuss the theory of fields in devices in two parts. In this section, we discuss the surface impedance of superconductors. The fields in the devices are discussed in Section IV under the subject of design and performance.

The theory of the behavior of superconductors in high-frequency fields for use in engineering and applied physics is not yet as well developed as for normal conductors. All of the information necessary to describe the surface resistance of superconductors due to the supercurrent can be taken from BCS theory, which was published in 1958 by Bardeen *et al.* (1957) (BCS). Only one year, later Mattis and Bardeen (1958), and independently Abrikosov *et al.* (1959), showed how it is applied in high-frequency fields. At lower temperature, losses caused by effects other than those connected with the supercurrent are also important. The various known loss mechanisms are discussed in Section III.

The formulas derived from BCS, however, are rather complicated, and several approximations have to be made to achieve analytical formulas for the surface impedance. For this reason, the phenomenological formulas of London (1934, 1961), Pippard (1947a, b, 1950), and Faber and Pippard (1955) are derived. Although they are not scientifically exact, they are very useful to introduce the quantum mechanical concepts. They continue to be used even though the BCS theory is well established. These empirical formulas are not valid over the full temperature range which is of practical interest, however, and better formulas can be given for special regions which are in good agreement with the BCS model. Moreover, it is possible to use computers to evaluate the integrals which are given by BCS. Therefore, the discussion of the surface resistance given in Section II.G is based on the BCS theory and should be used for the most exacting calculations.

According to the simplest model for superconductivity (London, 1934, 1940, 1961), electrons in a metal begin to condense into pairs (with opposite spin) when the temperature drops through the critical value T_c. These "Cooper pairs" move cooperatively through the lattice and do not suffer collisions as normal electrons do. This gives rise to two components of current, both of which obey the definition

$$J = nev \tag{3}$$

where J is current density, n is electron density, e is electronic charge, and v is the average velocity. They have individual differential equations, however:

$$\dot{J}_s = (n_s e^2/m)E \qquad \text{for the supercurrent} \tag{4}$$

$$\dot{J}_n + (1/\tau)J_n = (n_n e^2/m)E \qquad \text{for the normal component} \tag{5}$$

The normal component is collision limited with a momentum relaxation time τ.

In the absence of collisions, the n_s superelectrons are accelerated by the electric field E until a critical density is reached, whereupon the supercurrent switches back into the normal state. This phenomenological model is adequate to show how a superconductor will have zero resistance for a dc current but does have an ac resistance. The total current for a sinusoidal field becomes

$$J_0 = J_{0n} + J_{0s} = \left[\frac{n_n e^2 \tau(1 - j\omega\tau)}{m(1 + \omega^2\tau^2)^{1/2}} - j\frac{n_s e^2}{m\omega}\right]E_0 \tag{6}$$

Since $\omega^2\tau^2 \ll 1$ in most cases, the conductivity is

$$\sigma = \sigma_1 - j\sigma_2 = (n_n e^2 \tau/m) - j(e^2/m\omega)(n_s + n_n\omega\tau) \tag{7}$$

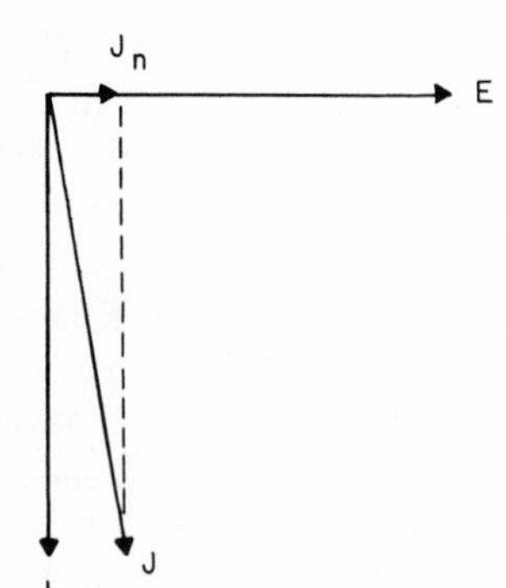

FIG. 1. Phase diagram for the normal and superconducting components of the total current density, J, with respect to the electric field, E.

The density of superelectrons, n_s, grows rapidly at the expense of the normal density, n_n, as the temperature $t = T/T_c$ drops below 1. London (1934) gives the ratio

$$n_n/n_s = t^4/(1 - t^4) = \phi(t) \tag{8}$$

In more practical terms, the total current has a normal component which is in phase with the voltage and causes a power loss. It also has a supercurrent component which lags by $\pi/2$ and produces no power loss (see Fig. 1). The power density is $EJ \cos \theta \simeq EJ_s\omega\tau\phi(t)$.

B. London Surface Impedance

Starting with the "two-fluid" model, a surface impedance can be derived. The four Maxwell equations are supplemented by two others:

$$\dot{J} = (1/\mu_0\lambda^2)E + \sigma_n\dot{E} \tag{9}$$

to introduce the nondissipative component of current, and

$$\nabla \times J = -(1/\lambda^2)H - \sigma_n\dot{H} \tag{10}$$

to account for the Meissner expulsion of magnetic flux. The penetration depth, λ, is defined by the relation

$$\mu_0\lambda^2 = m/n_se^2$$

in MKS units. When we set $\dot{H} = 0$ the dc H-field distribution is seen to be

$$H(x) = H(0)e^{-x/\lambda} \tag{11}$$

London shows that the displacement current is a negligible component and writes a set of electromagnetic equations, as follows:

$$\nabla^2 H = (1/\lambda^2)H + \mu_0\sigma_n\dot{H}$$

$$\nabla^2 E = (1/\lambda^2)E + \mu_0\sigma_n\dot{E}$$

$$\nabla^2 J = (1/\lambda^2)J + \mu_0\sigma_n\dot{J} \qquad (12)$$

The surface impedance

$$Z_s = E_{z0}\Big/\int_0^\infty J_z\,dx = E_{z0}\Big/\int_0^\infty [\mathbf{k}\cdot(\nabla\times H)]\,dx = (E_z/H_y)_{x=0} \qquad (13)$$

For a sinusoidal current,

$$\nabla \times E = -j\omega\mu_0 H_y = \frac{dE_z}{dx} \qquad (14)$$

Therefore

$$Z_s = \frac{E_{z0}}{(1/-j\omega\mu_0)\,(dE_z/dx)|_{x=0}} = j\omega\mu_0\lambda_e = j\omega\mu_0\left(\frac{2}{\omega\mu_0\sigma_e}\right)^{1/2} \qquad (15)$$

where λ_e is an effective current depth defined, in turn, by the effective conductivity from Eq. (7):

$$\sigma_e = \sigma_n + \frac{1}{j\omega\mu_0\lambda^2} = \frac{1 + j\omega\mu_0\lambda^2\sigma_n}{j\omega\mu_0\lambda^2}$$

Thus,

$$Z_s = R_s + jX_s = [-2\omega^2\mu_0^2\lambda^2/(1 + j\omega\mu_0\lambda^2\sigma_n)]^{1/2} \qquad (16)$$

Separating into real and imaginary components,

$$R_s = \left[\frac{\omega^2\mu_0^2\lambda^2}{1 + \omega^2\mu_0^2\lambda^4\sigma_n^2}[(1 + \omega^2\mu_0^2\lambda^4\sigma_n^2)^{1/2} - 1]\right]^{1/2} \qquad (17)$$

and

$$X_s = \left[\frac{\omega^2\mu_0^2\lambda^2}{1 + \omega^2\mu_0^2\lambda^4\sigma_n^2}[(1 + \omega^2\mu^2\lambda^4\sigma_n^2)^{1/2} + 1]\right]^{1/2} \qquad (18)$$

The London model is the first, and least exact, of several which are developed in the literature. It is useful at low frequencies, where $\omega\tau\phi(t) \ll 1$. For this case, assuming τ is independent of temperature,

$$R_s = \sqrt{\tfrac{1}{2}}\omega^2\mu_0\tau\lambda\phi = A_L\omega^2[t^4/(1 - t^4)^{3/2}] \qquad (19)$$

where $\lambda = \lambda(0)/(1 - t^4)^{1/2}$ and $\lambda(0)$ is the penetration depth at 0°K. R_s is proportional to ω^2, as in Eq. (19), because the losses vary as E^2, and E is proportional to the time derivative of the magnetic field. One frequently writes

$$r = \frac{R_s}{R_n} = \left\{\frac{\omega\tau\phi(t)}{1 + \omega^2\tau^2\phi^2(t)} \{[1 + \omega^2\tau^2\phi^2(t)]^{1/2} - 1\}\right\}^{1/2} \tag{20}$$

where $R_n = (\omega\mu_0/2\sigma_n)^{1/2}$ is the normal surface resistance at the same temperature. This reduces to

$$r \simeq [\omega\tau\phi(t)]^{3/2} = a_L\omega^{3/2}[t^6/(1 - t^4)^{3/2}] \tag{21}$$

The ratio R_s/R_n is used because the losses due to superconducting surface resistance are attributed to normal electrons inside the superconducting penetration depth. Thus λ is determined by n_s and is independent of frequency. R_n, on the other hand, is frequency dependent in a complicated way, depending upon the relative size of the various parameters such as l, δ, and even conductor thickness in the case of thin films.

The London model assumes a constant relaxation time, τ, and so the corresponding mean-free-path is $l = v_F\tau$, where v_F is the Fermi velocity of normal electrons. Equation (20) can be rewritten in terms of the classical skin depth, δ, since $\omega\mu_0^2\sigma_n = (2\lambda/\delta)^2$:

$$\frac{R_s}{R_n} = \left\{\frac{(2\lambda/\delta)^2}{1 + (2\lambda/\delta)^4} \left\{\left[1 + \left(\frac{2\lambda}{\delta}\right)^4\right]^{1/2} - 1\right\}\right\}^{1/2} \tag{22}$$

C. Anomalous Normal Surface Resistance

Resistance of normal metals is limited by size and skin effects. When the dimensions of a pure conductor become smaller than the mean-free-path, electron scattering off the surface will dominate. The reflection may be from spectral to diffuse, depending upon the surface smoothness. The effective mean-free-path will be shorter, and the resistivity will be greater than it is for large specimens. Thin films have a geometrical relation different from round wires.

As temperature is reduced, l increases and the size effect is more pronounced for a given size of conductor. At low temperatures, the phonon scattering process has less effect than impurities and other defects in determining the total scattering rate. Eventually, the dc resistivity becomes independent of temperature. This anomalous behavior also influences the ac surface resistance. The ac current is concentrated at the surface in the classical skin depth, δ, which is the penetration depth of the field and is, in this case, larger than the scattering length of the electrons. That means that on average the electrons do not leave the area where the field is concentrated before losing their energy by scattering processes.

The associated surface resistance is

$$R_n = [\pi f \mu_0/\sigma_n]^{1/2} \tag{23}$$

As temperature is reduced, the anomalous surface resistance has a radically different behavior, since $l \ll \delta$. Then the electrons which are accelerated in the penetrating electromagnetic field carry their energy out of the field area before losing it by scattering with lattice atoms. Therefore, the current-field relation in the metal is not given simply by the Maxwell equation. Following a suggestion of Pippard, Reuter and Sondheimer (1948) show the surface resistance for diffuse scattering in the anomalous limit as

$$R_n(\mathrm{AL}) = [3^{1/2}\pi\mu_0{}^2 l f^2/4\sigma_n]^{1/3} \tag{24}$$

The study of superconducting surface resistance is therefore complicated by a normal surface resistance frequency dependence that may lie between $\omega^{1/2}$ and $\omega^{2/3}$. The relative surface resistance ratio r will, therefore, have a frequency dependence between $\omega^{3/2}$ and $\omega^{4/3}$.

The London model gives a qualitative description that is valid for very pure metals at low frequencies and when $n_n\omega\tau \ll n_s$, which is below about $t = 0.90$. More sophisticated models have been derived which give good quantitative agreement with experiment over a wider range of parameters. For convenience in comparison with other models, Eqs. (19) and (21) are written as

$$R_s = A_L\omega^2 L(t) \qquad \text{and} \qquad r = a_L\omega^{3/2}\Lambda(t)$$

where A_L and a_L are constants.

D. Pippard Nonlocal Theory

London (1940, 1961) observed an anaomalous frequency dependence which he attributed to the normal skin depth becoming smaller than the mean-free-path as frequency was increased. Pippard (1947a, b, 1950), Faber and Pippard (1955), Fairbank (1949), and Maxwell *et al.* (1949) confirmed this experimentally with several metals in the normal and superconducting state. Pippard suggested in similarity to the anomalous limit that electrons in a superconductor may act coherently over distances, ξ, large compared to the penetration depth. There was considerable other evidence that current density could not be related to a point function of the local field. The nonlocal theory relates the current density to vector magnetic potential, $\bar{A}$, averaged over a region with radius, ξ, by the expression

$$\bar{J}_s = (\text{const}) \int [(\bar{A} \cdot \bar{r})\bar{r}/r^4]\epsilon^{-r/\xi} d\text{Vol} \tag{25}$$

ξ is related to l by $1/\xi \simeq 1/\xi_0 + 1/l$, and the constant is proportional to $n_s(T)/\xi_0$. If $\lambda \gg l$, then variations in A can be neglected, and J_s is similar to the London model. The range of coherence, $\xi_0 \sim 10^{-4}$ cm, is much larger than λ in the range of temperatures of interest.

The Pippard nonlocal model leads to a surface resistance ratio

$$r = A_P(\omega)[t^4(1 - t^2)/(1 - t^4)^2] + r_0 = A_P(\omega)f(t) + r_0 \tag{26}$$

where l, $\xi \gg \lambda$, and r_0 is a temperature-independent residual resistance. The separate frequency and temperature dependencies are a convenient feature of both the Pippard and London models.

The Pippard theory was used extensively on a wide variety of superconductors, with greatest success at microwave frequencies where the superconducting surface resistance could be made to dominate over the residual resistance.

E. Quantum Mechanical Model

The microscopic theory of superconductivity by Bardeen *et al.* in 1957 provided the basis for a quantum mechanical model of surface resistance. The original theory permitted calculations of the electrodynamics in a weak dc magnetic field. Mattis and Bardeen (MB) (1958) and Abrikosov *et al.* (1958) presented extensions for a superconductor in an ac magnetic field much less than the critical field. The MB treatment gives the surface resistance as a function of temperature, frequency, Fermi velocity, electron density of states, electron mean-free-path, the superconducting energy gap parameter, $\Delta(t)$, and the critical temperature. For microwave frequencies much less than $\Delta/\hbar$, and temperatures below $t \simeq \frac{1}{2}$, the superconducting surface resistance in simplified form becomes

$$R_s = A_{MB}\omega^2[1/T\ \epsilon^{-\Delta/kT}] \tag{27}$$

The experimental value of $2\Delta(0) \simeq 3.7kT_c$ and Δ does not vary rapidly until $t > \frac{1}{2}$, whence it decreases to zero at T_c (Mühlschlegel, 1959). The value of A_{MB} does not depend strongly upon frequency or temperature, but it must be evaluated to calculate the magnitude of R_s with precision. Methods of evaluating superconductor parameters are given below in Section II.F.

F. Superconductor Parameters

As shown by Bardeen *et al.* (1957), it turns out that the condensation of the conducting electrons into pairs is accompanied by the formation of an energy gap, which separates the superconducting pairs from the normal conducting electrons. The energy gap is temperature dependent and can be calculated from the BCS theory with the help of an implicit equation. As shown by Sheahen (1966), the solution of this equation is well fitted within the measurement accuracy by Eq. (28):

$$\Delta(t) = \Delta(0)\{\cos[\pi/2(T/T_c)^2]\}^{1/2} \tag{28}$$

The parameter $\Delta(0)$ is calculated from BCS theory to be $1.76kT_c$ but is found experimentally to be slightly different for different superconductors.

To split a Cooper pair and to put both electrons into the upper band, twice the gap energy, or $2\Delta(T)$, is required. Therefore, it is quite obvious that a superconductor shows a nearly normal conducting surface resistance if the quantum energy of the incident wave is larger than or equal to twice the gap energy. The second important parameter is the critical temperature. For temperatures above T_c a superconductor is normal, and the surface resistance has to be calculated from Reuter and Sondheimer (1948) theory.

Figure 2 shows the normal, nearly normal, and superconducting areas of a superconductor in a high-frequency field as a function of the gap energy and the critical temperature. By statistical mechanics it is possible that, with a certain probability, an electromagnetic wave can split Cooper pairs even if its quantum energy is lower than twice that of the gap energy. However, this has to happen in a time interval which is given by the Heisenberg uncertain relation,

$$\tau \cdot 2\Delta(T) = \hbar \tag{29}$$

τ is a characteristic time for the rate of change of the two energy states of the system. The traveling length of a Cooper pair during the time τ can be calculated with the help of the Fermi velocity. This length is called the the coherence length, and it gives a measure of the wavelength of an interacting electromagnetic wave, or wave component, which can interact with the Cooper pair. The coherence length is introduced into the literature with two different definitions which differ by a factor of $2/\pi$. We shall use the following definition:

$$\xi(T) = v_F\hbar/\pi\Delta \tag{30}$$

The coherence length increases with temperature.

$\Delta(0)$, the gap of the superconductor at 0°, and T_c, the transition tem-

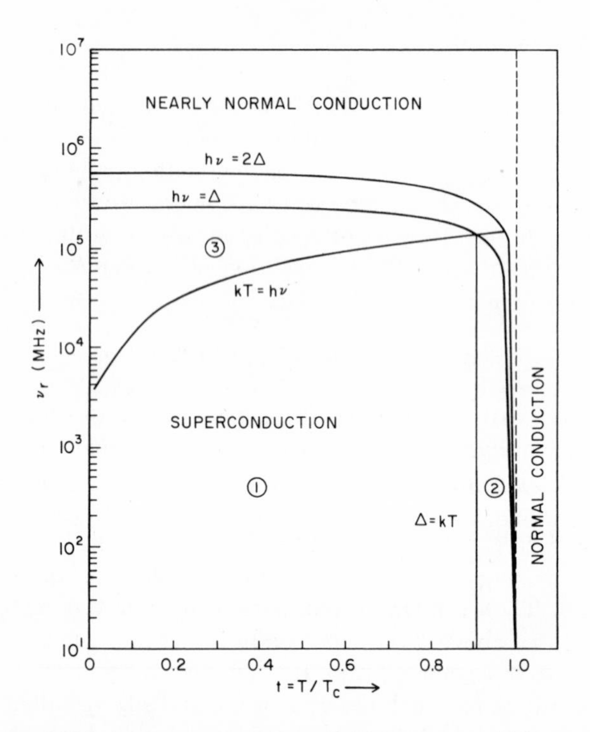

FIG. 2. Areas designate the temperature and frequency at which a superconductor is normal, nearly normal, and superconducting. The areas of possible approximation to calculate the integrals in Eqs. (39) and (40) are also shown. The frequency scale is calculated for lead. $\Delta(T)$ is calculated from Eq. (28). $\Delta(0)$ is assumed to be the theoretical value, $1.76kT_c$. For $T/T_c = 1$, the metal is a normal conductor. For $h\nu = 2\Delta(T)$, the quantum energy of the wave is large enough to split the Cooper pairs directly, and the metal acts nearly like a normal conductor. The approximate treatment of the metal as a normal conductor improves as the ratio $h\nu/2\Delta(T)$ increases. Important boundaries for approximations are $h\nu = kT$, $h\nu = \Delta(T)$, $h\nu = 2\Delta(T)$, and $\Delta(T) = kT$. The frequency scale for other superconductors can be calculated from $\nu_x = (T_{cx}/T_{c\text{Pb}})\nu_r$.

perature, are the only two basic parameters which describe the superconducting state. To describe the superconductor with respect to its behavior in high-frequency fields, we need to know four more parameters which are well known from the theory of the conductivity in normal metals. The four parameters are n_s, the density of conducting electrons; v_F, the Fermi velocity; m^*, the effective mass of the electrons; and l, the scattering length.

The six parameters given above are the only independent parameters which are necessary to describe a superconductor with the BCS theory. To give a better picture, or to simplify the expressions, some other parameters of the superconductor are used. One of these is the coherence length, ξ. The

TABLE I

BASIC MATERIAL PARAMETERS OF LEAD[a]

T_c (°K)	$\Delta(0)/kT_c$	l (Å)	m^*/m_0	n_s per atom	v_F (m/sec)	ξ_0 (Å)	δ_L (Å)	Reference
7.2	2.05	7100	(2.28)	(2.27)	6×10^5	970	308	Szecsi (1970, 1971)
7.2	2.17	912	2.32	2.09	(6.4×10^5)	990	308	Reichert and Hasse (1972), Halbritter (1969c, 1970d), Reichert (1971)
7.2	2.05	801	2.32	2.54	(6.8×10^5)	1114	280	Reichert and Hasse (1972), Hahn *et al.* (1968), Reichert (1971)

[a] These sets were found by Szecsi (1970, 1971), Reichert and Hasse (1972), and Reichert (1971) to fit their measurements best by using different theoretical formulas or computer programs. The values in parentheses are calculated by the authors of this article using the relation between the coherence length and the Fermi velocity and, in order to calculate m^*/m_0 and n, the formula derived from the quasi-free electron model, $m^*v_F = \hbar(3\pi^2 n)^{1/3}$.

London parameter is given by a combination

$$\delta_L = (m^*/q^2 n_s \mu_0)^{1/2} \tag{31}$$

This parameter is often called the London penetration depth, since it has the meaning of a penetration depth in the older London theory. However, for the behavior in high-frequency fields, it does not have the meaning of a penetration depth, nor does it distinguish whether the calculations of surface resistance should be performed by a London approximation or by a Pippard approximation. Therefore, we shall simply call it the London parameter. Another material parameter which is derived from the energy gap and the transition temperature is the critical magnetic field of a superconductor. The critical magnetic field is defined in Eq. (55), where its importance with respect to high electromagnetic fields is discussed.

TABLE II
THE SUPERCONDUCTING ENERGY GAP MEASURED FROM DIFFERENT EFFECTS OR CALCULATED FROM DIFFERENT EXPERIMENTS ON LEAD AND NIOBIUM

Taken from	Lead $2\Delta(0)/kT_c$	Niobium $2\Delta(0)/kT_c$
Microwave absorption	4.1 (Halbritter, 1969c, 1970d)	3.72 (Turneaure and Weissman, 1968)
	4.26 (Szecsi, 1970, 1971)	3.72 (Turneaure and Weissman, 1968)
	4.06 (Hahn *et al.*, 1968)	3.80 (Halbritter, 1969c, 1970d)
Infrared transmission	4.0 (Ginzberg and Tinkham, 1960)	
Infrared absorption	4.1 (Richards and Tinkham, 1960)	2.8 (Richards and Tinkham, 1960)
	4.37 (Leslie and Ginzberg, 1964)	
Tunneling	4.29 (Rowell *et al.*, 1963)	3.84 (Townsend and Sutton, 1962)
	4.38 (McMillan and Rowell, 1965)	3.6 (Giaever, 1963)
	4.18 (Douglass and Meservey, 1964)	3.59 (Sherrill and Edwards, 1961)
Ultrasound absorption	3.6 (Love and Shaw, 1964)	3.96 (Dobbs and Perz, 1963)
Thermodynamic measurements	3.95 (Meservey and Schwartz, 1969)	3.65 (Meservey and Schwartz, 1969)

TABLE III
MATERIAL PARAMETERS OF SUPERCONDUCTORS[a]

	T_c (°K)	H_c (G)	δ_L (nm)	ξ_0 (nm)	$\Delta(0)/kT_c$	ϕ (eV)
Niobium (Nb)	9.5	1944	39	38	1.80	4.0
Lead (Pb)	7.2	803	37	83	2.18	4.0
Vanadium (Va)	5.4	1370			1.75	
Tantalum (Ta)	4.5	830			1.80	
Mercury (Hg)	4.1	411			2.30	4.5
Tin (Sn)	3.7	309	34	230	1.75	4.4
Indium (In)	3.4	293	51	40	1.75	3.8

[a] The table gives rounded values as published in several textbooks. The gap parameter is taken from tunnel experiments, since this seems to be the most accurate value available at this time (see also Table II).

The six material parameters are not yet well known for all superconducting metals of interest, since they cannot be measured directly but only in combination. Moreover, for some parameters different values are measured in the different aspects of superconductivity. An example of a set of basic parameters is shown in Table I. This table gives the six parameters for Pb as measured and calculated by Reichert (1971), Reichert and Hasse (1972), and Szecsi (1970, 1971). Table II shows the gap height of lead and niobium at 0° as measured in different experiments and with different methods to estimate the gap height. Table III gives some of the important parameters, as usually published in the literature, for the metals of practical interest. The work function given in this table is used in Eq. (59).

G. QUANTUM MECHNICAL CALCULATION TECHNIQUES FOR THE SURFACE IMPEDANCE

The method used in quantum mechanics to calculate the surface resistance for normal conductors and for superconductors is, in principle, the same. First, electromagnetic plane wave equations are derived from Maxwell's equations of the vector potential which are valid inside the metal. These equations contain the current density produced by the conducting electrons of the metal. Consequently, the next step is to calculate the current density in terms of the material parameters and the generating field. This is done in a superconductor by using the wave functions of the Cooper pairs. Usually, the current density is given in its Fourier transformed representation in Gaussian CGS units used by most authors in the physics literature:

$$j(q) = (c/4\pi)K(q)A(q) \tag{32}$$

Here $A(q)$ is the Fourier transformed vector potential and $K(q)$ a complex function dependent on the material parameters of the metal. The function $K(q)$ was first derived by Mattis and Bardeen (1958) and independently by Abrikosov *et al.* (1959).

Both groups reached the same result by some different mathematical methods. We give a short representation of the results received by Mattis and Bardeen (1958), since it seems to show the physical meaning of the theory more directly. To use the quantum mechanical current density, Maxwell's equations are transformed also into their Fourier representation. This is done by a method described by Reuter and Sondheimer (1948) to introduce the boundary conditions given on the metal surface. It is done for the two possible cases of diffuse and specular reflection of the electrons in the metal surface.

The vector field $\bar{A}(z)$ then can be calculated by the inverted Fourier transformation. However, for most applications it is necessary only to calculate the surface impedance of a superconductor, and to do this only the fields at the surface ($z = 0+$) are needed. Therefore, it is possible to calculate the surface impedance directly as shown by Reuter and Sondheimer (1948) from Eq. (33) for specular reflection:

$$Z = \frac{1}{\pi}\int_{-\infty}^{+\infty} \frac{dq}{q^2 + \mu_0 K(q)} \tag{33}$$

and from Eq. (34) for diffuse reflection:

$$Z = \frac{\pi}{\displaystyle\int_0^{\infty} \ln[1 + K(q)/q^2]\, dq} \tag{34}$$

The integrals to calculate $K(q)$ are rather complicated, and an analytic solution can be found only after several approximations have been made. Moreover, the results in $K(q)$ are not, in any case, those of a simple dependence on q, and the integration over q cannot be performed analytically. Since the kind of approximation which is possible is dependent on the resulting penetration depth, the total evaluation procedure for the integrals is somewhat elaborate. At this time, there are no simple formulas which can be presented that allow an experimenter or engineer to calculate the surface impedance for every case.

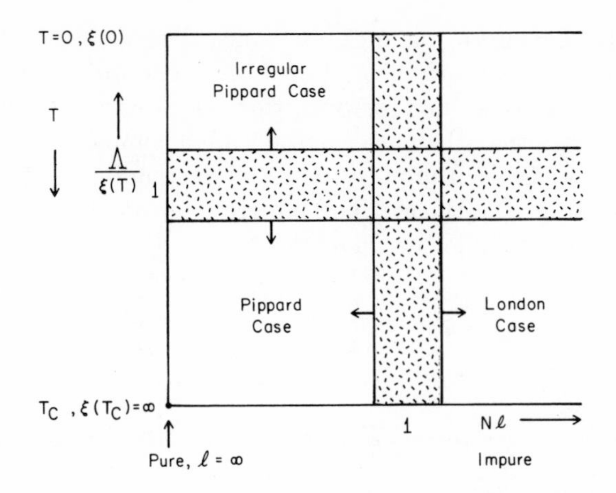

FIG. 3. Areas of possible approximations dependent upon the purity of a superconductor and its material parameters. The calculated penetration depth is given by Λ and is compared with the scattering length, l, and the coherence length, ξ, in the superconductor. Different approximations may be used depending upon the expected ratios, $\Lambda/\xi(T)$ and Λ/l.

To discuss $K(q)$ and its meaning, as well as the possible approximations, it is useful to write the function as a power series of q:

$$K(q) = a_{L0} + a_{L1}q + a_{L2}q^2 + \cdots + a_{P1}q^{-1} + a_{P2}q^{-2}\cdots \tag{35}$$

The coefficients are complex numbers which are dependent on the frequency, temperature, and the material parameters. The importance of the terms for the solution of Maxwell's equation is given, not only by the coefficient, but also by the ratios ξ_0/Λ and l/Λ. Here, Λ is the penetration depth which is finally calculated from the inversion of the Fourier transformation. The two important areas of possible approximations are the extreme London case and the extreme Pippard case. See also Fig. 3 for the following discussion.

The extreme London case is where the scattering length of the electrons in the metal is short compared to the penetration depth of the field into the metal. This is the case for dirty superconductors. In the extreme London case, only the term a_{L0} contributes to $K(q)$; the other terms are zero or can be neglected.

The extreme Pippard case is given for a pure metal if the scattering length in the metal is much longer than the penetration depth of the field into the metal, and if the coherence length is also large compared to the penetration depth. Then all coefficients except a_{P1} are small or zero and do not contribute to the solution.

There is also the intermediate case in which the coherence length and/or the scattering length is of the order of the penetration depth, and no approximations are possible. Moreover, there is the case that the scattering length is large and the coherence length is small compared to the penetration depth. This case is of interest for an understanding of the meaning of coherence length in the theory. The integrals to calculate the coefficients are dependent on the following functions:

$$h_{\pm} = \left| \frac{1}{ql} + \frac{i}{q\pi\xi(T)} \left[\frac{(E + h\nu)^2}{\Delta^2} - 1 \right]^{1/2} \pm \left[\left(\frac{E}{\Delta} \right)^2 - 1 \right]^{1/2} \right|$$

$$h_0 = \left| \frac{1}{ql} + \frac{1}{q\pi\xi(T)} \left[\left(\frac{E}{\Delta} \right)^2 - 1 \right]^{1/2} - i \left[\frac{(E + h\nu)^2}{\Delta^2} - 1 \right]^{1/2} \right| \tag{36}$$

The energy E is the integration variable of the integrals. E has to be integrated from Δ or $\Delta - h\nu$ to infinity. Up to a value where the functions equal unity, the integrals contribute to the Pippard coefficients. Those parts of the integrals for which the functions are larger than unity contribute to the London coefficients. The London case mentioned above will be given only if all three functions $h(E)$ are much larger than unity for the total range of integration over E. Therefore, the third possible region will be called the irregular Pippard case. The irregular Pippard case seems to be very unlikely in an extremely pure superconductor, $l = \infty$, since, as shown by Miller, this superconductor would show no losses in a high-frequency field if the quantum energy of the rf field is not high enough to split the Cooper pairs directly; that means if $h\nu < 2\Delta$. Miller pointed out that a free electron cannot absorb a quantum unless the velocity of the electron is greater than the phase velocity of the wave, $v_{\mathrm{f}} > v_{\mathrm{ph}}$. Since the field, penetrating into the metal, is given approximately by $\exp(z/\Lambda + j2\pi\nu t)$, the phase velocity of the most important wave component is given approximately by $\Lambda\nu$, and we obtain easily, using Eq. (30), the condition

$$\Lambda/\xi(T) > \pi\Delta/h\nu$$

for loss production in a pure superconductor.

The general result for $K(q)$ can be given in the following form:

$$K(q) = \frac{-3}{c^2 h v_{\mathrm{F}} \delta_{\mathrm{L}}^2} \int_0^\infty \int_{-1}^1 e^{iqRu} e^{-R/l} (1 - u^2) I(\nu, R, T)\, du\, dR \tag{37}$$

where $I(\nu, R, T)$ is a function containing two integrals over the energy which are quite long and are given here only in approximate form. The integration over u is the angular integration and can be performed for bulk material if the maximum length of the radius vector is independent of the angle. The resulting integral can be integrated over R. The result contains several integrals which, organized with respect to their dependence on q, give the coefficients a_L and a_P in Eq. (35). A more complete calculation of the coefficients is not yet published. Only the coefficients a_{P0} and a_{L0} are discussed in the literature.

The coefficient for the extreme Pippard case, that is, a coherence length which is long compared to the penetration depth and an infinitely long mean-free-path, is usually given in terms of a conductivity introduced by Glover and Tinkham (1957), and used by Mattis and Bardeen (1958).

Thus, for the Pippard case it is

$$K(q) = \frac{3\pi^2 \nu}{2v_F \delta_L^2} \frac{1}{q} \left(i \frac{\sigma_1}{\sigma_n} + \frac{\sigma_2}{\sigma_n} \right) \tag{38}$$

The ratio between the superconducting conductivity denoted with $i\sigma_1 + \sigma_2$ and the conductivity of the metal in the anomalous skin effect limit denoted by σ_n is given by the following integrals:

$$\sigma_1/\sigma_n = (2/h\nu) \int_{\Delta}^{\infty} [f(E) - f(E + h\nu)] g(E)\, dE$$
$$+ \gamma(h\nu - 2\Delta)(1/h\nu) \int_{\Delta - h\nu}^{-\Delta} [1 - 2f(E + h\nu)] g(E)\, dE \tag{39}$$

$$\sigma_2/\sigma_n = (1/h\nu) \int_{\Delta_g}^{\Delta} [1 - 2f(E + h\nu)] g(E)\, dE \tag{40}$$

Here $g(E) = (E^2 + \Delta^2 + h\nu E)/(|\Delta^2 - E^2|\, |(E + h\nu)^2 - \Delta^2|)^{1/2}$, and $f(E)$ is the Fermi distribution with the Fermi energy as a reference. The integration limit and the function Δ_g are defined by the relations

for $h\nu - 2\Delta < 0$, $\Delta_g = -h\nu$ and $\gamma(h\nu - 2\Delta) = 0$;
for $h\nu - 2\Delta > 0$, $\Delta_g = -\Delta$ and $\gamma(h\nu - 2\Delta) = 1$.

The surface impedance can be calculated for the case of diffuse reflection of the electrons on the surface for the extreme Pippard case to be

$$Z = [(\sigma_1 - i\sigma_2)/\sigma_n]^{-1/3} Z_n \tag{41}$$

Here Z_n is the impedance of the extreme anomalous limit in normal conduction.

A numerical evaluation of the formulas is difficult and not possible without further approximations. The integrals by which the superconducting conductivities in Eqs. (39) and (40) are given are elliptical integrals even with the approximations. The kind of approximation possible is strongly dependent on the temperature and frequency and its relation to the gap energy. The areas for possible approximations of the integrals are given in Fig. 2. To calculate the surface resistance of pure superconductors, Abrikosov *et al.* (1960) solved the integrals approximately in different areas given by Fig. 2. They obtained a set of seven solutions, each of which applies in a given area or under extreme conditions, for example, $T = 0$.

H. Experimental Verification of Surface Impedance Theories

The complex formalism described above for the Mattis and Bardeen theory which can be simplified only in the case of pure superconductors leads to Eq. (27) for a temperature and a frequency range given in area 1, Fig. 2. Fortunately, the matter has been improved by Turneaure (1967, 1968) and more recently by Halbritter (1969a, b, 1970) using computer solutions. Experiment verification was obtained only when superconducting cavity resonators were fabricated with vanishingly small residual resistance. The description of material preparation techniques to suppress the residual losses, and the origin of the various loss mechanisms, is discussed in the next sections.

Turneaure expressed MB in the form of a double integral which could be integrated numerically. Halbritter's calculations are based on a Green's function analysis used by Abrikosov *et al.* (1960) (see Figs. 4 and 5).

Both theories have been supported by measurements on several superconductors, including lead, tin, and niobium. Turneaure (1967, 1968), Pierce (1967), Flecher *et al.* (1969), and Kneisel *et al.* (1971a) electroplated TE_{011} cavities at S band and X band with highest purity tin and lead. The choice of mode assured minimum dielectric losses because the E fields are zero on the walls, and it permitted separable end plates because the H field is zero in the corners. The choice of frequencies insured that the superconducting surface resistance would dominate the losses and the residual and coupling losses could be separated out in the data reduction. Background magnetic fields were minimized to suppress trapped-flux losses.

The experimental and data analysis techniques are described by Turneaure (1967) to obtain superconducting surface resistance and reactance as a function of temperature. From Q and frequency change versus temperature at a given resonant frequency, each data point $[R_{si}\text{ (exptl)}, T_i]$ is a solution to the expression

$$R_{si}(\text{exptl}) = AR_{si}(\text{theor}) + R_0 \tag{42}$$

where $R_{si}(\text{theor})$ is the MB function of the energy gap parameter, Δ, the

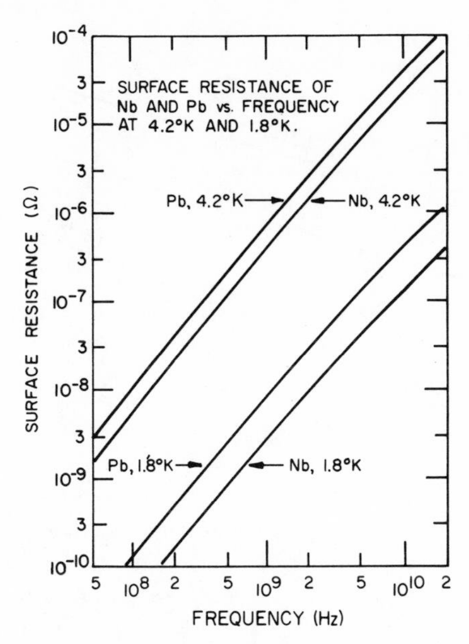

FIG. 4. Theoretical surface resistance of Pb and Nb from calculations by Turneaure based upon Mattis and Bardeen. (See Halbritter, 1969b,c, 1970c,d; Turneaure and Weissman, 1968; Turneaure, 1967; Schwettmann *et al.*, 1965; Pierce, 1967.)

London penetration depth, λ_L, the mean-free-path, l, and the Fermi velocity, v_F. The last three of these are known or determinable from simpler independent measurements. The determination of Δ can also be a part of the data analysis. The method of analysis consists of treating R_0, A, and Δ as free parameters and obtaining a best fit. The same type of procedure is used for surface reactance, X_s. The first use of this technique by Turneaure on Pb and Sn at 2.85 and 11.2 GHz and later on niobium verified the basic correctness of the MB surface impedance. The value of A obtained from the full calculation was nearly unity for niobium and for lead and tin at 11.2 GHz, but was less than unity in lead and tin at 2.85 GHz. The calculations made at the Pippard limit were in poorer agreement.

Turneaure reported that MB gives better agreement than all of the previous work based upon the Pippard and London models. The fact that A is not unity may be attributed to uncertainties in λ_L and v_F, or to subtle differences in the validity of MB for weak-coupling versus strong-coupling superconductors. Certainly, the cavity surface condition and crystal orientation on the surface will make various experiments differ in the best values of the normal and superconducting parameters. It is also likely that there are small superconducting surface resistance effects not anticipated by

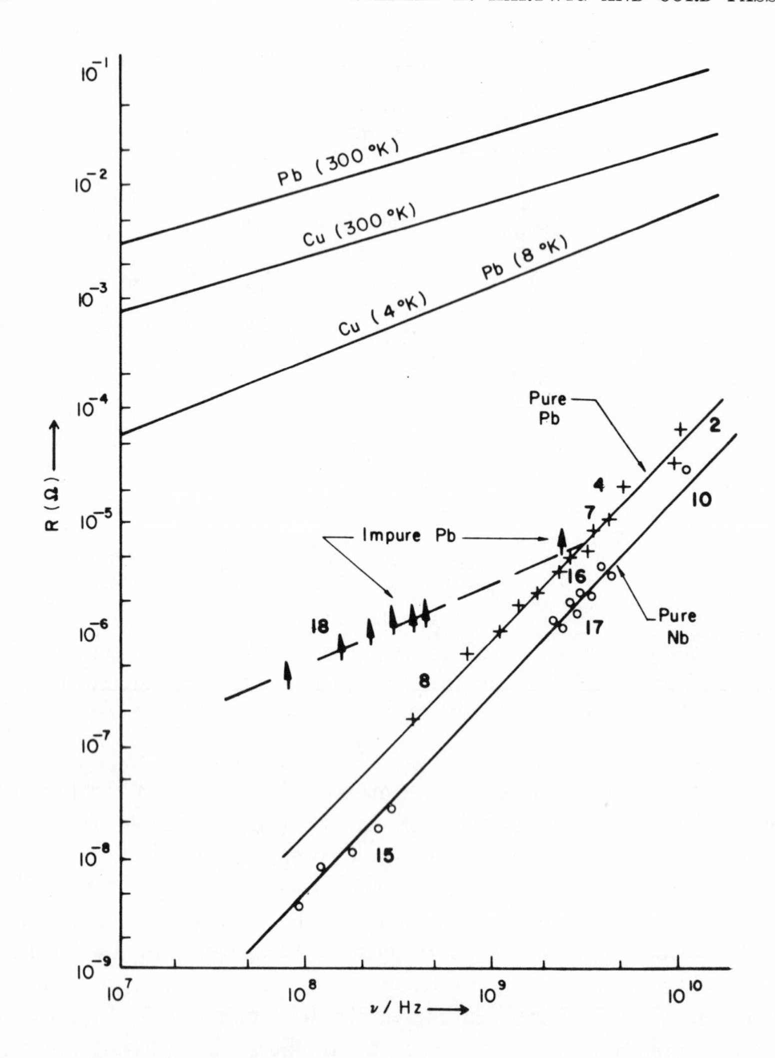

FIG. 5. Measured surface resistance of lead and niobium at 4.2°K. The values for pure material are well in agreement with the values predicted by theory as calculated by Halbritter, assuming the scattering length in lead to be 10^5 Å and in niobium to be 500 Å. Turneaure found agreement between theory and the measurements he performed on niobium at 10 GHz with $l = 10{,}000$ Å. For comparison, the room-temperature values for copper and lead, as well as the figures for clean normal conducting copper and lead at low temperatures, are also given. To show the importance of purity, measured values for lead films evaporated in poor vacuum are also given. The key for numbers appearing with the data points is: 2 (Pierce, 1967); 4 (Hahn *et al.*, 1968); 7 (Flecher, 1969); 8 (Szecsi, 1971); 10 (Turneaure and Weissman, 1968); 15 (Vetter *et al.*, 1970b); 16 (Hahn and Halama, 1970); 17 (Kneisel *et al.*, 1971a); 18 (Passow *et al.*, 1972).

MB. These will be masked by direct comparison of experimental data with any model using only one primary mechanism and only one residual mechanism.

Halbritter (1969c, 1970d) was able to refine the fit for tin, indium, lead, and niobium by selecting improved values of some parameters, particularly the London penetration depth and the coherence distance. He showed that the frequency dependence varies as ω^{α}, where α goes from 1.5 to 2.0 as mean-free-path decreases and λ_L/ξ decreases.

The assumption of an R_0 which is independent of temperature and, occasionally, of frequency) is, in general, oversimplified. In fairness to all of the excellent work done by many investigators in the past, we can now appreciate better the complex interactions of many residual phenomena. As early as 1959, Kaplan *et al.* reviewed the state of knowledge on superconducting surface resistance in the millimeter wavelength region. They observed, "A nonvanishing frequency-dependent r_0 has been observed in all the microwave experiments to date. However, since the origin of this effect is not understood, its inclusion as an additive constant is a hypothesis, albeit a reasonable one made in all the previous experiments." It appears clear to the present authors that, although the MB theory appears to be a correct one, the presence of other temperature- and frequency-dependent mechanisms makes a single r_0 an inadequate parameter, particularly at low frequencies.

In summary, the subject is obviously still an active research problem. Part of the solution will come from the results of better experimental methods. Whatever the outcome may be, it is the search for extremely small effects. The superconducting resonator will remain a vast improvement over its normal counterpart in Q, absolute stability and versatility.

III. Residual Resistance

A. Introduction

The Pippard theory was used extensively on a wide variety of superconductors up to the mid-1960s, with greatest success at microwave frequencies where the superconducting surface resistance could be made to dominate over the residual resistance. There was uncertainty as to the number and causes of the residual resistance, and, as a consequence, it was difficult to get comparison between experiments. It became increasingly clear that there were residual processes, such as trapped flux, which produced temperature- and frequency-dependent losses which were dominating the super-

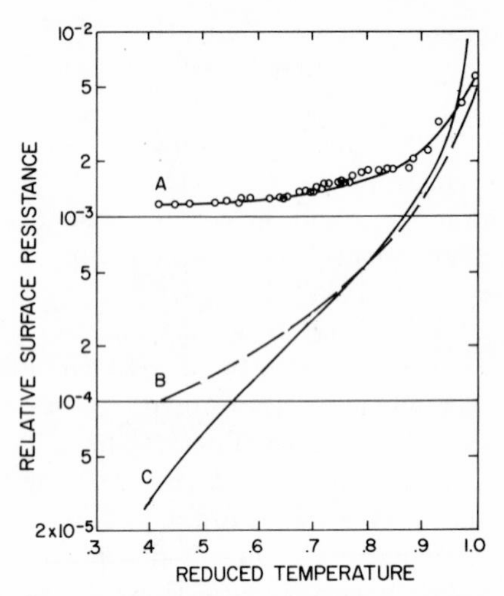

FIG. 6. Temperature dependence of relative surface resistance as measured for a 36-MHz *LC* circuit, 0.75Pb–0.25Sn. Curve (A) shows the residual resistance due to the Earth's magnetic field, curve (B) shows the data after subtraction of a constant r_0, and curve (C) is the Pippard temperature function fitted to curve (B) at 0.80. Note the failure of a temperature-independent loss to account for the observed data, indicating the presence of trapped flux loss.

conducting surface resistance. Figure 6 is an example of this situation as shown by Victor and Hartwig (VH) (1968). The observed relative surface resistance (curve A) cannot be made to agree with the Pippard $f(t)$ (curve C) by subtracting a constant r_0 (curve B). As recently as 1971, a major review of cryogenic applications by J. Clarke *et al.* (1971) commented that, "The origin of these (residual) losses is still an outstanding problem."

The situation is now greatly improved thanks to the contributions in three areas: better materials (particularly niobium and lead), the workable MB theory for superconducting surface resistance, and a number of important experiments which identify some of the dominant causes of residual loss. It now appears that several dissipating mechanisms can be separated and identified. This is vital to the progress in applied superconductivity, since these losses seem to dominate the BCS term below about $0.6T_c$.

B. SEPARATION OF LOSSES

The separation and exact determination of loss mechanisms in superconductors has been accomplished experimentally by measuring Q and resonant frequency changes in resonators. The techniques of measurement are described in Section IV.C.

The dissipated surface power density is

$$p_s = \tfrac{1}{2}R_sH_0^2 \tag{43}$$

where H_0 is the peak value of the magnetic field parallel to the surface. The surface resistance is related to the Q by the geometry factor, Γ, by

$$\Gamma = R_sQ_s \tag{44}$$

If one is measuring a homogeneous surface loss, it is convenient to measure Q_s and Q_n in the superconducting and normal state, respectively; then

$$R_s/R_n = Q_n/Q_s \tag{45}$$

If there are several loss mechanisms, one can write

$$Q_L^{-1} = \sum_i Q_i^{-1} \tag{46}$$

Each of the loss mechanisms is distinguishable, in principle, by its individual dependence upon such experimental variables as frequency, temperature, power level, surface treatment, and configuration.

Superconducting surface resistance varies approximately as ω^2 and exponentially with temperature, whereas the residual losses appear to vary more slowly and originate from numerous effects. Since not all of these can have an associated equivalent resistance and geometry factor, we simply expand the concept of loaded cavity Q to

$$\frac{1}{Q_L} = \frac{a}{Q_{MB}} + \frac{1}{Q_h} + \frac{1}{Q_c} + \frac{1}{Q_r} + \frac{1}{Q_d} + \frac{1}{Q_{em}} + \cdots \tag{47}$$

The loaded Q will reflect the well established losses due to dc magnetic fields (Q_h), coupling (Q_c), radiation (Q_r), and dielectric losses (Q_d). At high ac fields, there may be electron emission (Q_{em}) and magnetic breakdown, plus a number of losses due to imperfect surfaces, phonon generation, and momentum effects due to surface curvature.

Clearly, the number of ways one fails to observe the maximum possible Q seems limitless. We can take them in groups of increasing complexity. Q_{MB} is the value reached by a perfect superconducting resonator surface suffering only the theoretical MB surface resistance. If there are local effects which have the same frequency and temperature dependence, $a > 1$. Coupling losses are introduced by the external circuit and can be reduced until the output signal is just strong enough to be seen. Where coupling losses dominate all other losses, Q_L will increase as coupling is decreased, until coupling is no longer a significant source of loss.

Radiation losses are eliminated with a superconducting shield surrounding the circuit. A cavity is its own shield, and so the radiation loss becomes identical with the coupling loss.

Dielectric losses are caused by a variety of dissipating processes in coil forms, oxide films on the superconductor, and in dielectric tuning elements. Hartwig and Grissom (1965) show that the dielectrically loaded resonator has the relation

$$Q_{\mathrm{L}}^{-1} = Q_{\mathrm{u}}^{-1} + Q_{\mathrm{d}}^{-1} = Q_{\mathrm{u}}^{-1} + \alpha \tan \delta \tag{48}$$

where $\tan \delta$ is the dielectric loss tangent and α is the fraction of the electric field in the dielectric material. Q_{u} is the unloaded Q in the absence of dielectric material. The remaining dissipation processes are intrinsic to the resonator and its surface properties. Magnetic field effects and trapped flux losses are recognized and have considerable theoretical and experimental evidence to guide the designer. The remainder have been studied theoretically, and experimental verification is rapidly clearing up many previously unanswered questions.

C. Magnetic Field Effects

Dresselhaus and Dresselhaus (1960) developed an expression for the surface impedance of a superconductor, using the Boltzmann transport theory, the London model for the superconductor, and including the effect of a dc magnetic field, H_0. The change in relative surface resistance is

$$\Delta R_{\mathrm{s}}/R_{\mathrm{n}} = K(\omega, T)H_0^2 \tag{49}$$

which reduces to the Pippard expression at zero field.

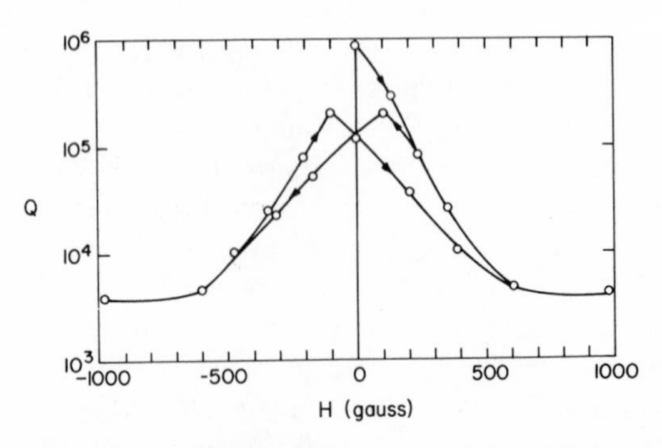

Fig. 7. Effect of a large dc magnetic field on the Q of a resonant circuit. Lumped parameter 30 MHz, lead plated; $T = 4.2°\mathrm{K}$.

There is a corresponding quadratic dependence in the surface reactance. These results have been confirmed theoretically and experimentally by many investigators (Richards, 1962; Hartwig, 1963a; Slay, 1963; Stone and Hartwig, 1968, 1972; Garfunkel, 1968). The effect of varying H_0 is not reversible as H_0 is taken to $\pm H_c$, as illustrated in Fig. 7.

1. *Stationary fluxoids*

Because of the Meissner effect, magnetic flux can be trapped when a resonator passes through the superconducting transition. Trapped flux losses due only to the background earth field may dominate the Q. This dissipating effect was studied by Haden *et al.* (1966) (see also Haden and Hartwig, 1965, 1966) and Pierce (1967; Schwettmann *et al.*, 1965), and shown to be confined to the normal regions in fluxoid cores within the penetration depth by Victor and Hartwig (1968).

Trapped flux had been suspected by numerous workers investigating superconducting surface impedance. The experiments of VH were conducted at low VHF (60–340 MHz) on rolled foils of Sn (and other soft superconductors) to suppress the superconducting surface resistance and insure an adequate density of flux-trapping sites. Relative and surface resistance measurements were made as a function of background dc magnetic field and temperature.

The theoretical model was for an additive term to Eq. (26):

$$r = A_{\mathrm{P}}(\omega)f(t) + r_{\mathrm{h}}(0)V(t) + r_0 \tag{50}$$

Here, $r_{\mathrm{h}} = r_{\mathrm{h}}(0)V(t)$ is the equivalent surface resistance having the temperature dependence of the volume of normal metal enclosed by the penetration depth and cross-sectional area of the fluxoid core. The term $r_{\mathrm{h}}(0) = \beta H/H_c$ is the fraction of flux trapped when the circuit is cooled through T_c in a background field H. Thus $V(t) = \lambda(t)h(t)$, and

$$r_{\mathrm{h}} = \beta H/H_c[(1-t^2)(1-t^4)^{1/2}]^{-1} \tag{51}$$

VH shows that the temperature functions $V(t)$ and $\lambda(t)$ are very similar in shape to the Pippard $f(t)$, and so flux trapping losses were masked along with superconducting surface resistance at higher frequencies. Only when $A_{\mathrm{P}}(\omega)$ was negligible could r_{h} plot a straight line versus the correct temperature function and yield a constant, positive value of r_0 (see Fig. 8).

When the rolled foils were annealed, the Pippard term and r_0 dropped significantly, but there was little reduction in the value of β. This indicates that flux-trapping site density may lie below any level reached by annealing and hence is an ever-present loss mechanism. After subtracting the trapped

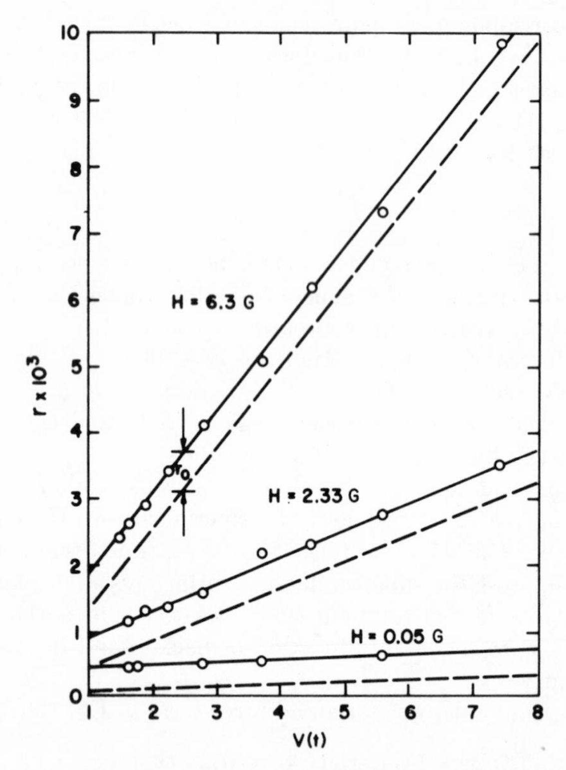

FIG. 8. Experimental resistance ratio of 66-MHz tinfoil LC circuit versus $V(t) = [(1-t^2)(1-t^4)^{1/2}]^{-1}$, showing agreement with flux-trapping model, and the presence of a temperature-independent r_0. (Victor and Hartwig, 1968.) ——: experimental data; - -: $[\Delta r/\Delta V(t)]V(t)$.

flux and residual resistance term, VH found that the magnitude $A_P(\omega)$ was close to the theoretical value.

2. *Nonstationary fluxoids*

Trapped fluxoids can also generate losses by oscillating around their pinning center or moving between different pinning centers while being driven by rf field.

The effect was investigated first by Gittleman and Rosenblum (1966) and was later discussed with respect to its importance in cavities operated below H_c by Rabinowitz (1971a), in efforts to account for some of the observed residual losses.

The motion of the fluxoid is described in terms of its displacement from the pinning center by the following formula:

$$M\ddot{x} + \eta\dot{x} + p(x) = j\phi_0/c \tag{52}$$

Here M is the mass per unit length of the fluxoid, η the flow viscosity, and $p(x)$ is the pinning force. The driving force is the Lorentz force and is given by the local current density j produced by the rf field and the flux quantum ϕ.

The mass of the fluxoid was calculated by Bardeen and Stephen (1965), and later by Gittleman and Rosenblum (1968), who give the result for a dirty superconductor. Despite some differences in the dependence of their results on geometry factors and material parameters, in both cases the mass is proportional to $1/H$, where H is the magnetic field in the fluxoid. There are also different formulas for the flow viscosity discussed in the literature.

The pinning force can be linearized for small-amplitude oscillations, $p(x) = -p_0 x$, where p_0 is a constant which is dependent on the unknown pinning strength in most cases, and on the magnetic field in the fluxoid. For larger oscillations, the potential was assumed by Gittleman and Rosenblum (1966) to be periodic and could be described by $p(x) = p_1 \sin(2x/d)$. This potential is based on the model that the fluxoids form a rigid lattice, and/or that the average distance between pinning centers is constant. The parameter d then gives the distance between the fluxoids and/or the pinning centers.

Equation (52) can be solved to calculate the power loss due to fluxoid movements. The normalized results explain the experiments well. However, there are differences in the measured parameters for viscosity and mass.

A resonator may be shielded from the earth's field by surrounding the Dewar with a high-permeability cylindrical shell. This has the effect of aligning the field with the cylindrical axis, which then can be canceled with a Helmholz coil. Residual fields less than 1 mG are readily achieved. It has been assumed but not demonstrated, to the authors' knowledge, that persistent currents set up by thermoelectric forces during cooling may generate some trapped flux.

In the case of a large ac field, $H_c \cos \omega t$, Rabinowitz (1971a) also derives a critical power loss associated with local heat generation at the fluxoid (see also Section III.G). The development is consistent with experimental observations of Turneaure and Weissman (1968) and Turneaure and Viet (1970) in both TE_{011} and TM_{010} niobium cavities, in which a sudden drop in Q_0 occurred at a peak ac magnetic field, $H_p < H_c$.

D. Surface Resistance of Imperfect and Impure Conductors

It seems to be established by nearly all experiments that impure superconductors and conductors without carefully polished surfaces have a higher than normal residual surface resistance. This effect is discussed in general by nearly all authors, who tried to achieve an extremely low residual resistance. However, no systematical experiments have been reported which investigate the dependence of the residual resistance on the number and kind of impurity atoms or the smoothness of the surface, or that give an empirical formula to describe the effect of the impurity distribution on the resistance. There has been no theory developed to date which results in a formula to describe the residual surface resistance with respect to its temperature dependence and frequency dependence as a function of the kind and number of the impurity atoms, or as a function of parameters describing the surface conditions. It is interesting to note, however, that the scattering of charge carriers on impurity atoms, crystal boundaries, and surface irregularities is included in the BCS theory and is shown to produce vanishing losses for $T \to 0$.

E. Residual Loss Due to Acoustic Coupling

Halbritter (1970a, 1971) and Passow (1972a) have called attention to a high-frequency surface loss mechanism based upon a theoretical treatment of Alig (1969). Transverse sound waves, generated by the incident electromagnetic radiation and negligible in normal conductors, are shown to have quantitative agreement with the temperature and frequency dependence of several authors (Szecsi, 1970, 1971; Hahn *et al.*, 1968; Turneaure, 1967; Pierce, 1967; Flecher *et al.*, 1969) for data below $t = 0.33$. Passow's analysis of acoustic waves launched by tangential E fields interacting with ions gives a surface resistance (in Gaussian CGS units) of

$$R = \frac{4\pi}{c^2}\,\omega^2\Lambda\left[\rho + \frac{\beta v_s \Lambda^3}{(1 + \omega^2\Lambda^2/v_s^2)^2}\right] \tag{53}$$

Equation (53) contains the MB and AGK superconducting surface resistance at the first term, $\Lambda = (8\pi)^{1/2}(\lambda_L \xi_0^2/3)^{1/3}$, $\rho = (2h/kT)(\ln 4kT/\gamma\hbar\omega)$ $\exp(-\Delta/kT)$, and $\beta = (4\pi/c^2)(e^2\eta^2 N/M v_s^2)$, where M is the ion mass, N the ion density, v_s the sound velocity, and η the fraction of N which is ionized.

Using published values for λ_L, ξ_0, Δ, and v_s for Pb, he gives the curve shown in Fig. 9. Data points from references cited fall on or above the curve, which has a broad peak near 1 GHz at $T = 0$ and falls off asymptotic to ω^2 on either side.

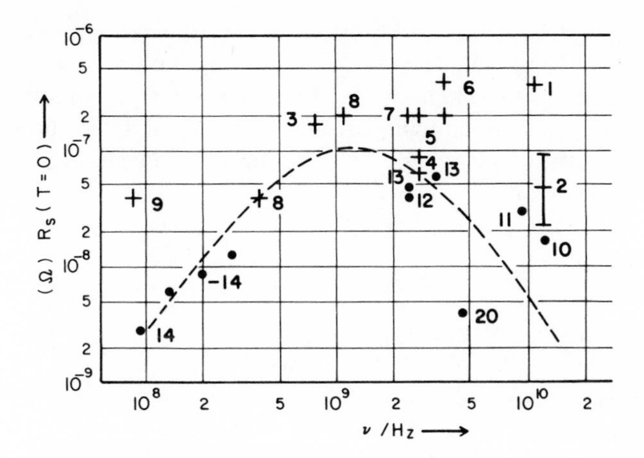

FIG. 9. Residual surface resistance measured in lead and niobium compared with theoretical acoustic loss. Trapped flux is absent, and the data points show the lowest low-field resistance calculated from measured values by subtracting the BCS term. The values for lower frequencies are measured in helix structures, and the values above 400 MHz are from cavity measurements. Lead values are crosses; niobium values are dots. The dashed curve shows the residual losses as calculated from Eq. (53) for lead by Passow (1972a). The calculations for niobium are more complicated, since acoustic standing waves are more probable in these cavities. The key for numbers appearing with the data points is: 1 (Turneaure, 1967); 2 (Pierce, 1967); 3 (Flecher, 1968a); 4 (Hahn *et al.*, 1968); 5 (Hietschold, 1968); 6 (Flecher *et al.*, 1969); 7 (Flecher, 1969); 8 (Szecsi, 1971); 9 (Fricke *et al.*, 1971); 10 (Turneaure and Weissman, 1968); 11 (Martens *et al.*, 1971); 12 (Kneisel *et al.*, 1971b); 13 (Halama, 1971); 14 (Fricke *et al.*, 1972a; Citron *et al.*, 1971); 20 (Kneisel *et al.*, 1972).

The calculations were performed in the Pippard limit for pure superconductors. This should be of sufficient accuracy if the scattering length of the electrons in the metal is large compared to the penetration depth, since then only the terms a_{P1} and a_{P2} of the kernel $H(q)$, Eq. (35), are contributing to the power loss, and the correction given by a_{P3} is small. The other terms of the series contribute only to the penetration depth, which depends on corrections given by a_{P2} and higher-order terms. Halbritter does not agree with this result and calculated considerably lower losses due to acoustic excitation. The question must be clarified by more definitive experiments, such as conducted recently by Goldstein and Zemel (1972) and by Abels (1967) and followed by a more detailed correlation between theory and physical behavior. Acoustic coupling is an intrinsic mechanism which is in addition to the superconducting surface resistance. It fixes a lower limit to the losses of a superconductor, and is a fundamental behavior in charge transport on surfaces.

Halbritter (1970a, 1971) also discusses phonon generation by the normal component of the E field, which may be much larger than the parallel component. The loss mechanism may be important at low frequencies, but poor electromechanical coupling between the field and the surface layers, plus the difficulty of specifying surface properties, makes it a speculative subject at this writing.

F. Losses from Nonideal Surfaces

Halbritter (1970a, 1971) analyzed several effects associated with nonideal surfaces and concluded that small fissures and joints would contribute important residual losses. The real surface will have fissures; microscopic ones at grain boundaries, strain cracks due to differences in expansion and contraction, machining grooves, and actual joints held together by pressure. Because of the greater chemical activity of the fissure surface and of the adsorption of O_2, H_2O, etc., the fissure will tend to fill shut with lossy metal oxides and other complexes. These regions will be efficient in phonon generation. They should display nearly temperature-independent losses, since there is no gap in their excitation spectrum at $T < 10°K$ (see Fig. 10).

The surface current is able to make an excursion around the fissure if its depth is less than λ. This creates a magnetic field along the fissure ($\nabla \times H_F = J$) and an electric field, E_F, across it due to the effect of the longitudinal field $E_{long} = ZH_F$. The current is partially converted to a displacement current, and a loss proportional to ω^2 results. Halbritter shows $E_F \gg Z_0 H(0)$ for TM modes in the fissure. In the case of a deep fissure or a superconducting joint, the current must cross as a transport and displacement current, so that $E_F/H(0)$ and the rf losses become frequency independent.

There is some experimental evidence to document the view that fissures and joints are sites for generating a temperature-independent residual resistance, with either a quadratic or zero frequency dependence. Szecsi

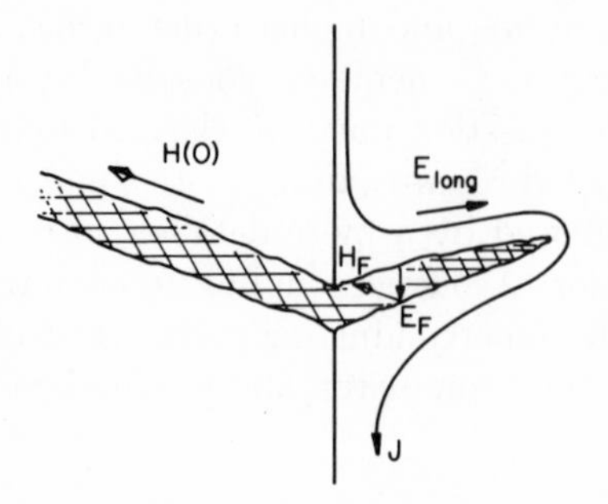

Fig. 10. Halbritter's model for the surface residual resistance caused by fields in an oxide-filled surface fissure set up by a perturbed surface current (Halbritter, 1970a, 1971).

(1970a, 1971) shows a larger $r_0(\omega^2) = 2 \times 10^{-7}$ (1 GHz) in a lead-plated copper cavity with TEM_{002n} modes. Grissom and Hartwig (1965) and Grissom (1965), for a Pb-plated helical resonator with a pressure joint, show a nearly frequency-independent Q_u from the fundamental at 48 MHz through each overtone to 1000 MHz. The Q_u dropped to about 70% of its initial value and remained essentially constant at each resonant frequency after repeated temperature cycling and reassembly. The equivalent residual resistance was $\sim 3 \times 10^{-7}\ \Omega$, or ~ 100 times the value from Turneaure (1972). The Q_u of a similar 132-MHz resonator (Hartwig and Grissom, 1965) rose from 10^7 to 2×10^7 when liquid He was allowed to immerse the helix and inner surface of the shield, showing the effect of better surface cooling.

G. High Magnetic Field Losses

There are losses associated with high rf field strength that limit the power rating of a superconducting resonator. A summary of experimental results is given in Table IV. The potential application of superconducting resonators to nuclear accelerators has stimulated efforts to understand and control the high field losses. The peak rf magnetic fields are proportional to $(Q_0 P_c)^{1/2}$, where P_c is the incident power. The voltage gain per foot is proportional to $(P_{dis} L r)^{1/2}$, where P_{dis} is the dissipated power, L is the length, and r is the shunt resistance per foot (Rabinowitz, 1970a, b, 1971b; Garwin, 1972). These losses are aggravated by local heating.

The critical magnetic field in a superconducting bulk sample can be calculated by the difference of the free energy between the superconducting and the normal conducting states. If the magnetic field energy is larger than the field energy at the critical field, then the superconductor becomes normal conducting. From BCS theory, the critical field can be obtained by setting the diamagnetic energy of the superconductor equal to the condensation energy:

$$H_c^2(0)/8\pi = \tfrac{1}{2}\eta(0)\,\Delta^2(0) \tag{54}$$

The critical fields, at $T = 0$, of several superconductors which are of possible interest for application in rf and microwave devices are given in Table IV.

The temperature dependence of the critical field can be derived from the two-fluid model, as well as from BCS theory. The following approximate formula is used:

$$H_c = H_c(0)[1 - (T/T_c)^{1/2}] \tag{55}$$

The formula is in rather good agreement with measurements made with dc fields. Even for lead, a rather strongly coupled superconductor, the error

TABLE IV

MEASURED FIELD STRENGTH IN CAVITIES[a]

Fabrication technique	T (°K)	f (GHz)	Q_0 10^9	Q_{max} at B 10^9	Q_{max} at B (Oe/G)	Q at B_{max} 10^9	Q at B_{max} (Oe/G)	E_{max} (MV/m)	Reference
Experimental cavities									
Fired	1.25	8.6	35			20	700 M		Turneaure and Viet (1970)
Fired	1.25	8.6				8	1080 M		Turneaure and Viet (1970)
Fired	1.4	2.4		267	357		554 M		Halama (1971)
Anodized	1.4	2.4		1.73	452		485 M		Halama (1971)
Anodized	1.4	2.4		3.35	280		442 M		Halama (1971)
Anodized		3.7					360 M		Kneisel *et al.* (1972)
Anodized		2.2					470 M		Kneisel *et al.* (1972)
Anodized		2.6					360 E		Kneisel *et al.* (1972)
Helix waveguides									
Anodized	1.6	0.08				0.12	990 E	24	Fricke *et al.* (1972a), Citron *et al.* (1971)
Separator structures									
Fired	1.8	2.9				1.4	470	17	Bauer *et al.* (1972)
	1.8	0.6				0.6	510	19	Bauer *et al.* (1972)
Accelerator structures									
Fired (pulsed)	1.8	1.3				2	250	16	McAshan *et al.* (1972)

[a] If available, the Q value measured at low fields, Q_0, the maximum Q value, Q_{max}, and the related field strength, B, as well as the maximum magnetic and electric field strength, B_{max} and E_{max}, together with the Q value at these fields, are given. M and E denote whether breakdown seems to have been caused by strong surface fields or field emission.

is less than 7% at the maximum. The critical field is also the natural upper limit for high-frequency fields on the surface of a superconductor. Since the typical time constants for the transition from the superconducting to the normal-conducting state is small compared to the period time of the alternating field, the peak value of the field has to be taken.

It was first shown by Easson *et al.* (1966) that the critical ac current in a lead bismuth eutectic ring was lower than the dc current, and was dependent on cooling. They operated the ring at 60 Hz and at several temperatures, and found a sharp rise of the critical current at the λ point of the liquid helium bath. Since the cooling effect of helium below the λ point is much better than above, they concluded that the current limitation for ac fields can be given, not only by the critical current, but also by the power loss inside the conductor leading to local heating above T_c. The effect was established by showing that the ac current limit was not changed if an additional dc persistent current was induced in the ring.

Rabinowitz (1971a) calculated the possible magnetic breakdown fields given by improperly cooled fluxoids, after Halbritter (1969a) had suggested that the findings of Easson *et al.* (1966) might also have importance for very high frequencies as used in radio and microwave cavities. Rabinowitz first calculated the temperature distribution in a fluxoid and its dependence on the peak magnetic field of the high-frequency wave by solving the heat-flow equations. He assumed that the fluxoid has cylindrical geometry and lies in a distance d parallel to and below the metal surface. Its diameter is denoted by a. The solution of the heat-flow equation results in the relation between the field and the temperature on the surface of the fluxoid. Rabinowitz assumed that the field on the surface of the fluxoid in the presence of an applied field H_a might exceed the critical magnetic field at the temperature calculated for the fluxoid surface. That means that the condition

$$FH_p + H_a = H_c = H_{c0}[1 - (T_a/T_c)^2] \tag{56}$$

would give a criterion for breakdown if the relation between the field in the fluxoid FH_p and the temperature T_a on its surface is inserted. F gives the fraction of the field which is trapped in the fluxoid. H_a is explained by Rabinowitz as an applied dc field and/or the field penetration from a fluxoid.

The criterion calculated from this equation is given in Eq. (57):

$$H_{pc} = gH_1\left[\left\{1 + \frac{2}{g}\frac{H_{c0}}{H_1}\left[1 - C_1 - \left(\frac{T_m}{T_c}\right)^2 - \left(\frac{H_a}{H_{c0}}\right)^2\right]\right\}^{1/2} - 1\right] \tag{57}$$

Here, $g = \lambda/a[\ln(a/b)]^{-1}$ and $H_1 = k_1T_c/2\rho_nH_{0c}$. Rabinowitz calculated the breakdown fields for three sets of parameters C_1 and K_1 describing the

TABLE V
ONE OF THREE SETS OF MAGNETIC BREAKDOWN FIELDS CALCULATED BY RABINOWITZ FOR THREE DIFFERENT VALUES OF THE THERMAL CONDUCTIVITY[a]

Bath temperature (°K)	H_{pc}	H_{c0} (Oe)	H_a	k_1 (W/cm °K²)	C_1 (W/cm)	T_a (°K)
1.0	690	4000	2400	0.045	0	4.5
1.8	660	4000	2400	0.045	0	4.6

[a] Here the thermal conductivity and its temperature dependence are taken from Styles and Weaver (1968). It is assumed that $\ln(b/a) = 12$.

heat conductivity of niobium. To calculate ρ_n, the normal electrical resistivity, he used the Wiedemann–Franz law, which gives ρ_n in terms of k_1. Some of the numerical results published by Rabinowitz are given in Table V. Rabinowitz pointed out that the observed breakdown fields are of the order of the fields measured in niobium cavities.

In a later paper, Rabinowitz (1970a,b, 1971b; Garwin, 1972) calculated also a similar criterion for breakdown caused by fluxoids perpendicular to the metal surface, as well as for superconductors of the second kind. The expressions are similar to Eq. (57). He also investigated the frequency dependence of the effect for the case of perpendicular fluxoids.

Halbritter (1969a) has given an alternative argument and believes that the nuclei for breakdown are normal droplets which are embedded in the Meissner phase and induce a growth of normal regions due to heating. Further on, he gives arguments to explain the order of magnitude for the typical breakdown time by pointing out that a thermal explosion could happen if the effect is somewhat advanced. However, he did not give a criterion which could be used to calculate the breakdown field nor a formula to calculate a time needed for the thermal breakdown.

Since the specific heat, C_p, is very low, ΔT can become so high that the superconductor will switch into the normal state. Thermal runaway comes from a local power dissipated, $R_s H^2/2$ per unit surface area, being balanced by $C_p\,\Delta T + K\,\Delta T$, or the local heat stored plus the heat conducted away. The steady-state temperature is reached when conduction losses balance generation. Power can also be generated by field and thermal electron emission, which is treated below.

Turneaure (1967) observed that Sn-plated cavities could be driven until the peak rf field, H_c^{ac}, approached the dc critical field, H_c, before normal surface resistance was encountered. Bruynseraede *et al.* (1971) observed

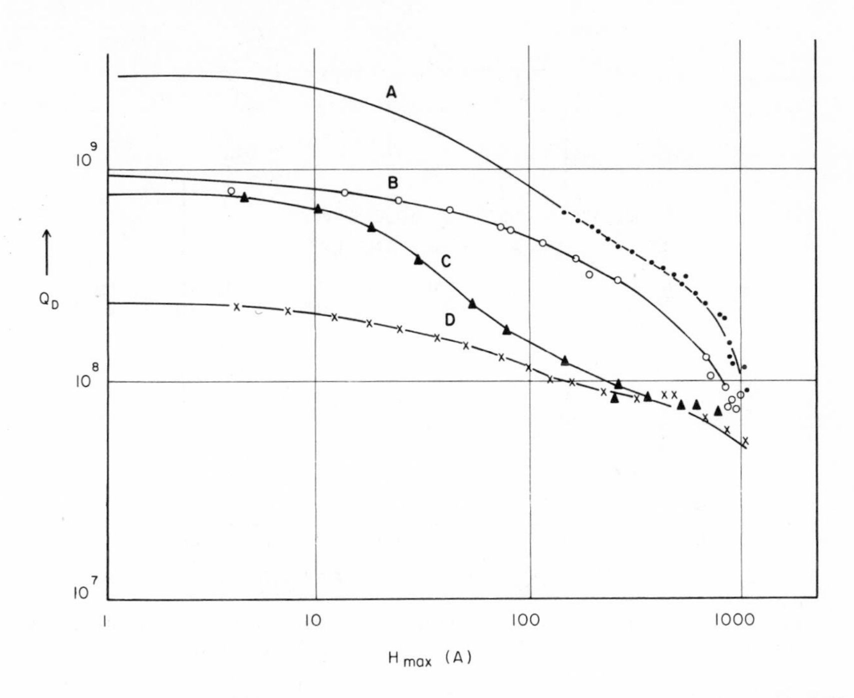

FIG. 11. Variations of measured Q in a helical waveguide dependent upon the field strength, measured before (see curves C and D) and after (A and B) treatment of the anodized niobium surface at 4.2°K (B and D) and 1.6°K (A and C). The breakdown and additional losses are attributed to field emission (Fricke *et al.* 1972a; Citron *et al.*, 1971).

an ac critical field as large as H_c (750 Oe at 1.8°K) in Pb. Although Nb (a Type-II superconductor) can achieve high Q's, Turneaure and Weissman (1968) showed that the critical rf magnetic field was well below H_{c1} for an X-band cavity. This was apparently associated with impurities at grain boundaries. In 1970, Turneaure and Viet produced several 8.6-GHz TM_{010} Nb cavities that measured $Q_0 \sim 10^{10}$ with peak surface fields of 1080 Oe and calculated electric fields of 70 MV/m. Kuntz is reported (Turneaure, 1970) to have reached about $2H_{c1}/3$ for Nb at 1.8°K. Rabinowitz, observing that L- and S-band cavities had lower Q and magnetic breakdown fields than X-band cavities, showed that his theoretical trapped flux model (Rabinowitz, 1971a) agreed with the observations of Turneaure (1971) at 1.36 GHz and Kneisel *et al.* (1971a) at S band. Metallurgical improvements and surface treatment have been responsible for the increase in H_c^{ac} for Nb cavities. See Figs. 11 and 12 for examples of experiments.

Martens *et al.* (1971), Kneisel *et al.* (1971a), and Garwin and Rabinowitz (1971) reported that Nb cavities anodized with Nb_2O_5 showed significant

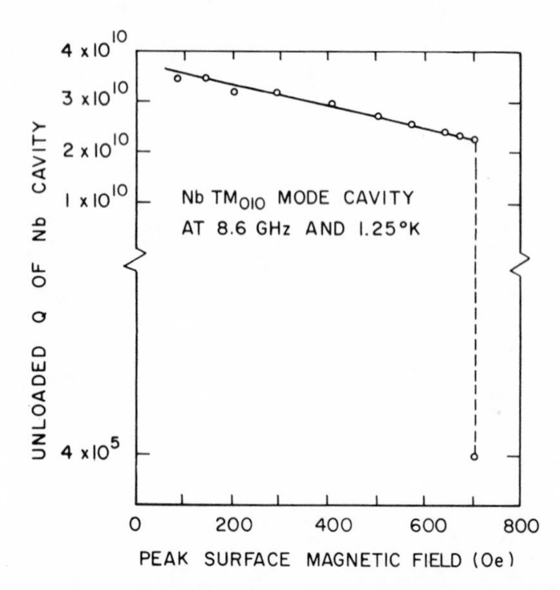

FIG. 12. Q value of a cavity as a function of the peak magnetic surface field. Breakdown is produced by the magnetic field (Turneaure and Viet, 1970).

improvements in their power-handling capability. Apparently, the intrinsic niobium oxide grows to about 60 Å in a way that does not improve the surface, but is influenced by lattice distortions and surface irregularities in a way that may actually enhance the residual resistance. The anodizing process, on the other hand, is driven by the external voltage instead of the local surface fields. This has the effect of converting surface Nb protrusions into Nb_2O_5, and also making the interface of metallic Nb and Nb_2O_5 a cleaner and purer one.

H. Field Emission

A large electrical field perpendicular to the metal surface may produce an electric breakdown in an electromagnetic device. Several different breakdown mechanisms are known; however, most of them are dependent on geometry, especially on the gap distance between two differently charged conductors. Field emission depends only on the surface field strength and introduces losses which degrade the performance.

Electrons may be drawn out of a metal surface by local fields of the order of 10^7–10^{10} V/m. These fields are enhanced at submicroscopic irregularities where the average applied field is very much lower.

TABLE VI
CALCULATED FIELD EMISSION CURRENT DENSITIES OF NIOBIUM AND LEAD (WORK FUNCTION 4.0 eV)[a]

E (V/m) $\times 10^{-9}$:	2.0	3.6	5.2	6.8	8.4	10.0
$\log j$ (A/cm^2):	−0.99	4.66	6.87	8.07	8.84	9.39

[a] From Dolan (1953).

An expression to calculate field emission current densities has been developed by Fowler and Nordheim (1928) on the basis of wave theory. They obtain the current density

$$J = 1.54 \times 10^{-6} \frac{E^2}{\phi} \frac{4(\mu\phi)^{1/2}}{\phi + \mu} \cdot E^2 \exp\left(-6.8 \times 10^7 \frac{\phi^{3/2}}{E}\right) \text{ A/cm}^{-2} \quad (58)$$

Here, E is in volts per centimeter, ϕ is the work function of the metal, and μ the energy of the Fermi level inside the metal relative to the bottom of the conduction band.

The theory has been subsequently modified by several authors, and a summary of their work is given in Good and Muller (1956). The more exact formula was evaluated numerically by Dolan (1953) for several different values of work function. The results for $\phi = 4.0$, which is the work function of lead and niobium, are given in Table VI.

The calculations show possible current densities up to 10^9 A/cm^2. Such a current, however, would be well above the critical current in a superconductor. This current can be calculated easily from the kinetic energy transfer which is necessary to break a Cooper pair (see, for example, Newhouse, 1964). The critical current in a superconductor is thus obtained to be

$$J_c = en\Delta/mv_F \quad (59)$$

The evaluation of this formula shows that the critical current is about 2×10^8 A/cm^2 for lead. The Fowler–Nordheim law is confirmed by experiment, but electron emission at protrusion spikes on "real" surfaces will be enhanced manyfold over emission rates calculated from average fields. This result means that particularly small microscopic peaks may become normal conducting even if the average field emission current is rather small. Since no relation is known which can be used to calculate the possible peak current as a function of the average current, experimental proof of the effect must be accepted on the basis of indirect evidence.

Secondary emission and multipactoring (resonant secondary emission multiplication) are observed in cavities which reach high E-field levels, −10 MV/m. Secondary electrons are emitted by a primary electron impact in the range of 3–10 eV, and, for energies above ~200 eV, the ratio of secondaries per primary is two or more. Real surfaces show enhanced secondary emission, e.g., niobium with adsorbed gas (Bruining, 1954; Dammertz, 1971). Multipactoring occurs when the time of flight of an emitted electron to another surface is in synchronism with frequency. Depending upon the mode, frequency, and field intensity, multipactoring can cause a large loss or it can be avoided with good design. All of these effects of electron loading will degenerate the Q and stimulate magnetic breakdown at the areas of the cavity surface where the impact energy is converted to heat.

I. Nonlinear Effects

The surface impedance of a superconductor and the numerous residual loss mechanisms discussed above are generally nonlinear or parametric. The high-frequency performance of superconducting devices is subject to their interaction in ways not observable at room temperature. For example, the change in penetration depth with temperature will introduce a change in surface reactance. This, in turn, will shift the resonant frequency of a superconducting cavity by a large amount compared to the cavity bandwidth. Similarly, free electrons from field emission at a wall will perturb the dielectric properties of the cavity and also shift the resonant frequency. The impact of these nonlinear effects on device performance are discussed in the next section.

IV. Design and Performance of Superconducting Devices

A. Design Considerations

A high-frequency superconducting device is usually referred to as a resonator or cavity in this chapter. Nonresonant structures are designed to have the optimum performance in some frequency range, and so some of what is said about design of resonators may apply to transmission lines and other devices.

1. *Configuration*

The geometry of a resonator determines the frequency and modes. Bulk circuits, consisting of helical inductors and plate capacitors, resemble conventional LC circuits. They can be conveniently fabricated for about

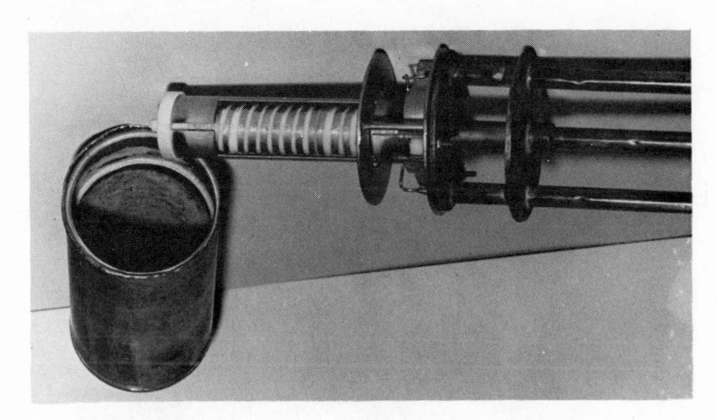

FIG. 13. *LC* circuit used by Victor and Hartwig to measure flux trapping (Victor and Hartwig, 1968).

100 MHz and below, and below 10 MHz they become very complex. Inductors with many turns, and even more than one layer of turns, and capacitors with many interleaved plates require considerable mechanical support by dielectric materials. Design of an *LC* circuit becomes a compromise between mechanical stability, *Q*, and tunability. *Q* is sacrificed as the dissipation in dielectric materials becomes the dominant loss mechanism.

Figure 13 is an *LC* circuit used by Victor and Hartwig (1968) to measure trapped flux losses in various superconductors. The test specimen was a rolled foil used to form the inductor. The capacitor plates and shield were electroplated with pure lead to tin, and the coil form was machined from Teflon. The entire circuit could be disassembled quickly to change samples. The resonant frequency range was from 60 to 70 MHz depending upon the extended length of the inductor helix.

Above about 100 MHz, the most convenient geometry is the helical resonator. The TEM fields are rather complicated to calculate (Hallen, 1930; Sauter, 1971), but the resonant frequency can be selected by nomograms developed by Macalpine and Schildknecht (1959) as a function of the geometric parameters. Using normal surface resistivity values for copper, it is possible to calculate a geometry factor, Eq. (44), and thereby calculate or estimate the superconducting *Q* limit set by the Mattis–Bardeen theory. Figure 14 is a helical resonator used by Hartwig and Grissom (1965), Grissom and Hartwig (1965), and Grissom (1965) to operate at 40 MHz and odd harmonics up to 1 GHz. The resonator was used to measure dielectric dissipation of materials which capacitively terminated the helix.

At about 700 MHz, the helical resonator degenerates into a quarter-wave transmission-line cavity. This configuration is useful above 1 GHz and at

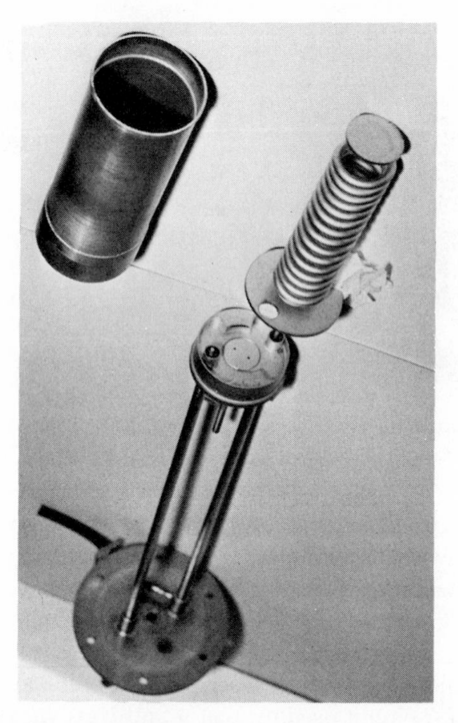

FIG. 14. Helical resonator used by Grissom and Hartwig to measure dielectric dissipation (Hartwig and Grissom, 1965; Grissom and Hartwig, 1965; Grissom, 1965).

any frequency is physically smaller than a standard cavity. Figure 15 is a sectional view of an 870-MHz re-entrant cavity used by Stone *et al.* (1969a) to demonstrate photodielectric tuning of the resonant frequency.

Microwave cavities that operate in TE and TM modes can be designed using standard calculations (Ramo *et al.*, 1965). This configuration becomes smaller in proportion to the wavelength for a given mode. At X band and above, the volume of a single cavity is small, and it is convenient to cool.

The geometry factor, defined in Eq. (44) in terms of Q and R_s, becomes

$$\Gamma = \mu_0\omega \int H^2 \, d\text{Vol} \Big/ \oint H^2 \, dS \tag{60}$$

Ramo *et al.* (1965) show theoretical methods to calculate Γ for rectangular, cylindrical, and spherical resonators. The field configurations are simplest, and their geometry factors are largest. Pierce (1967) and Turneaure (1967) give a value of $\Gamma = 752\ \Omega$ for a 2.85-GHz cavity in the TE_{011} mode. Turneaure and Weissman (1968) also used a TE_{011} cavity at 11.2 GHz, which

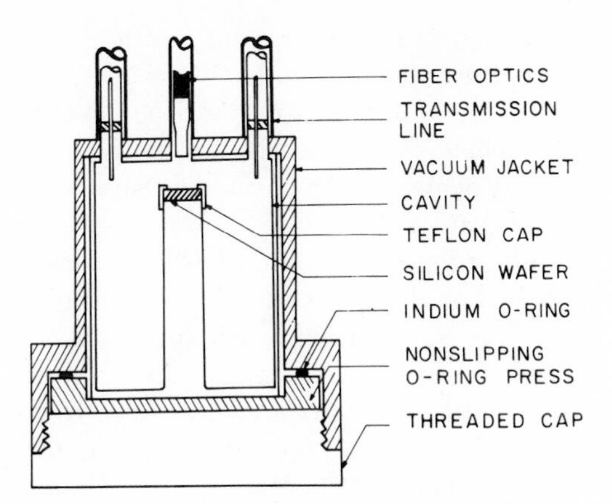

FIG. 15. Re-entrant cavity sealable for evacuation. Silicon wafer alters the resonant frequency due to its photodielectric response (Stone *et al.*, 1969a).

resonated at 8.4 GHz in the TM_{010} mode, and had Γ values of 752 and 301 Ω, respectively. Macalpine and Schildknecht (1959) give nomographs for helical TEM resonators from which a geometry factor can be calculated. For example, a copper resonator at 100 MHz, with a four-turn helix in a 3-in. shield can, has a room-temperature Q of 2200. Using $2.61 \times 10^{-3}\ \Omega$ for the surface resistance, the corresponding value for Γ is 5.75 Ω. An LC tank circuit will have an even lower value, since its field distribution is most unfavorable.

Using the published data for a given superconductor and the solutions to the MB surface resistance, a designer can estimate quite closely the maximum Q that could be observed for a given configuration and temperature.

2. *Materials survey*

Plenty of space is devoted to niobium cavities in current papers on rf applications of superconductivity, since the critical temperature, critical field, and surface resistivity make it a good material for simple microwave cavity structures. The disadvantages are its high initial cost, high reactivity to oxygen, need for elaborate fabrication facilities, and its relatively uncooperative metallurgy, which finds it naturally rich in tantalum impurities, very willing to absorb carbon, and hard in the sense of machinability. All of these problems can be suppressed at some cost.

Pure tin, indium, and lead are commonly used superconductors. The theoretical surface resistance has been observed to agree with experiment, as discussed in Section II. Lead, tin, and indium are rather soft metals and are used mostly as films deposited on copper or other metal supports or on a dielectric substrate. Pure lead and tin have also been successfully electroplated. The surface of these metals is rather delicate, and special care has to be taken to avoid chemical and mechanical deterioration. Lead wire is available commercially which could be used in superconducting devices; however, they are also rather delicate.

On the other hand, ordinary lead–tin solder has a high T_c (7.2°K, if it is rich in Pb), is applied by flowing it on relatively cheap copper substrates of any shape at modest temperatures with simple fluxes, retains a good surface finish indefinitely, and makes fairly reliable superconducting pressure joints. The disadvantages are lower critical field (by less than an order of magnitude) and higher surface resistance (by one or more orders of magnitude at the same reduced temperature). The designer does not lack for choices within this spectrum. Tables VII and VIII give useful physical data.

Dielectric materials for surface protection, supporting circuit structures, and separating evaporated thin film superconductors are limited to those

TABLE VII
PHYSICAL CHARACTERISTICS AT LOW TEMPERATURE[a]

Material	Electrical resistance (Ω cm $\times 10^6$)	Thermal conductivity $\left(\frac{\text{W}}{\text{cm °K}}\right)$	Specific heat $\left(\frac{\text{W-sec}}{\text{g °K}} \times 10^4\right)$	Expansion $\left(\frac{L_{273} - L}{L_{273}} \times 10^5\right)$	Ultimate tensile strength (lb/in.^{2} $\times$ 10^{-3})
Aluminum	0.121 (20)	26	101 (20)	358 ($\sim$0)	43–67 (20)
Brass	4	0.01		346 ($\sim$0)	68–87 (20)
Tin	0.04	56	2.36	271 (90)	
Lead	0.8 (20)	18	580	715 ($\sim$0)	
Niobium		0.08	3.15	560 (61)	
Stainless 304	50	0.002	46	263 ($\sim$0)	223–285
Inconel	103	0.005		210 ($\sim$0)	125–194 (20)
Copper	0.02 (20)	7	79.5	293 ($\sim$0)	18–76 (20)
Tantalum	0.5	0.23	1.8	140 (83)	112–208 (20)
Pyrex		0.001	230	48.5 ($\sim$0)	10.2 (20)
Fused quartz		0.5–4.0	251	−9.6	
Teflon		0.0004	766	1850	15 (77)

[a] The data are taken at 4.2°K unless followed by a number in parentheses indicating the temperature at which it has been taken. From "Superconductive Devices" by J. W. Bremer. Copyright 1962 by McGraw-Hill Book Company. Used with permission of McGraw-Hill Book Company.

TABLE VIII
MAXIMUM IMPURITY LIMITS (PPM) FOR NB

Al	20	Fe	50	Pb	20
B	2	H	10	Si	50
C	30	Hf	100	Sn	20
Ca	20	Mg	20	Ta	300
Cd	5	Mn	20	Ti	40
Co	20	Mo	50	V	20
Cr	20	N	60	W	100
Cu	40	Ni	20	Zr	100
		O	150		

with low rf losses and compatible properties in the superconducting device. These are discussed in Section IV.B.

Various tabulations of electrical, mechanical, and thermal properties of materials are found in other chapters in this book. Table IX is included here to develop several points of concern to a designer. Many materials

TABLE IX
TABLE OF LOSS TANGENTS OF DIELECTRIC MATERIALS[a]

Frequency (MHz)	Teflon	Quartz SiO_2	Ceramic Al_2O_3	Sapphire single-crystal Al_2O_3
34		2.0	1.2	0.35(0.5)
90	0.4(0.7)	1.5(2.0)	0.8	0.2(0.3)
119	1.2[b]			
136	0.9(1.2)	2.2(3.1)	0.7(0.8)	0.2(0.3)
185	0.6(1.5)		0.8(1.1)	
232	1.8(3.0)	1.8(3.0)	0.5(1.5)	
284		1.0(2.0)		
8,600	3.03[c]			
9,000	1.33			
14,000			1.4[d]	
Relative dielectric constant	2.0	4.4	9.0	9.0

[a] Given are the lowest measured values at 4.2°K. All values are multiplied by 10^6. The figures in parentheses give the highest measured values which are still in agreement with the error of the measurement. All values not indicated by a specific reference are measured by Mittag *et al.* (1970).

[b] Hartwig and Grissom (1965).

[c] Skizume and Vaher *et al.* (1962).

[d] Arams *et al.* (1967).

become more brittle when cooled due to a phase transition, but they also become much stronger. Specific heat and thermal expansion effects virtually disappear at very low temperatures. Teflon is an excellent dielectric at low temperatures, but its unusually high specific heat at room temperature means it requires hundreds of times more liquid He to cool it than an equal weight of many common metals. Although pure metals are usually good thermal conductors, stainless steel is so poor that it is a good material for reducing heat loads between the bath and room temperature. The tubes shown in Figs. 13 and 14 are thin-wall stainless-steel suspensions for the resonator and also serve as the outer conductor of the coaxial rf lines which couple the resonator outside the Dewar.

The values listed in Table IX are not applicable to an arbitrary sample of the material because of their dependence on individual defects, impurities, and stresses. Variations may be several orders of magnitude in the case of normal electrical resistivity.

3. *Power levels*

Since the peak energy stored at resonance is shifted from the electric field to the magnetic field twice each cycle, $\epsilon E_{\max}{}^2 = \mu H_{\max}{}^2$, how the cavity power level is limited will depend upon local tolerances to voltage breakdown, field emission, or magnetic breakdown. Theoretical limitations, discussed in Sections III.G, H, and I, are subject to considerable local variation from the average. A knowledge of the E and H distributions may improve the design, for example by selecting a mode which has no normal E field on the wall to minimize field emission.

4. *Assembly and closure*

The practical necessities of fabrication and assembly pose rigorous problems for the designer. The device must be machined in conveniently shaped pieces for assembly (and plating if that is the means to provide a superconducting surface). The wall must be thin enough to insure good cooling, and thick enough to damp temperature fluctuations and provide strength. The parting plane should not interrupt large surface currents, or losses will be introduced at the joints. If the resonator is to be evacuated, the seal should be reliable and not develop leaks in service. Input and output coupling should be designed to prevent cross-talk between the probes or apertures, or else there will be extraneous modes introduced and deterioration in performance. The design should permit normal maintenance, such as replating. If the resonator is to be used as an experiment chamber or optical detector, as described in Section VI, consideration should be given to convenience in changing samples.

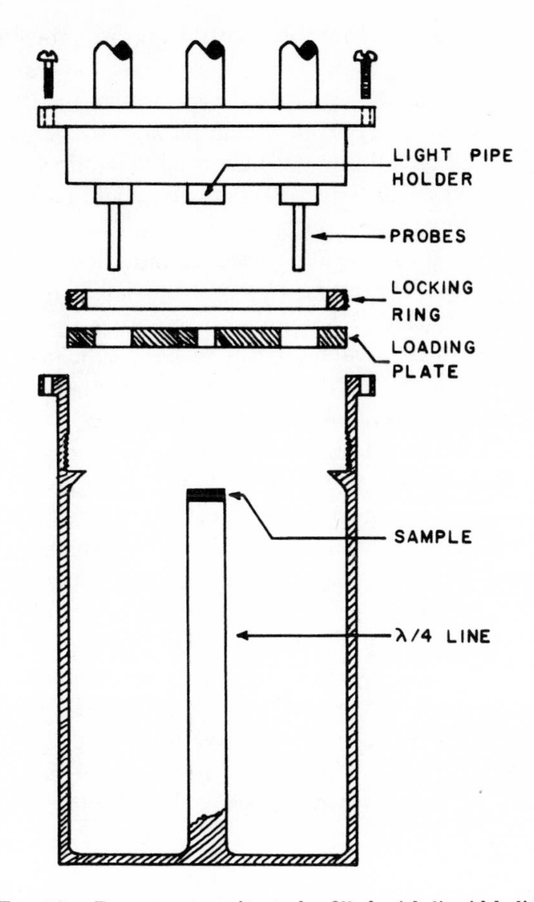

FIG. 16. Re-entrant cavity to be filled with liquid helium.

Figures 15 and 16 show two quarter-wave re-entrant cavities in the 700- to 900-MHz range used to observe and measure optical properties of semiconductor samples. There are several different design features incorporated in the two. Figure 15 shows an evacuated copper cavity made up of a cylindrical shell, two end plates, and a central post. The post was fastened to the bottom plate, and the top plate was fastened to the shell with high-melting-point solder. The cavity shell was then soldered to the bottom plate with medium-melting-point solder and the cavity filled with a low-melting-point alloy. The cavity was then sawed in half along a plane containing the cylindrical axis. The mating faces were milled flat and an indium foil gasket cut to replace the kerf and restore the cylindrical symmetry. The alloy was removed, and each half-cavity was then in a convenient configuration to be electroplated with a uniform thickness of superconductor.

There are several features to this design which justify the detailed description. By providing a parting plane parallel to the surface current direction, there would be a minimum loss in joints, and the sample, shown as a silicon wafer held in place by a Teflon cap, could be readily changed. Note that the holes for the coupling probes and fiber optic light pipe are very much smaller than the sample to reduce radiation losses. Evacuation is done through jackets shrouding the transmission lines and light pipe. The indium O-ring prevented helium from entering the cavity, and the gasket did not have to be a vacuum seal as a consequence. Thermal contact was made through the cavity top and bottom plates in contact with the bottom cap and vacuum jacket arrangement.

The cavity in Fig. 16 was designed to keep the sample thermally coupled to the helium bath by immersion. Helium liquid could flow into the cavity at the top, and helium vapor could escape the same way. The closure surfaces are the outer portion of the loading plate and a shoulder on the inside wall of the cavity. This is a region of nearly minimum current density. Uniform pressure was made against the entire closure surface with a locking ring. The presence of liquid helium in the cavity gave rise to a very slight bubble noise in the resonant frequency which could be controlled. This design has also given consideration to convenient disassembly and sample changing. Little change was observed in the loaded Q on reassembly because small variations in the resistance of the superconducting pressure joint were masked by the dielectric losses in the sample.

5. *Coupling*

Figure 17 is an equivalent circuit of a parallel LC resonator with a capacitive coupling probe. The probe couples energy in or out of the circuit and also has an effect on the Q and resonant frequency. If $x = C_2/C$ is the coupling factor, or ratio of the probe to circuit capacitance, the resonant frequency change is

$$(\omega_0 - \omega)/\omega_0 = \tfrac{1}{2}x \tag{61}$$

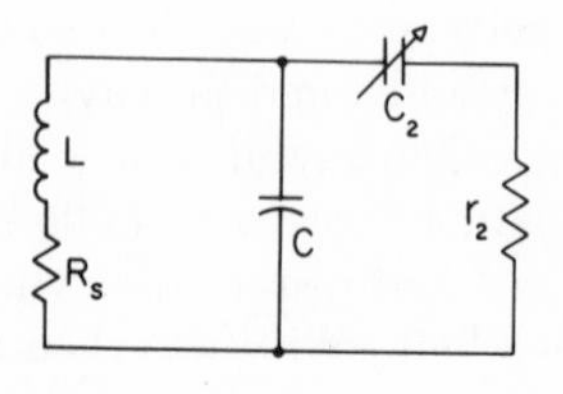

FIG. 17. Equivalent circuit of an LC superconducting resonator (Hartwig, 1963a,b).

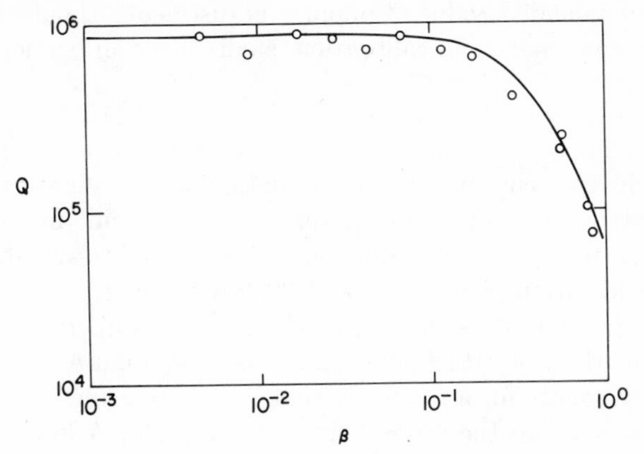

FIG. 18. Effect of coupling on Q of a 30-MHz LC circuit, electroplated with Pb, at 4.2°K. β is the voltage transfer ratio. o: experiment; —: theory (Stone and Hartwig, 1968, 1972).

if $x \ll 1$. The effect on loaded Q is

$$1/Q_L = (1/Q_0) + (r_2 x^2/\omega_0 L) \tag{62}$$

where $Q_0 = \omega_0 L/R_s$, and r_2 is the load impedance seen by the probe. A set of input and output probes which can vary C_2 by simply twisting the coaxial transmission lines is seen in Fig. 13. Inductive loop probes are also suitable, and a corresponding set of relations exists. One coupling loop probe and one capacitive probe make a convenient input-output pair because frequency changes can be cancelled as coupling is reduced on both probes simultaneously (Hartwig, 1963a,b).

The coupling must be largest at room temperature for convenience in testing while the Q is very low. Below T_c the coupling can be reduced until no coupling effects are seen and the signal is still adequate. Figure 18 is the confirmation of Eq. (62). It is noteworthy that, as coupling is reduced, the field intensity increases in the coupling-limited range; thus the output signal level remains virtually undiminished until the Q is no longer determined by the coupling alone. Figure 14 shows straight coupling probes which are adjusted by varying the distance of penetration into the resonator fields.

Coupling to cavities is by apertures between the cavity and the coupling waveguide. Variations in coupling can be achieved by an iris diaphragm, inductive loop, or capacitive probe. The reader is referred to the literature on microwave theory and techniques (Ramo *et al.*, 1965) for design information, since the subject becomes more complex as the dimensions approach the wavelength. When the cavity is critically coupled, the measured Q is one-half the unloaded value. Coupling adjustments should include this situation for convenience in calibration, as discussed in Section IV.C.

6. *Shielding*

Superconducting circuit losses by radiation will dominate those due to surface resistance at almost any frequency. The circuit in Fig. 13 had a room-temperature Q of about 200, and at $t = 0.8$ the Q was about 2×10^6 with the shield can in place and about 2000 without it.

Proper design considers the size of the shield, its apertures, seams and joints, and method of attachment to the header. Figure 16 shows screws which were adequate for a resonator that liquid He could enter freely. For an evacuated version, the screw holes were omitted. A low-melting-point solder insured that the plating would not be damaged by the heat. This same technique was used to close the helical resonator in Fig. 14. In this case, the thermal expansion coefficient of the solder was higher than the lead-plated brass shield. The superconducting pressure joint between the shoulder, seen on the inside of the shield can, and the loading plate at the top of the helix was, therefore, actually better as the system cooled.

This shield is a separate structure to an LC circuit, but will have the effect of a shorted turn coupled to the inductor. As a consequence, the resonant frequency will be influenced, since

$$\omega^2 = (LC)^{-1} = [(L_1 - M^2/L_s)C]^{-1} = [L_1(1 - k^2)C]^{-1} = \omega_0^2(1 - k^2)^{-1} \tag{63}$$

where L is the total circuit inductance, C the capacitance, L_1 the inductance without shield, M the mutual inductance between the shield and coil, L_s the surface inductance of the shield, and k the coupling coefficient. The shield inductance is the superconducting surface reactance parameter per unit frequency, or

$$L_s = X_s/\omega = \mu_0\gamma\lambda(t) = \mu_0\gamma\lambda(0)(1 - t^4)^{1/2} \tag{64}$$

where λ is the penetration depth and γ is the effective length/breadth ratio for shield currents. The coupling coefficient, k, will be dominated by the value of M which is a function of the separation between the circuit and the shield. The impact of temperature changes on frequency is discussed in Section IV.D.

The helical resonator has its own shield which functions as the outer conductor as the geometry approaches a coaxial cavity resonator at UHF. Shields of this type will have much lower current density than the inductor of the helix; hence, some liberties may be taken in their fabrication. The superconductor may be chosen for its durability, using lead–tin solder instead of electroplated lead, for example.

Resonant cavities are their own shields, and there are no currents except in the cavity walls. Full consideration must be given of the superconducting surface, therefore, as well as the location and size of apertures and parting planes.

7. *Cooling*

Several design parameters should be appreciated in relation to the cooling of a superconducting device. The refrigerant itself is convective liquid He or the cold plate of a helium refrigerator which cools by conduction. Below the λ point (2.1735°K), helium is a superfluid with a very high thermal conductivity. In any event, cooling is expensive and will place an economic constraint upon the design.

The cooling rates for a cavity will depend upon the thermal conductivity of the superconductor, which puts Nb at a great disadvantage compared to Cu or Pb, as seen in Table VII. The selection of materials will necessarily consider the power level, along with other factors in the design. For example, a niobium cavity will build up a larger temperature gradient per unit wall thickness for a given surface power density than will a leadplated copper cavity. The lower specific heat of Nb compared to lead or copper means greater temperature rise per unit density of thermal energy developed in the surface. The design of a niobium cavity for high power levels is clearly going to force a compromise between surface temperature and mechanical stability when the cavity wall thickness is set.

The cooling rate which is achievable with superfluid helium in helix pipes is limited by the possible temperature gradient along the pipe. This limitation leads to a maximum heat flux of about 2 W/cm^2. There is also a limitation for the heat flux from the metal to the superfluid helium; however, this limit will not be achieved in most practical applications (Passow *et al.*, 1970, 1971).

8. *Tunability*

A resonant circuit or cavity can be tuned in frequency by any method that alters the distribution of the electric or magnetic fields. When the Q is extremely high, the stability of the tuning method becomes as important as the method itself. The discussion of stability is taken up in Section IV.D.

Superconducting LC circuits can be tuned by the familiar techniques of changing capacitor plate separation or altering the mutual inductance between coils. The Meissner effect can be used to expel the H field from the inductor. The frequency shift, in that case, can be calculated by Eqs. (63) and (64), with suitable consideration of the position and size of a solid superconducting cylinder along the axis of the inductor.

Arams *et al.* (1967a) show that a series resonant *LC* circuit is preferred over a parallel configuration to maintain constant bandwidth over the tuning range and to minimize the detuning effects when adjusting bandwidth.

Helical resonators can be tuned by adjusting the terminating capacitance on the helix, as shown in Fig. 14, for instance. The adjustment can be geometrical or by changing the effective dielectric constant. Smaller effects, such as adjustment of the coupling probe position or elastically deforming the outer conductor, have been used. Grossly changing the resonant field structure, by inserting a Meissner slug into the helix, may generate spurious modes which deteriorate the performance.

Cavities are tuned by moving an end wall in the simplest case, but such a discontinuity would severely limit the Q if the wall were to be a sliding plunger. Flexible superconducting bellows can be used as a compromise between reduced tuning range and lower surface losses. The most used techniques for adjustment of a few bandwidths include various forms of mechanical deformation. High tuning rates can be achieved by the photodielectric technique described in Section VI.

B. Materials and Fabrication Techniques

The design of superconducting devices is much more limited in the choice of materials than the corresponding room-temperature device. This is true for the superconductor itself, as well as for any dielectric materials needed for support or electrical insulation. Both classes of materials should be selected for minimum loss at the operating temperature and frequency. In addition, they should have physical and chemical properties that give them a long useful life.

1. *Metal properties*

The selection of the conductor metal has to be done with respect to its superconducting properties, the feasibility of processing it, and its resistivity against chemical and mechanical aging. The transition temperature is its most important parameter. Lower surface resistances can be expected for greater T_c, since the losses are determined by a power of the ratio between the operation temperature and the transition temperature. The cryotechnique is simplest for an operating temperature at the boiling temperature of liquid helium, or about 4.2°K. Metals with a critical temperature well above this are preferred for devices in which power savings are important. The cryotechnique gets more expensive if lower bath temperatures are required, and it has to be estimated for each special application, whether or not the advantages given by power savings might compensate the higher cooling costs.

Niobium, lead, and technetium are the pure superconductors with the highest transition temperatures. Pure metals which could be used in devices based on the inductive behavior of superconductors and which have to operate close to its transition point are tantalum and mercury. Tin and indium, which are often used in experiments, have a transition temperature which is some below 4.2°K and, therefore, need a more complicated cryostat and pumping system, which might be a disadvantage in practical applications. Alloys and compounds can be made with nearly any transition temperature below about 15°K; however, there is little experience on how these materials perform in high-frequency fields.

Thin films usually show a slightly lower transition temperature than the bulk material. Technetium can be ruled out for possible application, since it is very expensive and radioactive. Mercury is a liquid at room temperature and, therefore, of little use for building devices. Lead, niobium, tin, indium, and lead–tin solder have been discussed already in Section IV.A.2.

a. Bulk niobium. Niobium is commonly used in nuclear reactors and therefore is commercially available in the form of solid ingots, sheets, square and round bars, extruded and drawn tubing, and forged cups. Nb production processes make it available in recrystallized fully wrought, arc-melted, or electron-beam-melted forms. The fabrication of a high-frequency device out of the raw material is not too difficult, since niobium can be rolled and machined, and tubes can be bent if some special care is taken. Parts of the device can also be joined by electron-beam weld. The primary producers are Fansteel, Kawecki, and Wah-Chang Albany. In most cases the specifications are the impurity limits given in Table VIII, and the stipulation that it be free of voids and inclusions as well as annealed.

Surfaces which are smooth enough and pure enough to show a really low surface resistance are more difficult to achieve. It seems necessary, even during the fabrication, to handle the pieces of the device very carefully. Several special treatments of the surface have been discussed and measurements made to show what kinds of surface have the lowest surface resistance. Turneaure and Weissmann (1968) first investigated niobium cavities and described possible surface treatments. Later several other groups published measurements and procedures for surface treatments of niobium. Halama (1971) published a review of the methods, and compared some of them with respect to their effects on the surface resistance and breakdown field.

High-temperature vacuum outgasing was proposed (1968) for niobium and found to give good results (Meservey and Schwartz, 1969; Rowell *et al.*, 1963) in Europe and at the Stanford High Energy Physics Laboratory (HEPL). A temperature above 1800°C is recommended to give a polish to the surface and to evaporate NbO. Heating times between 3 and 47 h are reported by various workers. Halama found an influence of the heating time on cavity performance not found by other authors.

Chemical polishing is another method to achieve smooth surfaces. Halama (1971) used a polishing solution consisting of 40%HF and 60% HNO at 0–5°C. After the desired amount of Nb is removed, the acid is displaced by water and then flushed with alcohol. The procedure is done without exposing the cavity to air. Finally, the cavity is evaporated dry and tested without exposure to air.

The third possible surface treatment is electropolishing. A method which reduces boundary etching was developed by Diepers and Schmidt (see Kneisel *et al.*, 1971a). A solution of H_2SO_4 + HF at 9–15 V is used.

Anodic oxidation is a fourth method discussed in literature. Martens *et al.* (1971) measured an improvement of the surface resistance, as well as of the breakdown fields, with this technique (also see Flecher *et al.*, 1969). Anodic oxidation is basically the electrolysis of water at the surface to be treated. The electrical conductivity of the water is increased by an NH_3 solution at 20°C. The current density was 2 mA/cm^2, and the resulting film thickness was 0.1 μ.

It has also been proposed to protect the Nb surface by reacting it with tin or nitrogen (Dickey *et al.*, 1971) to form Nb_3Sn or NbN. Before starting the treatment, the niobium surface must be cleaned carefully. This can be done with chemicals usually used to clean metals, such as alcohol, acetone, and trichlorethylene.

b. Bulk lead and alloys. For applications with lower requirements on the surface resistance, cavities and lines can be built from bulk lead. This is much cheaper, since lead can be machined very easily. However, Pb is very soft, and dimensional stability problems are severe. For low surface resistances, pure lead seems to be necessary, which rules out most of the commercially available lead and lead products. Bruynseraede *et al.* (1971) tested bulk indium–lead alloys in high-frequency fields and reported moderately low surface resistance.

c. Films and plated surfaces. i. SUBSTRATES. Since the current in a superconductor flows only a few penetration depths, or a few thousand angstrom units, below the surface, the superconductor thickness need not be any greater. It is not necessary, therefore, to fabricate the device entirely out of superconducting metal. This can result in such design improvements as lower cost, easier fabrication, better heat transfer, greater mechanical stability, and improved performance characteristics.

Several different methods are available to produce thick or thin superconducting metal films on a base metal or insulating substrate (Berry *et al.*, 1968). The requirements of a good substrate include compatibility with the intended superconductor and ease of fabrication into the desired geometry. It must be capable of reaching a high polish if the best surface is sought for the superconductor. Copper, brass, and aluminum are common metal substrates. Dielectric substrates, which are also used as supporting structures for superconducting wires, must be chosen for their low-loss characteristics if they are to be placed in an electric field region.

ii. ELECTROPLATING. Superconducting cavities and resonators made by electroplating Pb on copper substrates first produced near-theoretical Q and rf field levels at Stanford (Turneaure, 1967; Pierce, 1967), at Brookhaven (Hahn *et al.*, 1968), and at Karlsruhe (Passow, 1967a,b) at S band. Turneaure (1972) points out that a number of factors must be considered to produce plated Pb resonators with low surface resistance, including copper purity (OFHC is recommended by nearly everyone to simplify the surface treatment of the copper), purity of the lead fluoborate bath and the lead electrodes (99.999% purity is recommended), additives (such as animal glue) to control the surface smoothness (by enhancing the "throwing power" of the bath), electroplating current density, and the postplating rinse technique. (The comments in parentheses are the author's—WHH.) More work on electroplating has been reported by Carne *et al.* (1971).

The film thickness is usually several microns thick to exceed the penetration depth. Others have reported successful plated cavities with tin and indium deposited from the corresponding fluoborate bath.

2. *Evaporation and sputtering*

Vacuum evaporation and sputtering are well established techniques for depositing thin films of metals on relatively uncomplicated shapes. The practical problems of coating resonant circuits are very great for low-frequency LC circuits, but become simpler for cylindrical cavities at microwave frequencies. Metal atoms evaporated from a resistance heat source travel in straight lines and so tend to fall unevenly on targets that do not have significant radial symmetry with respect to the source.

Evaporated Pb and In films were investigated by Flecher (1968b, 1969, 1970) as a means to achieve low resistance at high frequencies. In a vacuum of 10^{-5}–10^{-6} Torr, the resulting surface resistance was from two to ten times the BCS value. The vacuum, as shown by Flecher, must be of the order of 10^{-9} Torr to produce a satisfactory superconducting film.

Niobium and tantalum, having much lower vapor pressures and higher melting temperatures, require electron-beam melting in higher vacuum before they can be successfully evaporated. The apparatus discussed by Gerstenberg and Hall (1964; Fowler, 1963) operates at 10^{-9} Torr before and less than 10^{-6} Torr during evaporation. The high chemical reactivity of these metals, particularly to oxygen, make these techniques generally unattractive. Sputtering is a related technique (Gerstenberg and Hall, 1964; Frerichs and Kircher, 1963) which has not produced low-resistance superconducting films (Frerichs and Kircher, 1963).

Evaporation techniques have produced some interesting effects where films have been made from two or more different metal layers. Hauser and Theurer (1965) have shown that the transition temperature of a lead–aluminum film sandwich varies from 1.2 to 7.2°K depending upon the thickness of the lead film. Others (Dickey *et al.*, 1971) have proposed Nb_3Sn on niobium. The dependence of the layer thickness is due to the proximity effect and is theoretically well understood.

Alloys and compounds of superconductors might also be used to produce films, and there is already some experience with their behavior in high-frequency fields. Flecher (1969, 1970) measured the surface resistance of a Pb–Bi alloy. His measurements found to be in agreement with the theory (Halbritter, 1969c, 1970d), Mason (1971) measured the behavior of a thin-film indium–tin alloy strip lines for tin concentrations between 0 and 23%. This is still a research area in many respects, but, for a good surface resistance with along useful life, Pb–Sn solder is an excellent and familiar alloy found around the laboratory. One author (WHH) has found it very satisfactory for applications where the BCS surface resistance is much less than residual losses introduced into the cavity by virtue of the application.

3. *Dielectric materials*

Dielectric materials in superconducting devices are used as substrates or supports for a conductor such as a coil form, for control of the wave velocity on a line or in a waveguide, and to protect or passivate a superconductor surface.

a. Dielectric parameters. The dielectric constant and loss tangent are the parameters most vital to the electromagnetic performance. To be useful in a superconducting device, however, a dielectric material must be compatible with other materials with which it is used. For that reason, the thermal expansion coefficient and other mechanical properties should be known at low temperature and should introduce no special problems when the device is brought to room temperature.

The state of knowledge, at this writing, is not yet sufficient for precise design, but enough is known about enough materials to permit good choices to be made. Even though the loss tangent of every useful dielectric as a function of temperature and frequency is not yet available, if the designer limits the volume of dielectric material in regions of high electric field, the added losses need not be excessive.

Table IX gives a summary of loss tangents measured by several authors (Hartwig and Grissom, 1963; Shizume and Vaher, 1962; Arams *et al.*, 1967b; Mittag *et al.*, 1970), at 4.2°K in the radio and microwave frequency range. The materials are Teflon (polytetrafluorethylene), quartz (single-crystal SiO_2), ceramic Al_2O_3, and single-crystal sapphire (Al_2O_3). These materials are probably the least lossy of those commonly recommended for use. Shizume and Vaher (1962) measured loss tangents of 4×10^{-5} and 3×10^{-5} for polystyrene and styrene, respectively, at 8.8 GHz and 4.2°K. Mylar was reported by Gottlieb and Garbuny (1964) to have a value of 1.8×10^{-4} at 500 Hz and 4.2°K. The loss tangent of liquid helium was found to be as low as 10^{-10} (Hartwig and Grissom, 1965; Mittag *et al.*, 1970).

The tabulations, such as given in Table IX, should be taken to have order-of-magnitude precision, particularly if material quality is not known to be close to that for which measurements have been published. Grissom and Hartwig (1966) show that there are large temperature and frequency dependences for the loss tangent of dielectric materials at low temperature. Fused quartz (glassy SiO_2), for example, is an excellent material for dimentional stability and strength, and much less expensive than single-crystal quartz. The loss-tangent–frequency–temperature surface, shown in Fig. 19, reveals considerable structure. There is clearly a peak in tan δ near 200 MHz and an even larger one below 40 MHz. The measurement techniques and theoretical considerations are discussed in Section VI.F.

b. Techniques for fabricating dielectrics. Most of the materials listed in Table IX are commercial products. Their surface must be cleaned and polished very carefully before depositing superconducting films. However, it might be a problem to get rather large sheets of the material which have a low average loss tangent. It is also difficult to get very thin sheets which can be used as a gap material in strip lines. Teflon sheets are available down to a thickness of 2.5×10^{-5} m. These sheets can have a surface treatment to improve the adhesion of metal film and then have a loss tangent of about 10^{-5}.

Thin films of dielectric material can be fabricated also by anodic oxida-

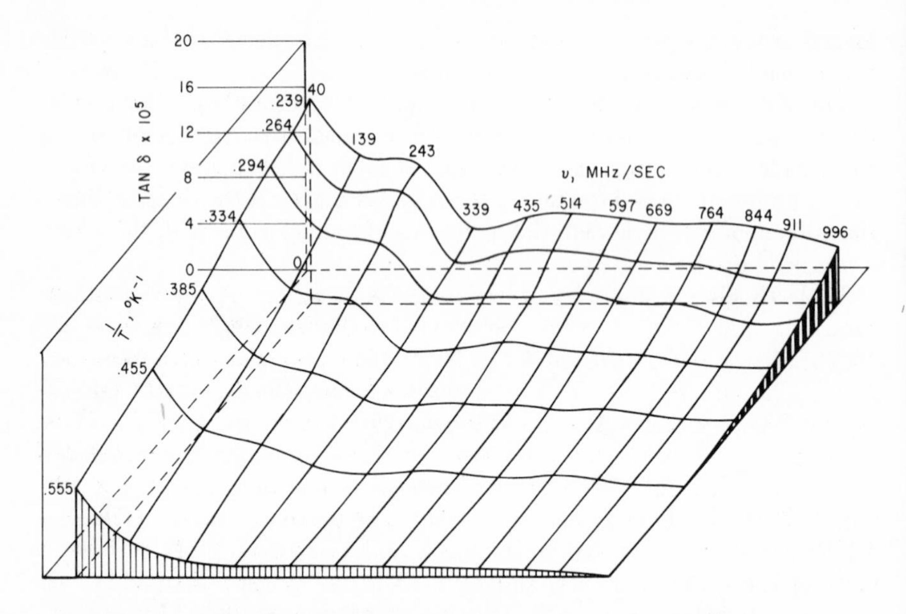

FIG. 19. The loss-tangent–frequency–temperature surface of fused quartz.

tion, evaporation, and chemical vapor deposition. Anodic oxidation to protect niobium cavity surfaces is discussed more fully below. Here the layers are too thin to produce measurable losses. Mason and Gould (1969) anodized tantalum sheets to form an insulator between the ground plane and the conductor of strip lines. However, the loss tangent is not known. The dielectric constant of Ni_2O_5 = 41 and of Ta_2O_5 = 22; therefore, it can be used very well to slow down waves on a strip line. SiO is a dielectric material which can be evaporated very easily and is often used as a protective layer. SiO_2 and Al_2O_3 films have been evaporated with an electron gun (see, for example, Maissel and Glang, 1970). It is not known whether the loss tangent measured in bulk material in Table IX also will be found in films. Chemical vapor deposition results in films with greater crystalite orientation than with usual evaporation techniques. However, the larger stress in crystals causes difficulties due to the temperature change suffered by a device operated at low temperatures.

The generation of geometrical patterns in films to form the desired geometry of devices, made by strip line techniques, can be done either by masking during deposition to permit only selected substrate areas to receive the film, or by uniformly coating the substrate with film, and then selectively removing the metal to generate the desired pattern. The mask necessary to perform the first method can be made either mechanically or by photolithographic techniques. In the second method, a photolithographic processing method is used to form the pattern. It is not known at this time whether the photoetching process deteriorates the superconducting surface resistance; therefore, it might be better to use masks if required dimensions are not too small. For very small dimensions, however, photoetching is necessary, and strip lines and devices down to a width of 3 μ can be processed without much difficulty.

4. *Fabrication techniques for niobium cavities*

Since rf components are fabricated in many shapes and sizes, and from several materials, it is not worthwhile to describe the state of the art in detail. The niobium cavity has been given so much development effort in so many laboratories, however, that a short summary is essential. The authors draw upon a survey by Turneaure (1972).

The early work (Turneaure and Weissman, 1968; Turneaure and Viet, 1970) on small research cavities consisted of machining cavity sections from solid Nb bar stock and joining them with demountable seals or electron-beam (EB) welds. This method becomes prohibitively expensive for large cavities due to the waste of raw material.

The Nb cavities for the superconducting electron linear accelerator at HEPL (Stanford) (McAshan *et al.*, 1972; Schwettmann *et al.*, 1967) have a diameter of 20 cm and an assembled total length of about 140 m. Raw material for the individual cavity sections is in the form of sheets about 5 mm thick. They are hydroformed into cups and stress-relieved at 975°C for 1 hr. They are then machined to final dimensions and joined by EB welding, as shown in Fig. 20.

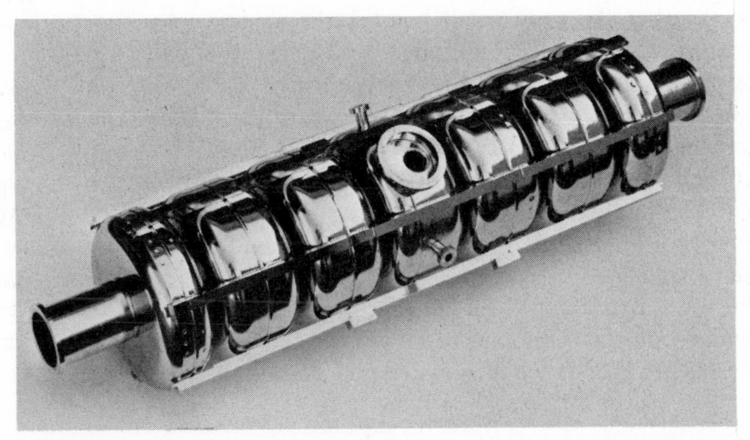

FIG. 20. L-band niobium cavity system for electron linac at Stanford.

A similar accelerator for the Physics Research Laboratory (Illinois) (Allen *et al.*, 1970) is being fabricated using forged Nb cups. S-band cavities at the Stanford Linear Accelerator Center (SLAC) are made by high-pressure coining and at HEPL by die forming.

At lower frequencies, helical resonators have been made by winding Nb tubing on a mandrel at Karlsruhe (Citron *et al.*, 1971) and winding niobium wire on a Teflon coil form at Union Carbide.

C. Measurement of Performance

The performance of a superconducting circuit or cavity is principally measured in terms of its Q and frequency of resonance as a function of various external variables, such as temperature and magnetic field.

1. *Measurement of Q*

The exact determination of Q for a superconducting resonator is much more complicated than the equivalent measurement on a normal one. There are four common methods: measurement of bandwidth, measurement of phase shift, transient measurements of stored energy decay, and comparison of reflected to incident steady-state power using directional-coupler separation. The latter is useful only at microwave frequencies.

Bandwidth measurements consist of observing the transmitted or absorbed signal and calculating Q as given by $Q = f_0/\Delta f$, where f_0 is the resonant frequency and Δf is the separation of half-power points. The phase shift of a signal at either half-power frequency is 45°, and this is the basis of another method of measuring Q. These simple techniques are useless for measuring the very small bandwidths achieved by superconducting resonators. Commercial electronic signal generators have insufficient stability or adjustment to produce a frequency signal with the resolution required, except the quartz crystal oscillator, which cannot be tuned.

Lerner and Wheeler (1960) describe a technique of bandwidth measurement for microwave resonators by observing the phase shift of a modulating signal. The method is reported to work well for Q's as high as 25,000 and avoids the requirement of stability better than 1 ppm of the signal frequency.

The most satisfactory technique to measure Q of superconducting helical resonators and LC circuits is to record the transient decay of rf energy. If the losses are proportional to stored energy, the decay is exponential, and

$$W = W_0 \exp(-t\omega/Q_\mathrm{L}) \tag{65}$$

If the measuring circuit includes a square-law detector, the value of Q_L is

$$Q_L = 2\pi f_0 \, \Delta t \tag{66}$$

where Δt is the time for the output signal to decay to 36.8% of its peak value. If the measuring circuit includes a linear detector, the output signal is proportional to the field amplitude in volts, and

$$Q_L = \pi f_0 \, \Delta t \tag{67}$$

The procedure is to drive the circuit at its resonant frequency and gate the carrier off abruptly. This method does not require an ultrastable frequency generator. Once the circuit has absorbed some rf energy and the carrier is switched off, the output signal from the resonator will be its instantaneous resonant frequency. The only measurement to be made is of the elapsed time interval, which can be recorded as a decay envelope on an oscilloscope. Hartwig (1963b) describes several versions of circuits used to observe the decay.

Microwave measurements of cavity Q are also made by the decay transient technique, but the quantitative measurement of field intensity is more convenient by steady-state power reflection techniques. Turneaure (1967) discusses both methods as they have been applied to microwave cavities.

Cavities coupled with waveguides have a relation

$$Q_0 = (1 + \beta) Q_L \tag{68}$$

where β is the coupling coefficient, $\beta = P_{rad}/P_{dis}$, and P_{rad} is the power radiated out of the cavity to the external detector, and P_{dis} is the power lost inside the cavity. At critical coupling, $\beta = 1$ and $Q_0 = 2Q_L$. This is a very useful relation, since it can be used to calibrate the absolute power levels and field strengths. Care must be taken in analysis of data over wide ranges of temperature and frequency because β becomes $\beta(\omega, T)$.

The steady-state measurement requires an ultrastable oscillator. Figure 21 is a block diagram of the microwave circuit used by Pierce (1967).

2. *Measurement of frequency*

Quartz crystal oscillators are the basic laboratory frequency standards. Electronic counters, using a quartz reference oscillator, can measure frequency by counting cycles for 1 sec with an error of plus or minus one count. This means that a frequency of 10 GHz can be measured with a resolution of 10 significant figures.

Resonant frequency of a superconducting circuit can be determined by

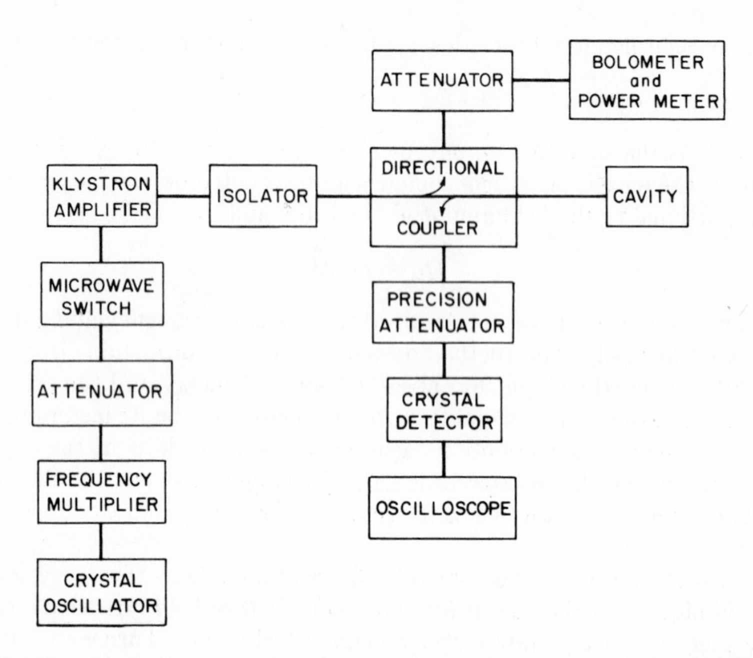

FIG. 21. Block diagram of the microwave cavity circuit used by Pierce (1967).

measuring the frequency of a reference oscillator driving the circuit. An alternate method is to incorporate the superconducting resonator into an oscillator by using an amplifier. The latter is a very accurate method for observing small deviations in frequency.

D. Frequency Stability

Superconducting resonators have extremely narrow bandwidths and clean resonant modes. As a result, a variety of small effects can influence the resonant frequency which would be unimportant in a normal resonator. Because the Q is high, the effect of a change in Q on the resonant frequency, however, is negligible. The center frequency can be approximated by

$$\omega_0 = [1/(LC)^{1/2}][1 + (1/2Q^2)] \tag{69}$$

A change of a factor of 2 in the quality factor having an initial value of 10^8 causes a frequency change of only one part in 10^{16}.

Frequency stability and spectral purity are related through the Fourier transform, since spectral purity in the frequency domain is equivalent to frequency stability in the time domain. Thus, frequency stability, for some considerations, is defined for the short term by

$$S_f = \omega_0 \frac{d\phi}{d\omega} \tag{70}$$

where ϕ is the phase. For an LC equivalent circuit, $S_f = -2Q$. This is not the context in which we wish to discuss frequency stability. Instead we are concerned with a number of long-term effects.

1. *Temperature effects*

The effect of thermal contraction of metals at low temperature is very small. For example, the thermal coefficient of linear expansion of copper is virtually undetectable below 20°K, but is about 30 ppm from room temperature to 20°. This means that a lead-plated copper cavity would have a frequency increase of about 10^{-3}%/°K as it was cooled, but the rate would become negligible below 20°.

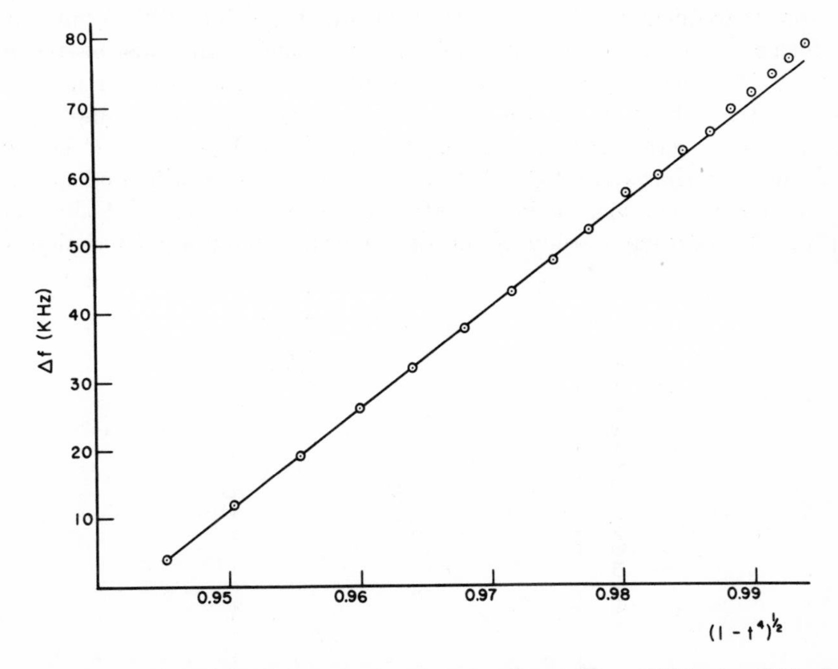

FIG. 22. Resonant frequency change due to temperature: $t = T/T_c$ where T is the temperature and T_c the critical temperature; $2.3°K \leq T \leq 4.2°K$; and $T_c = 7.2°K$ (lead). Abscissa is plotted according to the temperature dependence of the penetration depth, showing that the change is due to surface reactance effects (Stone and Hartwig, 1968, 1972).

At the other extreme, Stone and Hartwig (1968, 1972) observed the frequency of a 30-MHz circuit to decrease linearly at 0.4%/°K due to shrinkage of Teflon supports.

The dielectric constant of all insulating material is temperature sensitive to some extent. Liquid helium as a dielectric material will cause the frequency to decrease about 0.35%/°K below 4.2°K. In terms of atmospheric pressure, this means about 5 Hz/MHz/Torr. Normal atmospheric pressure variations are several millimeters per day.

The penetration depth determines the surface inductance, as given by Eq. (64), and there will be a corresponding change in resonant frequency. Figure 22 is the frequency shift of a 30-MHz Pb-plated LC circuit due to the temperature variation in the surface reactance of the shield can, in agreement with Eq. (63).

2. *Nonlinear effects*

The various loss mechanisms have corresponding effects on the cavity resonant frequency. When a 1-GHz resonator has a Q of 10^9, any process which generates frequency shifts of 1 Hz (the bandwidth) must be understood. DC magnetic fields produce a quadratic shift in surface reactance due to their effect on the penetration depth. Figure 23 is an example of a change of several kilohertz out of 30 MHz for a Pb-plated LC circuit (Stone and Hartwig, 1968, 1972). AC magnetic fields will produce the related nonlinear change in the eigenstate of the resonator. Halbritter (1970b) has made a study of the theory using perturbation techniques,

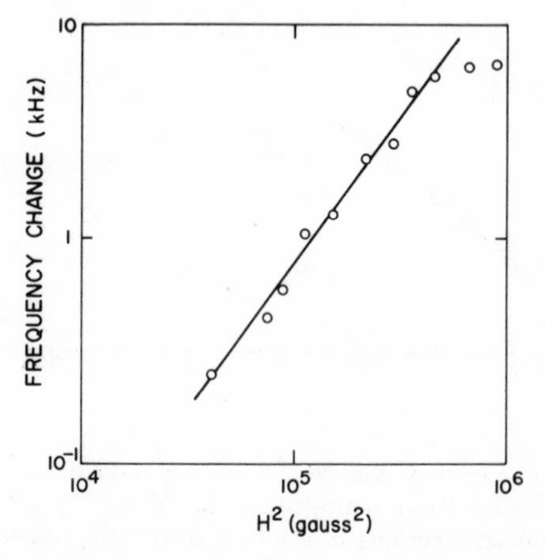

FIG. 23. Resonant frequency change due to external dc magnetic field. Lumped parameter 30 MHz, lead plated; $T = 4.2°K$. (Stone and Hartwig, 1968, 1972).

which accounts for this phenomenon plus the changes due to radiation pressure on the cavity walls and to emitted electrons influencing the E-field distribution. The frequency shift due to field emission has been used to calibrate the field intensity and electron density at high power levels (Fricke *et al.*, unpublished).

3. *Mechanical vibration*

The effect of dimensional distortion due to vibration is to modulate the resonant frequency at the natural frequency of vibration. The excursion in resonant frequency will depend upon the amplitude of vibration.

4. *Frequency compensation*

It is possible to reduce the temperature-induced drift in resonant frequency by introducing compensating effects. For example, as temperature is reduced, the frequency will decrease due to changes in the dielectric constant of liquid helium. It will simultaneously increase due to changes in the penetration depth and dimensions. With careful design, these effects can have an approximately equal and opposite effect over a substantial temperature range.

V. Devices for Particle Acceleration and Deflection

A. Introduction

High-frequency fields are more advantageous than dc fields for particle acceleration or deflection (Lapostolle and Septier, 1970). The traveling wave in a waveguide can be made to have its phase velocity less than the velocity of light. In addition, the mode and phase velocity can be made to vary along the waveguide, and so a charged particle can experience a constant field strength along its trajectory or can be steered, deflected, or decelerated as necessary. The high-frequency field strength and total voltage are also less limited than in the dc case for various practical reasons. For example, it was possible to build a machine with a total accelerating voltage of 20,000 MV at Stanford using traveling waves, whereas only about 20 MV is achievable with a dc field. To insure that the particle beam deviates a minimum amount from the beam axis and from the designed phase along the beam, focusing forces must be exerted along the trajectory. These forces are usually produced by static magnetic fields, but ac field components are also useful.

B. Advantages and Problems of Using Superconductors

A severe disadvantage of ac beam optical systems is their high installation and operation costs. The power loss in the waveguide walls of accelerators are so high that the installation of the necessary power supply and the cooling system, as well as its operation cost, becomes extremely expensive. Usually, an accelerator is operated with a duty cycle well below unity to reduce the average power. To achieve a high average beam intensity during the short operating pulse, as many particles as possible are injected into the beam.

Banford and Stafford (1961; Banford, 1965) first proposed to use superconducting waveguides to reduce the installation and operating cost of linear accelerators in 1961. It can be shown that there is a chance that the reduction of power losses in waveguides using high-transition-temperature superconductors would lead to overall savings in the installation in spite of the fact that rather large refrigerators and cryostats are necessary to achieve the low operating temperature. The savings are expected to be so high that the designers of superconducting accelerator projects have proposed a duty cycle which is one or close to one. This would be an advantage for nuclear physics experiments. However, a large duty cycle is also necessary to keep superconducting accelerators competitive with high-intensity normal conducting accelerators, since there are no savings possible by using superconductors if the power dissipated into the beam is of the same order of magnitude as the power usually dissipated in the walls of a normal conducting accelerator. We, therefore, can expect to build cheaper only those accelerators using superconductors which are designed for medium high currents and long duty cycles. The superconducting accelerator will show one more advantage, however, and this is a possible higher energy definition and beam stability. Because of the small ratio between power loss and stored energy, the decay time of the fields is much longer, and, since the accelerator is not pulsed, it will be simpler to design systems for accurately controlling the field strength.

Now, 14 years after the paper of Banford and Stafford was published, it seems to be clear that building superconducting accelerators and beam optical systems will be more difficult than expected. This is because it is extremely difficult to achieve the surface resistances predicted by the BCS theory in devices of large scale. Joints in the waveguides cannot be avoided, and the surface treatment of superconductors is difficult. Moreover, the accelerator techniques developed for normal conducting accelerators have to be revised completely to take those special problems into account which are posed by the use of superconductors.

Finally, it turns out that the advantage of the superconducting waveguide becomes effective only if the accelerator does not become too long, since the cost per length of the structure and cryostat is rather high. The energy gain per unit length in the accelerator has to be rather large. This requires rather high electrical field strength, and, up to this time, all efforts have failed to show that the necessary field strength can be achieved in large-scale devices under operating conditions.

The waveguide, the part of the accelerator to be built from superconducting metals, has the function of slowing the waves down. Depending on whether the particle has a velocity close to the velocity of light, or the same order, or an order of magnitude smaller than light velocity, we have to use different waveguide structures. We therefore divide the possible superconducting waveguides into two groups: those for particles with a velocity $v \geq 0.2\ c$, and those for particles with lower velocities.

C. Accelerators for Fast Particles

1. *High-energy accelerators*

The first superconducting accelerator project was started by a group of scientists in Stanford in 1962. Since that time, they have investigated the problems connected with the building of an electron accelerator. The important design criteria were worked out by Schwettman and co-workers (1967, Wilson *et al.*, 1964). They proposed an accelerator with a final energy in the 1000-MeV range. The design parameters have changed a little during the years of development but were kept in the same order of magnitude as given in Table X. The waveguide is a cavity chain which seems to be optimal to slow the wave down to a little less than the velocity of light. It was proposed to be operated with standing waves, using only its traveling-wave component in one direction in order to avoid complicating phasing problems.

They designed the accelerator for an operating temperature of about 1.8° after showing that the reduction in temperature from 4.2 to 1.8°K promised larger economical savings and would make cooling the cavities easier, since the heat conductivity of superfluid helium can be used. The group started the investigation with lead cavities, but, after finding that the lead surfaces are very delicate, they proposed the use of niobium. Besides the problems with the superconductor, electrical problems also had to be solved. For example, the power dissipated into the walls of a superconducting electron accelerator is about 100 times smaller than the power fed into the proposed beam current. Therefore, the rf generator has to be coupled to a load which varies abruptly by a factor of 100 when the beam is injected (Wilson and Schwettmann, 1965). Moreover, the small band-

TABLE X
PARAMETERS OF SUPERCONDUCTING ELECTRON ACCELERATORS[a]

	HEPL[b] (Stanford)	URBANA[c] (Microtron)	SLAC[d] (Stanford)	
Energy	2	20×0.03	100	GeV
Intensity (average)	100	200	3–48	μA
Duty cycle	1	1	$\frac{1}{16}$	
Acceleration	17	8	32.8	MeV/m
Maximum field strength	ca. 40	ca. 18.5	55	MV/m
Power loss in waveguide	2[e]	13	4	W/m
Operation temperature	1.85	1.85	1.85	°K
Frequency	1.3	1.3	2.856	GHz
Required R_s	2.5[e]	61	61	$n\Omega$

[a] The data were collected and calculated from published parameters by Bathow and Freytag (1971). The parameters have changed only a little during the past few years and are not consistent. They give, however, an indication of which improvements are necessary to build economical accelerators.

[b] Schwettmann *et al.* (1967); Wilson and Schwettman (1965); and Wilson *et al.* (1964).

[c] Allen *et al.* (1970).

[d] Stanford Linear Accelerator Center (1969); and Wilson *et al.* (1970).

[e] The power loss for HEPL is found under the assumption that the cooling power of 300 W will be sufficient for the accelerator.

width causes severe tuning and phasing problems which must be solved by the construction of fast control systems (Suelzle, 1968).

The Stanford group was joined in 1964 by a group in Karlsruhe which worked out a proposal for a 7-GeV proton accelerator (Passow, 1967a,b). They started to investigate accelerator structures for particles with a velocity of about half of the velocity of light. Waves traveling with this velocity can be maintained in cavity chains also, but the cavities have to be more heavily loaded, and the rf power cannot be transmitted through the beam aperture only (Giordano, 1964). Therefore, more complicated cavity shapes had to be considered, and the Karlsruhe group was especially interested to find an optimal superconducting structure. A model was built to investigate the excitation of higher modes in the cavity by the beam, and their effect on the beam stability (Heller *et al.*, 1970; Mittag *et al.*, 1969).

A third group interested in the field was established at Illinois to investigate the possibility of a microtron with a superconducting accelerating component (Allen *et al.*, 1970). Another group at SLAC was investigating the possibility of increasing the final energy of the existing normal conducting electron accelerator from the 20 to 40 GeV now achievable to a possible 100 GeV (Stanford Linear Accelerator Center, 1969; Wilson *et al.*, 1970) by installing a superconducting waveguide. Their concern, therefore, was to achieve a high-energy gain per meter. They proposed a traveling waveguide structure with a separate waveguide to feed the power back into the injection region. Thus the maximum peak fields in the accelerator are only given by the accelerating field and not by the vector sum of the two field components of a standing wave traveling in opposite direction.

2. *Low-energy accelerators for electron microscopes*

The use of electron linear accelerators with a final energy of 1 to 4 MeV as a voltage source for electron microscopes was proposed first by Klema (1966), Rymer (1966), and Hanson (1966) independently in 1966. In order to replace the very large and expensive electrostatic generators, high-energy microscopes could have a higher resolving power, and could be used to investigate thicker objects when compared to current instruments. Since electrons already have relativistic velocities, even at rather low energy, cavity chains can be used to build the accelerating waveguide structure. The most severe problem is to achieve a voltage stability of 3×10^{-5} to 5×10^{-6}. This can possibly be done by a superconducting spectrometer and a control device to stabilize the rf field amplitude (Suelzle, 1968; Passow, 1969a, 1972b). To supply a high current, it may be necessary to build up two waves in the waveguide where the frequency of the second wave is a higher harmonic of the ground wave, with a phase and an amplitude chosen in a way that the field around the maximum is constant for a larger phase angle (Passow, 1969a, 1972b). Thus more particles distributed over a larger phase will be accelerated up to the same final energy. The high Q value of superconducting cavities will be helpful to build the control system, since the accelerator can be operated with a duty cycle of one, and the field decay time is long.

If a high-frequency beam is used to photograph an object, it will be possible to build a beam optical device to reduce the chromatic aberrations of the lens system, or to select out of the beam the inelastic scattered electrons with low energy. Moreover, high-frequency fields may also be useful to build up the high field strength required for field emission cathodes. This kind of cathode promises to give greater brightness and higher beam intensities (Passow, 1969a, 1972b).

D. Accelerators for Particles with Slow Velocities

Since the particles cannot be injected with mathematical exactness to be coincident with the accelerator axis, and since this axis cannot be built to be an exact straight line, the particle beam in every accelerator has to be focused along the beam path. This is usually done by magnets and is not too difficult for fast particles even in superconducting accelerators where care must be taken that necessarily high magnetic fields do not deteriorate the superconductivity. However, for particles with slow velocities, and especially for protons and ions where the available particle sources are not as efficient as for electrons, the focusing turns out to be an especially difficult problem in superconducting accelerators (Passow, 1969b). Focusing problems can be solved best by using waves of lower frequencies to accelerate the particles. Passow called attention to this point after investigating the necessary conditions for focusing particles in superconducting accelerators. Comparing the known structures (cavity chains, Alvarez structures, helices, and single cavities) with respect to the feasibility of building them out of superconducting material, he came to the conclusion that the helix structure is most practical, since its dimensions are small enough to make easy processing possible, even if it is operated at very low frequencies (Passow, 1969c, 1970). The slightly smaller shunt impedance and Q value of the helix, compared with the Alvarez structure, which is used to build normal conducting accelerators for low particle velocities, is not a severe disadvantage for application in superconducting accelerators.

The fact that the surface resistance of superconductors is decreasing with the second power of the frequency instead of the square root in case of normal conducting metals leads to large power savings and high Q values in low-frequency structures. The Q value and shunt impedance are therefore larger in a helix than an Alvarez structure, which has to be operated at a higher frequency in order to keep its size small. The system can be cooled easily by feeding superfluid helium through the helix pipes (Passow *et al.*, 1970, 1971).

Technical problems connected with the helix structure are its sensitivity to vibration and the change of the helix length due to radiation pressure. Even if this change is only of the order 1 μ, this results in a frequency shift which is larger than the bandwidth of the structure (Schulze, 1971a,b). Superconducting helical waveguides are under investigation at Karlsruhe (Vetter *et al.*, 1970, unpublished; Fricke *et al.*, 1972), Oak Ridge (Jones *et al.*, 1972), Cal Tech (Sierk *et al.*, 1971), and Argonne (Benaroya *et al.*, 1972).

Single electrically separated re-entrant cavities, which are heavily loaded to get an acceptable cavity size, even if it is operated at 300 or 400 MHz, can be used to accelerate ions or protons if the required beam intensity is not too high, and the particle velocity is not too low.

E. Beam Optical Devices. Separators

To build a beam optical system for application other than acceleration, the same kinds of problems must be solved. Only the phase or the wave mode has to be different to get a decelerator, a deflector, or a quadrupole lens instead of an accelerator. It was proposed first by Montague (1963) to build a separator waveguide out of superconducting material. A separator is used in nuclear physics to separate different kinds of particles, for example, K mesons and π mesons (Lapostolle and Septier, 1970). For large particle momenta (larger than about 5 GeV/c), this can be done only with rf fields, since, like an accelerator, the total possible voltage in an electrostatic separator is limited. A separator is a waveguide in which the wave is slowed to the velocity of one kind of particle. These particles will then travel in a deflecting field which is constant in their coordinate system, and will be deflected.

The second kind of particle in the beam has the same momentum, but a different mass and therefore a different velocity. In their coordinate system, the electromagnetic field component will change sign periodically, the resulting average force will be zero, and the particle will not be deflected. The arguments given for building superconducting separator waveguides are the same as given for accelerators. However, economical considerations show that, in the case of the separator, the superconducting solution is more favored than it is for the accelerator (Citron and Schopper, 1970). Therefore, it is not necessary to achieve extremely high-quality superconducting cavities for a separator. This is because the particles which are to be separated are usually not stable, and decay after traveling only a short distance. Consequently, the necessary deflecting voltage has to be applied over a short distance, which makes high field strength necessary. A normal conducting waveguide has its economical optimum at rather low field strength and has a long waveguide length. Used as a separator, it has to be operated far away from its economical optimum. The superconducting waveguide, however, has its economical optimum about at the desired field strength. Moreover, for an experiment where particle counters are needed, a large duty cycle of the separator is required.

F. Applications and Projects Under Way

1. *High-energy electron accelerators*

High-energy electron accelerators (Wilson, 1970) are used only for nuclear research up to this time. At Stanford, a group in HEPL is building superconducting electron accelerator with a final energy of 2 GeV, with parameters given in Table X (Schwettmann *et al.*, 1967; Wilson *et al.*, 1964; Wilson and Schwettmann, 1965). Its long duty cycle and high energy make possible the performance of experiments which need a higher energy resolution and/or a small pulse current, but also need high average current

TABLE XI
PERFORMANCE DATA OF THE FIRST TEST OF THE SCA AT STANFORD[a]

	Measured value	Design goal
Capture section		
Energy gradient (cw)	2.7	9 MV/m
Q_0	2×10^9	10^{10}
Preaccelerator		
Energy gradient (cw)	1.8	9 MV/m
Q_0	1×10^9	10^{10}
Maximum cw current	10	100 uA
Energy spectrum ($\Delta E/E$)	0.1%	0.3%

[a] From McAshan *et al.* (1972).

to suppress the relative number of acidentally counted particles which are not involved in the same scattering or production process. A part of this accelerator 16 ft long is already in operation (McAshen *et al.*, 1972).

The measured parameters are compared with the design goal in Table XI. It shows the difficulties in obtaining the necessary Q values and energy gain per length in real accelerator structures. At SLAC, a project was underway to investigate the possibility of using superconducting traveling-wave accelerator structures in the future instead of the presently installed normal conducting waveguide (Stanford Linear Accelerator Center, 1969; Wilson *et al.*, 1970). It was hoped that the final energy could then be increased from the 20- to 40-GeV range up to 100 GeV (see Table X). The project has been closed down, however, in favor of an improvement program based on normal conducting accelerator and beam optical devices.

A project is underway which aims at building a microtron with a superconducting accelerator part (Allen *et al.*, 1970). In a microtron, each particle passes through the accelerator several times. The parameter list of the Illinois project given in Table X shows that the energy gain per turn is 30 MeV and that the particles will be returned by magnets 20 times through the structure before leaving the accelerator. Similar investigations, but performed in a smaller scale, are underway in Hamburg at Deutches Electronen–Synchrotron.

2. *Proton and ion linear accelerators*

Proton linear accelerators are useful mainly in nuclear research. Superconducting accelerators are supposed to give a cheap particle factory to

TABLE XII

PARAMETERS OF A TEST HELIX AND THE FIRST SECTION OF A SUPERCONDUCTING PROTON ACCELERATOR AT KARLSRUHE[a]

	Test helix	Accelerator
Helix radius	3.2	3.7–4.2 cm
Radius of outer conductor	6.4	20 cm
Pitch	1.0	1.0 cm
Particle velocity divided by the velocity of light	0.06	0.04
Maximum field strength	26	15[b] MV/m
Field on axis	3.4	1.15[b] MV/m
Q value	1.7×10^8	3.3×10^7
Frequency	91	91 MHz

[a] Data from Citron *et al.* (1971) and Brandelik *et al.* (1972).
[b] Design goals achieved.

produce K-meson and/or π-meson beams with a long duty cycle (Citron and Schopper, 1970). Its application as a radiation source in medicine, for example, is still doubtful, since a very high intensity will be required. This radiation can also be produced in short pulses, and that means that the advantages of superconducting accelerators would not be fully used.

Heavy ion accelerators are used only for nuclear research at this time. Further applications of these machines might also be possible. For each particular application, the question must be answered whether the normal conducting version or a superconducting version will be more economical. A comparison of the cost of a superconducting ion accelerator and a normal conducting version is given by Wilson (1970). A short superconducting proton test accelerator is under construction at Karlsruhe (Brandelik *et al.*, 1972); the parameters of the helix test structure are given in Table XII. The machine uses a helix for the low-energy part and a cavity for the higher energy part. The final energy of the accelerator is about 60 MeV.

A proposal for a K-meson factory and/or a π-meson factory was prepared in 1967 (Passow, 1967a,b). This paper gives still valid considerations about the high-energy part and the possible facilities of the accelerator, but not the low-energy parts. A possible modification is discussed very briefly by Passow (1969c, 1970). The Cal Tech group proposed a superconducting heavy ion accelerator based on a helix structure (Sierk *et al.*, 1971). The Stanford group proposed to use very heavily loaded reentrant cavities and worked out a proposal for a superconducting heavy ion accelerator based on a cavity structure (Glavish *et al.*, 1971). Some measurements on the cavities have already been performed.

3. *Particle separators*

Separator systems usually consist of one or more separator tanks which are used in combination with magnetic lenses to build the beam transport system in which only one kind of particle is directed to the target while the other particles produced simultaneously are trapped in shielding material.

A two-tank separator is under construction at Karlsruhe (Bauer *et al.*, 1972); its parameters are given in Table XIII. Each tank consists of a cavity chain in which a deflecting electromagnetic standing wave is built up.

TABLE XIII
PARAMETERS OF A TWO-TANK SEPARATOR TO SEPARATE PARTICLES WITH MOMENTA BETWEEN 5 AND 15 GEV[a]

Material	Niobium
Operating frequency	2.855 GHz
Operating temperatures	1.8°K
Q value	10^9
Peak electric field	11 MV/m
Deflecting field	2 MV/m
Distance between two tanks	20 m
Length of one tank	2.73 m

[a] From Bauer *et al.* (1972).

The distance between the two tanks is used to build up magnetic focusing and defocusing lenses and increase the phase difference between the different particles. The separator is designed to be used at the European Center for Nuclear Research (CERN) 28-GeV proton accelerator. Extensive work on the development of superconducting separators and investigations of their efficiency has been done at Brookhaven (Brown *et al.*, 1971); measurements on lead separator cavities have been performed at Rutherford Laboratories (Carne *et al.*, 1971). The group at the Rutherford Institute plans to build a separator for π mesons with a momentum of 1 GeV/c to reduce the beam length compared with an electrostatic separator.

VI. Applications of Low-Power Superconducting Resonators

A. INTRODUCTION

Superconducting resonant circuits have superior performance characteristics compared to room-temperature circuits. In a recent survey, Hartwig

(1973) has summarized the many contributions to electronic systems for communication, control, and instrumentation. An adequate body of knowledge now exists for the engineer to design and build the superconducting devices.

The most profound feature of the superconductor is its spectacular reduction in surface resistance below T_c. At 1 GHz, for example, one can achieve an improvement of 5 or 6 orders of magnitude. Superconductors also display a variety of temperature dependences and nonlinearities not associated with normal conductors. These features have been exploited by several investigators to add value to the technology. The Josephson junction and its high-frequency device applications represent a unique and very important area of applied superconductivity which extends the variety and range of high-frequency devices. It is based upon quite different mechanical phenomena, however, and is therefore not appropriate in this chapter.

Since the superconducting system operates in a liquid helium bath or with a closed-cycle refrigerator, most experimental work has been done with size as an economic parameter. As a consequence, the components and systems tend to operate above 10 MHz. Superconducting systems have been designed, and many have been built and demonstrated which are compact and portable. The existence of highly reliable helium refrigerators makes it possible to achieve the improved performance of superconducting components in a practical way.

B. Superconducting Filters and Tuners

Superconducting tank circuits for radio communication applications were investigated by several groups since 1961 (Stone and Hartwig, 1968, 1972; Arams *et al.*, 1967; Hartwig, 1961–1963, 1963c; Markard and Perlman, 1966; Stone, 1964; DiNardo *et al.*, 1971; Minet and Combet, date unknown; Vig and Gikow, 1973). Although the references are probably not exhaustive, the work shows several areas where useful applications might be made. Hartwig (1961–1963, 1963c, 1973) built *LC* tanks with Meissner tuning slugs that expelled flux from the inductor, which were tunable over the range 21–24 MHz, and had a $Q = 2 \times 10^5$. Markard and Perlman (1966) pointed out that the ultra-low-noise microwave masers and parametric amplifiers had stimulated the development of cryoelectric microwave components. The over-all receiver noise is dependent upon the insertion losses of any diplexer, circulator, or filter introduced into the transmission system between the antenna and the low-noise amplifier. A superconducting preselector might have its noise temperature below that of the corresponding preamplifier

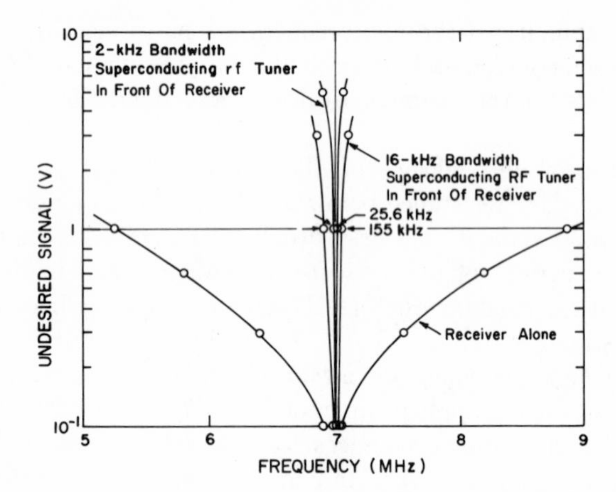

FIG. 24. Results of cross-modulation measurements of an rf receiver using a superconducting preselector filter. Undesired modulation 20 dB below desired signal modulation reference: desired signal, 100 μV; modulation, 400 Hz, 30% (Arams *et al.*, 1967a).

in use. The high isolation between transmitter and receiver made possible by a diplexer suggested a superconducting high-power transmitter filter and low-noise receiver filter.

Arams and associates (1967a) applied superconductivity to the problem of reducing rf interference. They built a low-loss, very narrow-band preselector that was tunable from 6.3 to 21 MHz. It had a bandwidth adjustable from 60 Hz to 50 kHz, and Q_L from 3.5 to 6×10^5. They demonstrated reduction in intermodulation interference susceptibility of 40 dB and 23 dB of improvement in teletype signals. Measurements were made to determine cross-modulation of an interfering signal on a desired signal as a function of power level and frequency separation from the desired signal. The interfering signal power levels recorded were those generating, on the desired cw signal, a modulation that was 20 dB down from the 30% modulated signal. Figure 24 shows some of the results of cross-modulation measurements on a high-frequency receiver with and without the superconducting filter as a preselector. This receiver was of the double-conversion type. For the receiver without the superconducting filter, a 1-V interfering signal produced 10% cross-modulation ± 3 MHz away from the desired 100-MV signal. With the superconducting tuner in front of the receiver and the tuner 3-dB bandwidth adjusted to 2 kHz at the same signal power levels, the interfering signal had to be within ± 13 kHz of the desired signal to produce the same interference.

DiNardo *et al.* (1971) constructed superconducting miniature microstrip resonators, operating at X band, using vacuum-deposited lead on low-loss

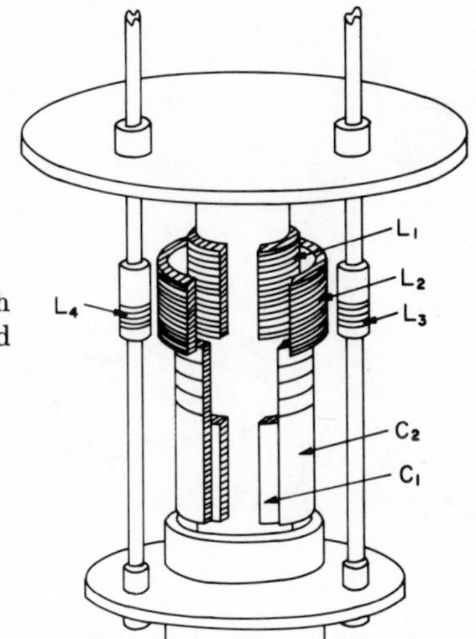

FIG. 25. Superconducting tuner with broad tuning range, as built by Vig and Gikow (Vig and Gikow, 1973).

sintered alumina substrates. They measured unloaded Q's as high as $2\text{–}5 \times 10^5$ at 14.3 GHz between 4.2 and 1.8°K. Stone and Hartwig (1968, 1972) and Stone (1964) examined effects of temperature and dc magnetic fields on the frequency stability of a 30-MHz tank circuit to be used in an oscillator, and verified the theoretical performance. Vig and Gikow (1973) demonstrated a superconducting tuner with a Q above 10^5 which is adequate for voice communication bandwidth, and which had a tuning range from 1.3 to 23 MHz (see Fig. 25). Significantly, the superconductors used were Nb_3Sn and NbTi. These are Type II superconducting alloys valued for their high T_c (18 and 10°K) and high H_c, but have been given little consideration as high-frequency materials. It must be emphasized that the improvement in Q is adequate for the application, which establishes these materials as competitors to niobium and lead.

C. OSCILLATORS AND DETECTORS

Several efforts to build ultrastable oscillators have been reported which take advantage of the ultrahigh Q of superconducting resonators (Stone and Hartwig, 1968, 1972; Turneaure, 1972; Stone, 1964; Stone *et al.*, 1969b; Biquard, 1970; Benard *et al.*, 1972). Stone and Hartwig (1968, 1972) and Stone (1964) reported a 30-MHz oscillator using a Pb-plated LC circuit with a Q of $\sim 10^6$, and having a 20-dB insertion loss due to light coupling. A GR1216A IF amplifier provided the required gain in the for-

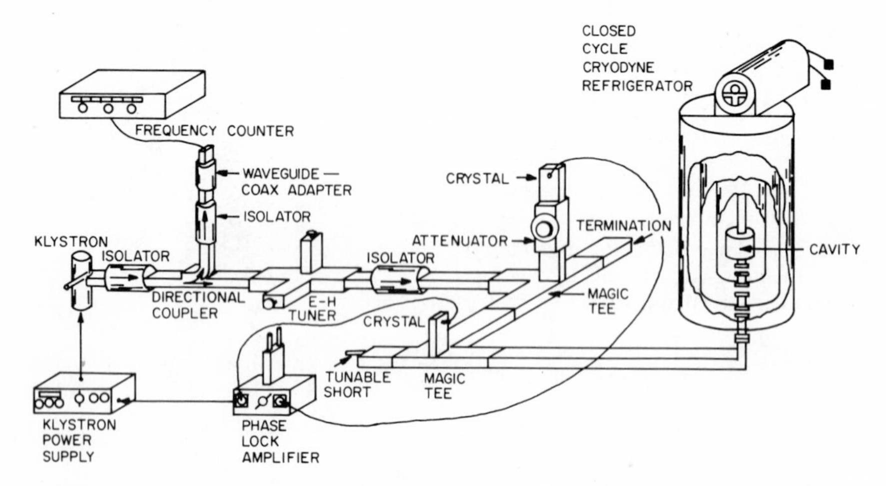

FIG. 26. X-band oscillator with superconducting cavity and closed-cycle refrigerator. Frequency stability at 6.7°K (T_c = 7.2°K) was 10^{-8} (Stone *et al.*, 1969b).

ward loop. Frequency stability was measured against a crystal-controlled HP 524C counter to be about the same as the counter stability, or about 10^{-7} for a 1-sec averaging time at constant bath temperature. Temperature effects on frequency were attributed to changes in surface reactance with penetration depth, $\lambda(t)$. Since the dielectric constant of LHe is also temperature dependent, it is possible to compensate somewhat for temperature changes over a reasonable range if the resonator is immersed.

Stone *et al.* (1969b) reported on an X-band experimental oscillator using a Pb-plated cavity cooled to 6.7°K with a CRYODYNE closed-cycle refrigerator (see Fig. 26). The frequency stability of 10^{-7} with a room-temperature Q of 2×10^4 increased to 10^{-8} with a Q (6.7°) of 2×10^5. The system operated for several days in the laboratory and was subsequently displayed at the 1969 Physics Show in New York, courtesy of Arthur D. Little, Inc.

More recently, Biquard (1970) has achieved a stability of 10^{-10} using a monotron oscillator; Benard *et al.* (1972) reached 3×10^{-12} for 10-sec periods using a reflex klystron stabilized with a superconducting cavity. Turneaure (1972) reports that he and Stein have reached 1.2×10^{-13} for 10 sec using two Gunn-effect oscillators, each stabilized with Nb cavities, and he predicted that someone would reach $\sigma(\Delta f/f) = 10^{-15}$ within a year. It would be about two orders of magnitude better than the current state of the art. The applications to improve frequency standards and clocks are justifiably optimistic, with all that this implies for communication, navigation, geodesy, and scientific research.

The nonlinear electromagnetic response of the ac Josephson junction (Josephson, 1962; Shapiro, 1963) has been exploited by Silver and Zimmerman (1967) and others for simultaneous rf generation and detection. They used these voltage-biased superconducting weak-link contacts to observe resonance of *LC* circuits, coaxial line resonance modes, and the zero-field nuclear magnetic resonance (NMR) in ^{59}Co at 218 MHz. Since then, Eck and Ulrich (1967) demonstrated an electronically tunable far-infrared and millimeter wave detector, and Ulrich (1970) successfully applied it to astronomical observations of the sun, moon, and Venus at 1.2 mm. Subsequently, Ulrich and Kluth (1972) coupled a 1-mm source out of the Dewar and self-detected the radiation with an estimated noise equivalent power (NEP) of 10^{-16} W/Hz with a signal/noise of 10^3 over 1 sec.

The theory and applications of the superconducting weak-link junction are covered in detail elsewhere in this book, but are included here for continuity with the subject of oscillators and detectors.

D. Mixers and Amplifiers

Superconducting mixers have been built using the superconducting/normal transition and others using the Josephson junction. Feucht and Woodford (1961) built a device consisting of an input loop and output loop of Pb wire separated by a Sn film (see Fig. 27). Coupling between the input and output was prevented by the inability of flux to penetrate the superconducting tin. The film could be driven into the normal state at the current peaks in a third coil which surrounded the loops and film. In other words, the switching signal frequency is half of the modulation frequency, f_0. The output loop had a voltage induced in it composed of the input frequency, f_1, and the sums and differences of harmonics of the switching signal frequency, or $f_2 = f_1 \pm nf_0$. The output circuit could be tuned to pass only one frequency, say $f_1 - f_0$. The input circuit could be tuned to f_1 to prevent feeding back undesired signals to the signal source.

Microwave superconducting mixers operating at 10–13 GHz utilizing

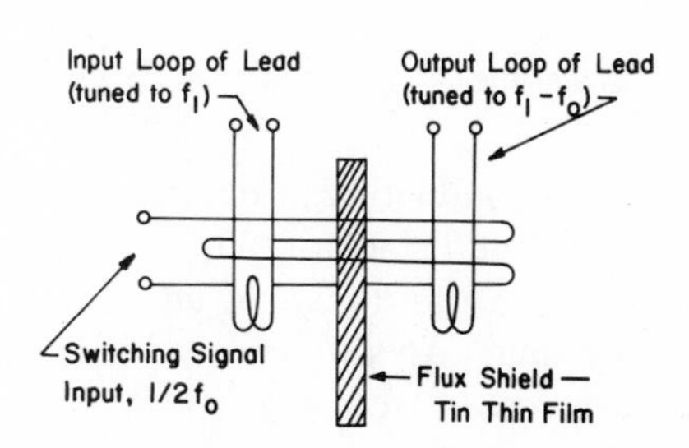

FIG. 27. Superconducting mixer of Feucht and Woodford (1961).

Josephson junctions have been investigated analytically and experimentally by DiNardo and Sard (1971). Reported sensitivity was −90 dBm/MHz, including mismatch losses in the waveguide structure. A local oscillator 60 MHz below the signal frequency was used to excite the junction. Mixing was observed to occur between the constant-voltage steps of the junction dc I–V characteristic.

Amplification by superconducting devices is in the laboratory stage, although Ulrich and Kluth (1973) has shown theoretically that the Josephson junction coupled to a superconducting cavity is capable of high parametric gain without an external oscillator.

Miles *et al.* (1963) also investigated the possibility of using the voltage-current characteristic of the tunnel junction between two superconductors for amplification.

Clorfeine (1964) has demonstrated microwave amplification of 11 dB at 6.06 GHz with a "modified dielectric resonator" consisting of rutile and quartz blocks separated by a tin film and mounted in a waveguide. Bura (1966) obtained 27 dB gain with a parametric amplifier consisting of superconducting films on mica substrates, inserted radially between the inner and outer conductors of a coaxial line. The nonlinear variation of film magnetization with the magnetic field set up in the line is shown to be responsible for the parametric amplification.

Another class of amplifiers not incorporating resonant circuitry is the linear cryotron amplifier of Newhouse and Edwards (1964). Originally thought to be useful in cryotron computer systems and bolometers, Radhakrishan and Newhouse (1971) show advantages when the CFC amplifier is compared to the maser. The analysis is made based on the noise characteristics in terms of the product of the smallest energy measurable and the smallest possible response time, called the "action factor." They conclude that, for low-frequency sensitive measuring devices, the CFC amplifier is potentially superior. A harmonic generator for frequency multiplication was built by Zimmer (1965). He used a field emission source in a superconducting cavity, and generated 0.6 mW output power at 34 GHz with an input power of 1 W at 8.5 GHz.

E. Antennas and Output Tanks

A superconducting antenna and matching output tank have been shown by Walker and Haden (1969) and Ramer and Walker (1969) to offer unique solutions to the design of efficient systems that are physically and electrically small. An antenna equivalent circuit, including its matching resonator, is represented in Fig. 28. R_R is the radiation resistance, or the equiva-

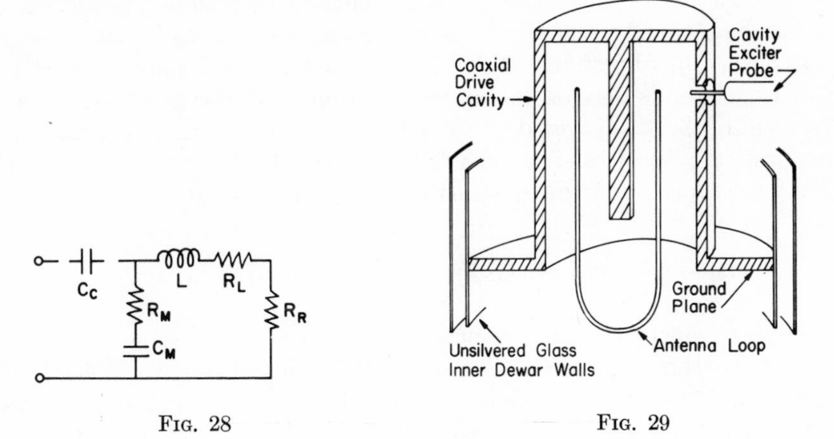

FIG. 28 FIG. 29

FIG. 28. Equivalent circuit of an antenna (Walker and Haden, 1969; Ramer and Walker, 1969).

FIG. 29. Superconducting antenna configuration (Walker and Haden, 1969, Ramer and Walker, 1969).

lent circuit component accounting for the power radiation. R_R is a function of size, and, for a loop in free space,

$$R_R = 3.12 \times 10^4 (A_L/\Lambda^2)^2 \tag{71}$$

where A_L is the loop area and Λ is the wavelength. The ohmic resistance of the normal wire loop is

$$R_L = (\omega\mu/2\sigma)^{1/2}(r/r_0) \tag{72}$$

where r/r_0 is the loop radius over wire radius. The efficiency of the antenna system depends upon the relative size of R_L and R_R, or

$$\gamma = 100\%/(1 + R_L/R_R) \tag{73}$$

and R_L should include the small resistance reflected in from the driving circuit R_M. The antenna configuration shown in Fig. 29 is a 1-cm² loop, suspended through a ground plane and capacitively coupled to a coaxial transmission line cavity. The system was excited by a capacitive probe with freedom to cancel the reactive component of field and match the impedance of the driving system. The performance showed an increase of 27 dB in the received signal from 300 to 4.2°K, an increase in Q of 100 times, and an efficiency of nearly 100%.

The success of this experiment is matched by another technological advantage, namely, the ability to construct *superdirective* arrays of small antennas. Superdirectivity (1969) implies very narrow beamwidths, which, in turn, depends only on the number and phase of the loops and is independent of the area covered by the array. Ohmic losses, narrow bandwidth, and critical tolerances of antenna parameters are the limiting factors, which gives the superconducting antenna array a significant advantage.

F. Superconducting Resonators for Materials Research

High Q superconducting resonators have been designed as experiment chambers (Hartwig and Grissom, 1965; Grissom and Hartwig, 1965, 1966; Grissom, 1965; Hartwig, 1966a; Kung, 1969; Hartwig and Hinds, 1969; Hinds and Hartwig, 1971; Johnson, 1972) for research on low-temperature properties of materials. These applications take advantage of the precision in measuring frequency and the ability to suppress cavity losses.

The complex permittivity of a material can be written as

$$\epsilon^* = \epsilon_0 K^* = \epsilon_0(K_1 - jK_2)$$

where K_1 and K_2 are the real and imaginary components of the dielectric constant. The loss tangent is defined as $\tan \delta = K_2/K_1$ and is usually very small for most dielectric solids. If a dielectric sample is placed in a superconducting cavity, its losses will dominate, however, because the unloaded cavity losses can be made comparatively negligible.

In such a situation Eq. (48) can be written (Hartwig and Grissom, 1965; Grissom and Hartwig, 1966) as

$$Q_D^{-1} = Q_u^{-1} + \alpha \tan \delta \simeq \alpha \tan \delta \tag{74}$$

Tan δ is the loss tangent of a dielectric sample in the resonator, and α is its filling factor evaluated from its dielectric constant and the shift in frequency associated with its presence in the resonator. Very sensitive measurements of defects in dielectric solids can be made by observing the effects of temperature and frequency on the loss tangent.

The relation between the sample's loss tangent and the measured change in Q is

$$\tan \delta = \omega/(2\,\Delta\omega)[(K_1 - K_0)/Q_D] \tag{75}$$

where the resonant angular frequency with the sample in the cavity is ω. $\Delta\omega$ is the change when the sample is removed. K_0 has the free space value of unity for an evacuated cavity or a value of 1.049 if the cavity is filled with liquid He at 4.2°K. Grissom and Hartwig (1966) show that the

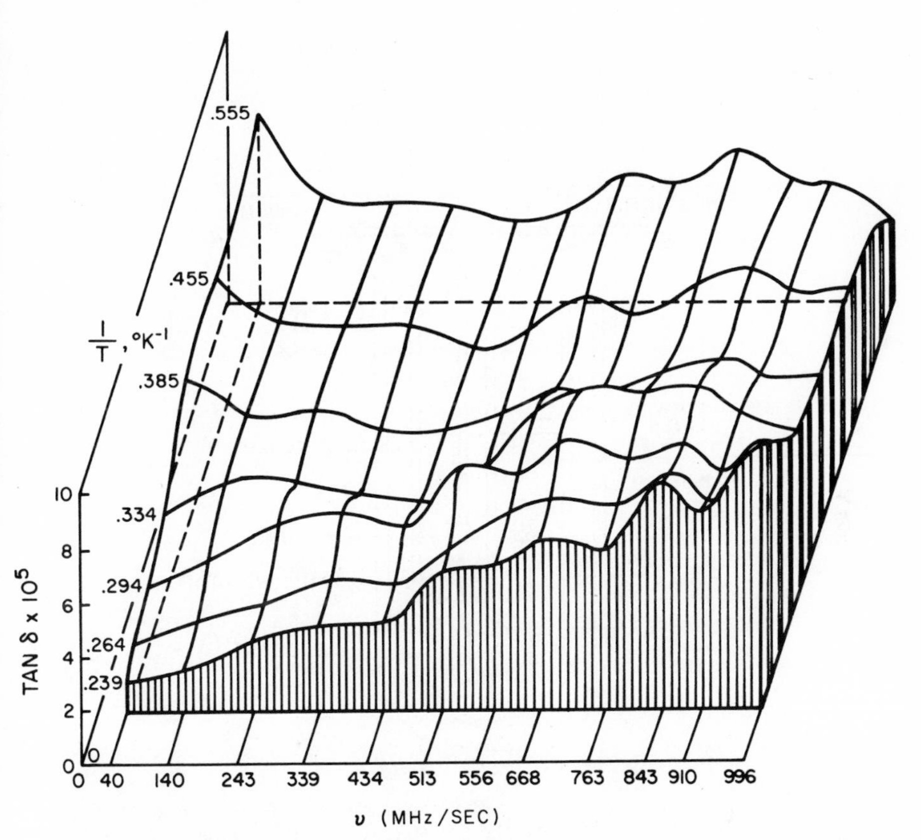

FIG. 30. Loss-tangent–frequency–temperature surface of single-crystal NaCl (Grissom and Hartwig, 1966).

analysis of a polarizable defect in a material with a dominant Debye–Breckenridge relaxation loss mechanism is made with the expression

$$\tan \delta = (np^2/9\epsilon_0 kT)[(K_1 + 2)^2/K_1][\omega\tau/(1 + \omega^2\tau^2)] \tag{76}$$

where τ is the relaxation time, n is the density of defects, and p is their dipole moment.

Figure 30 is the frequency-temperature-tan δ surface for a single crystal of NaCl from 1.8 to 4.2°K and from 40 to 1000 MHz. The data can be analyzed to isolate a Debye relaxation peak produced by an absorption process with an activation energy of 1.1 meV at one lattice site in 10^{10}. At room temperature, this mechanism is unobservable, being submerged in the background of the sum of all dissipation mechanisms. This same technique used by the authors (Hartwig and Grissom, 1965) shows that the loss

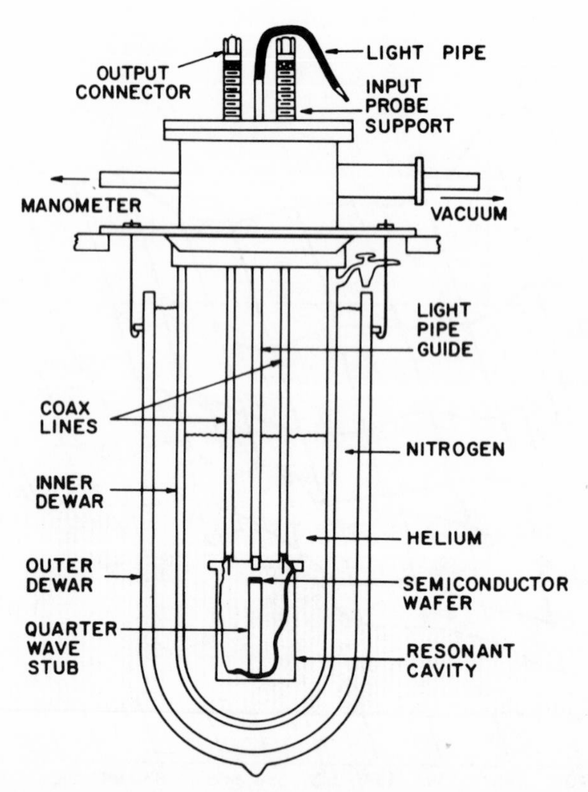

FIG. 31. Experiment to observe the photodielectric effect in semiconductors (Kung 1969; Hartwig and Hinds, 1969; Hinds and Hartwig, 1971; Johnson, 1972; Hartwig, 1966, 1970; Arndt *et al.*, 1968; Albanese *et al.*, 1972a,b; Mesa Instruments, 1968).

tangent of liquid helium is less than 10^{-9}, and its dielectric constant is measurable to six or more significant figures.

Using superconducting resonators, Hartwig and Arndt (1966a) observed the photodielectric effect in several semiconductors illuminated with band-gap light. The change in the real part of complex dielectric constant, caused by free carriers, shifted the resonant frequency several kilohertz at 787 MHz. Figure 31 is a schematic drawing of the experiment. The frequency shift is given by

$$\Delta\omega = \omega_0 G \epsilon_\psi / \epsilon_1^2 \tag{77}$$

where G is the filling factor, ϵ_1 is the lattice dielectric constant, and ϵ_ψ is the optical component contributed by the inertia of free carriers. Figure 32 is a plot of frequency change versus light intensity for Si and Ge.

Other experiments showed the change in recombination lifetime with Cu

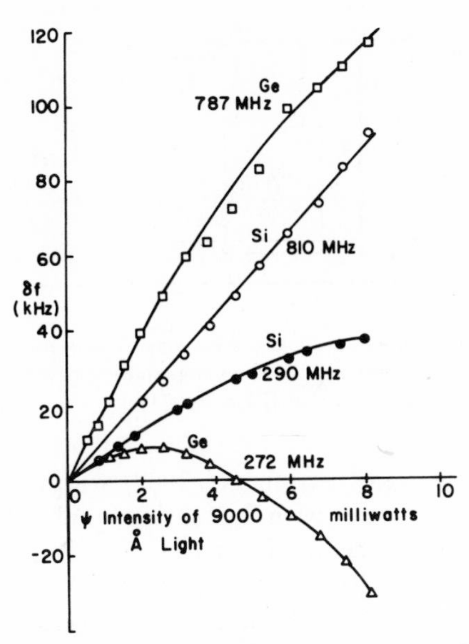

FIG. 32. Frequency shift of resonance caused by the photodielectric effect (Arndt *et al.*, 1968).

impurities in silicon (Kung, 1969), and the trapping of electrons in shallow states in CdS and other II–VI compounds (Hartwig and Hinds, 1969; Hinds and Hartwig, 1971). Johnson demonstrated the effect of a distributed plasma frequency on the photodielectric response of InSb (Johnson, 1972). These experiments are capable of measuring properties of dielectric and semiconductor materials which cannot be seen by other techniques, including those materials to which ohmic contacts cannot be made.

G. RADIATION DETECTION WITH LOADED SUPERCONDUCTING RESONATORS

The superconducting cavity loaded with a semiconductor, as in Fig. 31, is also a unique photodetector (Stone *et al.*, 1969a; Hartwig, 1966b, 1970; Arndt *et al.*, 1968). By proper selection of semiconductor and cavity resonant frequency, detectors can be built for infrared, optical, and even nuclear (Alworth and Haden, 1971) radiation. Albanese (1972a,b) has shown that the photodielectric detector can be much more sensitive than an ac photoconductive detector. By using optical feedback (Stone *et al.*, 1969a; Mesa Instruments, 1968), it is possible to build a carrier-tracking preselector for a secure radio communications receiver, and to add a micro-

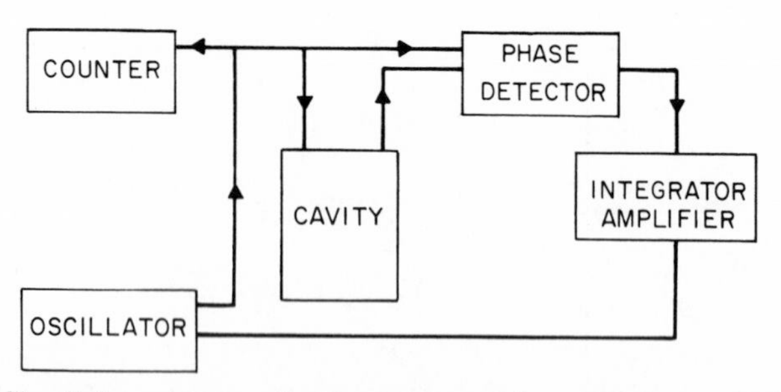

FIG. 33. Circuit for a superconducting cavity used as a nuclear radiation detector. The resonant frequency changes due to the change in dielectric constant of an irradiated crystal inside the cavity (Alworth and Haden, 1971).

wave power leveler (Mesa Instruments, 1968) with low insertion loss in a line from a microwave source with fluctuating amplitude.

Optical and infrared detectors, using the photodielectric effect, combine a semiconductor crystal with a bandgap energy, E_g, in a superconducting resonator. Photons with greater energy create electron-hole pairs in the crystal when they are absorbed in the crystal. The resulting change in the real part of the complex dielectric constant, ϵ_ψ, shifts the resonant frequency. In this way, a detector can be sensitive to any wavelength by selecting an appropriate semiconductor. The optimum design criteria for sensitivity dictate that the sample have a high resistivity in the dark, that the cavity frequency should be equal to the plasma frequency in the crystal, that the cavity should be coupled to the external circuit at the critical coupling position, that the sample thickness be only great enough to absorb all of the light, that the material should have a high quantum efficiency at the desired wavelength, and that cavity resistive losses will be negligible compared to the sample losses.

A nuclear radiation detector has been demonstrated by Alworth and Haden (1971) which measures the permanent change in crystal polarization due to radiation damage to the semiconductor crystal. Using the loaded cavity to control the frequency of a voltage variable oscillator, shown in Fig. 33, they observed 4-kHz/min changes in the resonant frequency of 380 MHz when a CdS crystal was exposed to a ^{14}C source of 1 μCi. This is a very sensitive detector for low levels of radiation.

VII. Transmission and Delay Lines

In 1947, Pippard (1947a,b, 1950; Faber and Pippard, 1955) recognized that a wave on a superconducting transmission line would be slowed because of the penetration of the H field. This led to numerous research investigations on penetration depth behavior using thin-film transmission lines, represented in Fig. 34. Swihart (1961) used the London two-fluid model to derive the slow mode propagation velocity of a strip line which is determined by the ratio of dielectric and superconductor film thickness to the penetration depth. He showed that the low-temperature, low-frequency velocity is dispersionless, even though there is a component of the E field in the direction of propagation.

The circuits and cavities described above confine the electromagnetic fields into geometrical distributions that determine a resonant frequency. Transmission lines guide traveling waves, and their time and frequency relations are governed by the wavelength and velocity of propagation along the line. The science and applications of superconductivity have been enriched by studies made with the many configurations of transmission lines.

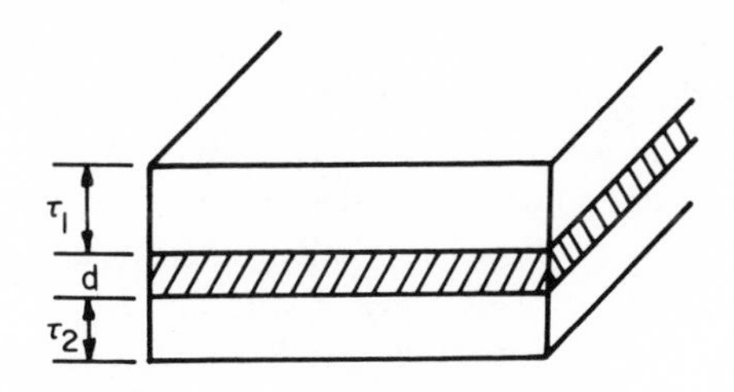

FIG. 34. Representation of a thin-film superconducting transmission line.

The strip line above a ground plane was examined by Sass (1964) using the London model, and the superconducting-to-normal transitions were studied experimentally and theoretically by Sass and Friedlaender (1964) using the Pippard formalism. They showed that the solution to the moving boundary value problem depends upon the current propagation on the line, the temperature, and the penetration depth. Emphasis was on the application to cryotron computer switching dynamics, but the results are pertinent to all superconducting-to-normal interface propagation.

Soderman and Rose (1968) examined the linear and nonlinear microwave transmission characteristics of superconducting tin films with the BCS theory. They showed that residual reactances, which had a substantial effect on the nonlinear transmission, were related to incident power and temperature, particularly near T_c.

Mason and Gould (1969) extended Swihart (1961) to show velocity dependence on the coherence length, mean-free-path, surface reflectance, as well as temperature and magnetic field. Both Swihart (1961) and Sass (1964) show that the effective penetration depth is a modification of the penetration depth into bulk material, λ_B, given by

$$\lambda(t) = \lambda_B \coth(\tau/\lambda_B) \tag{78}$$

where τ is the strip line thickness (see Fig. 34). The phase velocity of the dominant slow-wave E mode is

$$v_\phi \simeq c_d[1 + \lambda_1/d + \lambda_2/d]^{-1/2} \tag{79}$$

where c_d is the velocity of light in the dielectric, d is the dielectric film thickness, and λ_1 and λ_2 are the $\lambda(\tau)$'s in the superconducting films. Mason (1971) showed that the effect of reducing the mean-free-path with impurities increases the delay per unit length without increasing attentuation or degrading the pulse shape significantly.

Miniature 50-Ω coaxial superconducting lines were built and analyzed by Nahman and associates (Nahman and Gooch, 1960; Allen and Nahman, 1964; McCaa and Nahman, 1969; Ekstrom *et al.*, 1971) and Shizume and Vaher (1962). The lines investigated by Nahman and co-workers had a niobium inner conductor, Teflon dielectric, and a lead outer conductor. The 100-ft line was able to operate relatively losslessly and to transmit nanosecond pulses with no apparent distortion. Under lossless conditions, the line must be near perfect geometrically, since reflections within the line are not attenuated. The result is a reduction in signal-to-noise ratio. Analysis with the Mattis–Bardeen (1958) surface impedance improved the calculations, but dielectric losses complicated the match between theory and experiment. A miniature superconducting coaxial transmission line 278 m long showed a swept frequency (0.1–12 GHz) attenuation which was a quasi-periodic function of frequency (Ekstrom *et al.*, 1971). This indicates that the nonuniformity of the line was significant. Cummings and Wilson (1966) showed that similar lines fabricated from both pure and alloyed metals were capable of transmitting 15-kV 4-μsec pulses with nanosecond rise times.

The advantages of high-frequency superconducting meander lines have been demonstrated by Gandolfo *et al.* (1968) and Passow *et al.* (1972). Meander lines are made of series-connected lines, as shown in Fig. 35. The wave velocity is slowed by the increased length due to folding, by the lower propagation velocity in the dielectric, and dependent on the geometry more or less by the electromagnetic delay in the penetration depth (Swihart, 1961), and the wave interaction between adjacent lines. Since the wave interaction produces a slightly nonlinear dispersion, one advantage is that both linear and dispersive performance are possible. Other advantages (Gandolfo *et al.*, 1968) are the large packing density possible with microcircuit fabrication techniques, and the ease of interconnection.

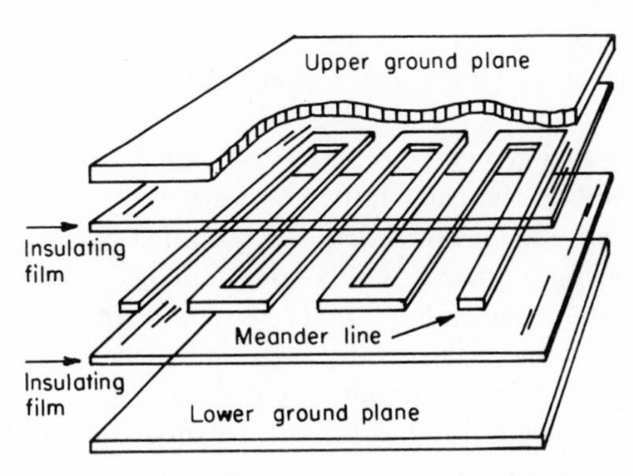

FIG. 35. Exploded view of a meander strip line.

Linear operation is observed for linewidths much less than line separation, and nonlinear (or dispersive) operation is obtained when the ratio is near unity. The line described by Gandolfo *et al.* (1968) was of the nonlinear type that slowed the center-line frequency by a factor of ~ 2. This geometry is useful as the dispersive delay element in a pulse compression filter. The slowing factor, v_ϕ/c_d, approached 10^{-2}; the insertion loss was only a few decibels and at least 20 dB less than a copper line of identical geometry and temperature.

Passow *et al.* (1972) described three devices: a nondispersive line for filtering, storage, and signal delay; an electrically tunable line that can also be used as a switch; and a strongly dispersive line for pulse compression. The designs were all ultracompact, easy to fabricate, and rugged. The nondispersive line was built to have high Q and thus exhibit maximum storage rather than maximum slowness. A rugged structure, it consisted of a lead meander line evaporated on one side of a Teflon sheet. To achieve high Q, lead ground planes were placed above and below the line to reduce radiation losses. See Fig. 35 for an exploded view. At all frequencies from 75 to 450 MHz, the Q was 2–4 times the value with only one ground plane. Compared with acoustic delay lines, the superconducting line has low insertion loss, can be more easily fabricated for frequencies above 100 MHz, and, although the wavelength on an acoustic line is several orders of magnitude smaller, it cannot be made in thin-film form.

The electrically tunable line takes advantage of the surface inductance, $\mu_0\lambda$, and switches its phase velocity or resonant frequency by thermally adjusting the penetration depth. The temperature is controlled by a thin-film heater deposited on an alumina substrate and mounted against the line. The separation of 0.25 mm was optimum for control and avoiding losses coupled to the line from the heater. A frequency change of $\simeq 6\%$ was achieved over a 0.6°K temperature swing, with $Q \sim 400$ and a switching time of 3 msec.

The dispersive line geometry requires that the capacitance between lines be large compared to the capacitance between the line and ground plane. On the other hand, mode conversion and radiation can be reduced and the impedance matching improved with a ground plane. The line built (Passow *et al.*, 1972) had a ground plane under the input and output region. The performance of the line was thereby improved; particularly, the insertion losses were suppressed at the upper end of its frequency range.

In conclusion, the superior performance of superconducting transmission line devices over the normal metal structures has been as impressive as that demonstrated for resonators. The added advantages of size reduction makes them competitive on a per-unit-volume performance basis, including the penalty of the Dewar. This is even more true when the delay line devices are used in conjunction with masers that also operate in a cryogenic mode.

References

Abels, B. (1967). Electromagnetic excitation of transverse waves in metals. *Phys. Rev. Lett.* **19**, 1181.

Abrikosov, A. A., Gor'kov, L. P., and Khalatnikov, I. M. (1958). A superconductor in a high frequency field. *Sov. Phys.-JETP* **8**, 182.

Abrikosov, A. A., Gor'kov, L. P., and Khalatnikov, I. M. (1960). Analysis of experimental data relating to the surface impedance of superconductors. *Sov. Phys. JETP* **37**, 132.

Albanese, A. (1972a). M. S. Thesis, Univ. of Texas at Austin. Albanese, A. (1972b). Optical Properties of Materials at Low Temperature and Their Application to Optical Detection. Final Rep., NGR 44-012-104, Electron. Mater. Res. Lab., Dept. of Elec. Eng. Univ. of Texas at Austin.

Alig, R. C. (1969). Direct electromagnetic generation of transverse acoustic waves in metals. *Phys. Rev.* **178**, 1050–1058.

Allen, R. J., and Nahman N. S. (1964). Analysis and performance of superconductive coaxial transmission lines. *Proc. I.R.E.* **52**, 1147–1154.

Allen, J. S., Axel, P., Hanson, A. O., Harlan, J. R., Hoffswell, R. A., Jamnik, D., Robinson, C. S., Staples, J. W., and Sutton, D. C. (1970). Design of a 600 MeV microtron using a superconducting linac. *Particle Accelerators* **1**, 239.

Alworth, C. W., and Haden, C. R. (1971). Nuclear radiation detection using a superconducting resonant cavity. *J. Appl. Phys.* **42**, 166–169.

Arams, F. R., Siegel, K., and Domchick, R. (1967a). Superconducting ultrahigh Q tunable RF preselector. *IEEE Trans.*, **EMC-9**, 110–123.

Arams, F. R., Domchick, R., Levinson, D. S., Seigel, K., Seven, R., Taub, J. J., and Worontzoff, N. (1967b). Final Rep. ECOM 01435-F 9. Airborne Instrum. Lab., Long Island, New York.

Arndt, G. D., Hartwig, W. H., and Stone, J. L. (1968). Photodielectric detector using a superconducting cavity. *J. Appl. Phys.* **39,** 2653–2656.

Banford, A. P. (1965). *Advan. Cryogen. Eng.* **10,** 80–87.

Banford, A. P., and Stafford, G. H. (1961). Feasibility of a superconducting proton linear accelerator. *Plasma Phys.* **3,** 287.

Bardeen, J. and Stephen, M. J. (1965). Theory of the motion of votices in superconductors. *Phys. Rev.* **140,** A1197.

Bardeen, J., Cooper, L. N., and Schrieffer, J. R. (1957). Theory of superconductivity. *Phys. Rev.* **108,** 1175–1204.

Bathow, G., and Frytag, E. (1971). Interner Bericht, DESY, Hamburg, B3/2 (unpublished).

Bauer, W., Citron, A., Dammertz, G., Eschelbacher, H. C., Jüngst, W., Lengler, H., Miller, H., Rathgeber, E., and Diepers, H. (1972). Investigations of niobium deflection cavity models for use in a superconducting RF particle separator. IEEE Conf. Rec. 72 CHO 682-5 TABSC.

Benard, J., Jimenez, J. J., Sudraud, P., and Viet, N. T. (1972). Frequency stability improvements in a klystron stabilized by a superconducting cavity. *Electron. Lett.* **8,** 117.

Benaroya, R. *et al.* (1972). Test on superconducting helix resonator. *Appl. Phys. Lett.* **21,** 235.

Berry, R. W., Hall P. M., and Harris, M. T. (1968). "Thin Film Technology." Van Nostrand-Reinhold, Princeton, New Jersey.

Biquard, F. (1970) Etude d'un oscillateur de type monotron a cavité superconductrice presentant une très haute stabilité de frequence. *Rev. Phys. Appl.* **5,** 705.

Brandelik, A. *et al.* (1972). Accelerating tests on the first section of the Karlsruhe superconducting proton accelerator. Externer Bericht 3/72-9, Ges. f. Kernforschung, Karlsruhe.

Bremer, J. W. (1962). "Superconductive Devices." McGraw-Hill, New York.

Brown, H. N., Foelsche, H. W., Halama, H. J., and Lazarus, D. M. (1971). Long duty-cycle beam separation developments. *Proc. Int. Conf. High Energy Accelerators, 7th* p. 262. CERN, Geneva.

Bruining, H. (1954). "Physics and Applications of Secondary Electron Emission." Pergamon, Oxford.

Bruynseraede, Y., Gorle, D., Leroy, D., and Morignot, P. (1971). Surface resistance measurements in TE_{011} mode cavities of superconducting indium, lead, and an indium-lead alloy and high magnetic fields. *Physica* **54.**

Bura, P. (1966). Parametric amplification with superconducting films. *Proc. IEEE* **54,** 687–688.

Carne, A., Bendall, R. G., Brady, B. G., Sidlow, R., and Kuston, R. L. (1971). High fields in a model superconducting RF separator at 1.3 GHz. *Proc. Int. Conf. High Energy Accelerators, 8th* p. 248. CERN, Geneva.

Citron, A., and Schopper, H. (1970). *In* "Linear Accelerators" (P. M. Lapostelle and A. L. Septier, eds.), p. 1141. North-Holland Publ., Amsterdam and Wiley, New York.

Citron, A. J., *et al.* (1971). Measurements on superconducting helices for the first section of the superconducting proton accerator at Karlsruhe. *Proc. Int. Conf. High Energy Accelerators, 8th, CERN, Geneva.*

Clorfeine, A. S. (1964). Microwave amplification with superconductors. *Proc. IEEE* **52,** 844–845.

Clarke, J., *et al.* (1971). "Device Applications of Cryogenics, Part I, Superconducting Electronics." Res. Advisory Inst., Newport Beach, California.

Cummings, A. J., and Wilson, A. R. (1966). High voltage pulse characteristics of superconducting transmission lines. *J. Appl. Phys.* **37**, 3297–3300.

Dammertz, G. (1971). The effects of electrons on superconducting cavities in the GHz region. *IEEE Trans.* **NS-18**, 153–158.

Dickey, J. M., Strongin, M., and Kammerer, O. F. (1971). Studies of thin films of Nb_3Sn on Nb. *J. Appl. Phys.* **42**, 5808.

DiNardo, A. J., and Sard, E. (1971). Superconductive microwave mixers utilizing Josephson junctions. *J. Appl. Phys.* (*abst.*) **42**, 105.

DiNardo, A. J., Smith, J. G., and Arams, F. R. (1971). Superconducting microstrip high Q microwave resonators. *J. Appl. Phys.* **42**, 186–189.

Dobbs, E. R., and Perz, J. M. (1963). Ultrasonic Determination of the energy gap in superconducting niobium. *Proc. Int. Conf. Low Temp. Phys., 8th* (R. Davies, ed.). Butterworths, London and Washington, D.C.

Dolan, W. W. (1953). Current density tables for field emission theory. *Phys. Rev.* **91**, 510.

Douglass, D. K., Jr., and Meservey, R. H. (1964). Measurements of the superconducting energy gap in lead and aluminum films. *Proc. Int. Conf. Low Temp. Phys., 8th* (R. Davies, ed.). Butterworths, London and Washington, D.C.

Dresselhaus, D., and Dresselhaus, M. S. (1960). Surface impedance of a superconductor in a magnetic field. *Phys. Rev.* **118**, 77.

Easson, R. M., Hlawiczka, P., and Ross, J. M. (1966). Thermal nature of the AC phase transition in Type-II superconductors. *Phys. Lett.* **20**, 465.

Eck, R. E., and Ulrich, B. T. (1967). Paper presented at the *Conf. Electron Device Res., Montreal.*

Ekstrom, M. P., McCaa, W. D., and Nahman, N. S. (1971). The measured time and frequency response of a miniature superconducting coaxial line. *IEEE Trans. Nucl. Sci.* **NS-18**, No. 5, 18.

Faber, T. E., and Pippard, A. B. (1955). *Proc. Roy. Soc. Ser. A* **231**, 336.

Fairbank, W. M. (1949). High frequency surface resistivity of tin in the normal and superconducting states. *Phys. Rev.* **76**, 1106.

Feucht, D. L., and Woodford, J. B. (1961). Superconducting-transition switching time measurements using a superconductive radio frequency mixer. *J. Appl. Phys.* **32**, 10.

Flecher, P. (1968a). Q-measurements in superconducting lead cavity at 800 MHz. *Proc. Linear Accelerator Conf., Brookhaven.*

Flecher, P. (1968b). Hochfrequenzabsorbtion von aufgedampften und zerstäubten schichten. Externer Bericht 3/68-9, IEKP, Kernforschungszentrum, Karlsruhe.

Flecher, P. (1969). *Nucl. Instrum. Methods* **75**, 29.

Flecher, P. (1970). Oberflächenwiderstand in blei in abhängigkeit der mittleren freien weglänge. Dissertation, Univ. of Karlsruhe, and Externer Bericht 3/70-5, Kernforschungszentrum, Karlsruhe.

Flecher, P., Halbritter J., Hietschold, R., Kneisel, P., Kuhn, W., and Stolz, O. (1969). Measurements of the RF absorption of superconducting resonators. *IEEE Trans.* **NS-16**, no. 3, 1018.

Fowler, P. (1963). Superconducting films by vacuum deposition. *J. Appl. Phys.* **34**, 3538.

Fowler, R. H., and Nordheim, L. Electron emission in intense electric fields. *Proc. Roy. Soc. London Ser. A* **119**, 173.

Frerichs, R., and Kircher, C. J. (1963). Properties of superconducting niobium films made by asymmetric AC sputtering. *J. Appl. Phys.* **34**, 3541.

Fricke, J. L., Piosczk, B., and Vetter, J. E. (1972a). Messungen am Nb helix resonator. Notiz Nr. 183, IEKP, Karlsruhe (unpublished).

Fricke, J. L., Piosczyk, B., Vetter, J. E., and Klein, H. (1972b). Measurements on superconducting helical loaded resonators at high field strength. *Particle Accelerators* **3**, 35.

Fricke, J., Muench, N., Piosczyk, B., Vetter, J., and Westenfelder, G. (1971). Hf messungen an supraleitenden Nb und verbleiten Cu helices in verbleiten Cu aussentank. Interne Notiz IEKP, Kerforschungszentrum, Karlsruhe, No. 148, unpublished.

Gandolfo, D. A., Boornard, A., and Morris, L. C. (1968). Superconductive microwave meander lines. *J. Appl. Phys.* **39**, 2657–2660.

Garfunkel, M. P. (1968). Surface impedance of Type-I superconductors; calculation of the effect of a static magnetic field. *Phys. Rev.* **173**, 516–525.

Garwin, E. L., and Rabinowitz, M. Thin dielectric films in superconducting cavities. *Nuovo Cimento Lett.* **2**, 450.

Gerstenberger, D., and Hall, P. M. (1964). Superconducting thin films of niobium, tantalum, tantalum nitride, tantalum carbide, and niobium nitride. *J. Electrochem. Soc.* **111**, 936.

Giaever, I. (1963). Energy gaps of some hard superconductors. *Proc. Int. Conf. Low Temp. Phys., 8th* (R. Davies, ed.), p. 171. Butterworths, London and Washington, D.C.

Ginzberg, D. M., and Tinkman, M. (1960). Infrared transmission of superconducting films. *Phys. Rev.* **118**, 990.

Giordano, S. (1964). *Linear Accelerator Conf.* UC 28, TID 4500, S. 60. Nat. Bur. Std., Springfield, Virginia.

Gittleman, J. I., and Rosenblum, B. (1966). Radio frequency resistance in the mixed state of supercritical currents. *Phys. Rev. Lett.* **16**, 734.

Gittleman, J. I., and Rosenblum, B. (1968). The pinning potential and high frequency studies of Type-II superconductors. *J. Appl. Phys.* **39**, 2617.

Glavish, H. F., Hanna, S. S., and Schwettmann, H. A. (1971). Proposal to the Nat. Sci. Foundation, HEPL, Stanford Univ.

Glover, R. E., and Tinkham, M. (1957). Conductivity of superconducting films for phonon energies between 0.3 and 40 kT_c. *Phys. Rev.* **108**, 243.

Goldstein, Y., and Zemel, A. (1972). Electromagnetic generation of transverse waves in the gigahertz range using indium films. *Phys. Rev. Lett.* **28**, 147.

Good, R. H., Jr., and Müller, E. W. (1956). Field Emission. "Encyclopedia of Physics" (S. Flügge, ed.), Vol. XXI, Electron Emission Gas Discharges I. Springer-Verlag, Berlin, and New York.

Gottlieb, M., and Garbuny, R. (1964). Low frequency dielectric dissipation in mylar at liquid helium temperatures. *Rev. Sci. Inst.* **35**, 641–642.

Grissom, D. (1965). PhD Dissertation. Univ. of Texas at Austin.

Grissom, D., and Hartwig, W. H. (1965). Theory and Measurement of Dielectric Properties of Alkali Halide Crystals at Cryogenic Temperatures, Tech. Rep. No. 6. Lab. for Electron. and Related Sci. Res., Univ. of Texas, Austin.

Grissom, D., and Hartwig, W. H. (1966). Dielectric dissipation in NaCl and KCl below 4.2°K. *J. Appl. Phys.* **37**, 4784–4789.

Haden, C. R., and Hartwig, W. H. (1965). *Phys. Lett.* **17**, 106.

Haden, C. R., and Hartwig, W. H. (1966). *Phys. Rev.* **148**, 313.

Haden, C. R., Hartwig, W. H., and Victor, J. M. (1966). Magnetic losses in superconductors at high frequencies. *IEEE Trans.* **MAG-2**, 331–333.

Hahn, H., and Halama, H. J. (1970). Investigations of a superconducting niobium cavity at S-band. *Proc. Int. Conf. High Energy Accelerators, 1969, 7th* p. 674. Acad. Sci. Armenian SSR, Yerevan.

Hahn, H., Halama, H. J., and Forster, E. H. (1968). Measurement of the surface resistance of superconducting lead at 2.868 GHz. *J. Appl. Phys.* **39**, 2606–2609.

Halama, H. L. (1971). Effects of oxide films on surface resistance and peak fields of superconducting Niobium cavities. *Particle Accelerators* **2**, 335.

Halbritter, J. (1969a). The dependence of the RF field amplitude due to a finite thermal conductivity. Externer Bericht 3/69-6, Kernforschungszentrum, Karlsruhe.

Halbritter, J. (1969b). Exterior Bericht 3/69-2, Kernforschungszentrum, Karlsruhe.

Halbritter, J. (1969c). Paper presented at the *Joint Karlsruhe Symp. Superconduct. RF Separators, Karlsruhe, June 11–13.*

Halbritter, J. (1970a). Externer Bericht 3/70-3, Kerforschungszentrum, Karlsruhe.

Halbritter, J. (1970b). Change of eigenstate in a superconducting RF cavity due to a non-linear response. *J. Appl. Phys.* **41**, 4581–4588.

Halbritter, J. (1970c). Fortran program for the computation of the surface impedance of superconductors. Externer Bericht 3/70-6, Kernforschungzentrum, Karlsruhe.

Halbritter, J. (1970d). Comparison between measured and calculated RF losses in the superconducting state. *Z. Phys.* **238**, 466–476.

Halbritter, J. (1971). Surface residual resistance of high-Q superconducting resonators. *J. Appl. Phys.* **42**, 82–87.

Hallen, E. (1930). Über die elektrischen schwingungen in drahtförmigen leitern. Uppsala Univ. Årsskrift, A. B. Lundequistka Bokhandeln.

Hansen, A. O. (1966). *Proc. AmuAnl Workshop High Voltage Electron Microsc.*, ANL 7275, Argonne Nat. Lab., June 13–July 15.

Hartwig, W. H. (1961–1963). Superconducting Frequency Control Devices, Rep. No. 111–123. Electron. Mater. Res. Lab., Univ. of Texas at Austin.

Hartwig, W. H. (1963a). Superconducting resonant circuits. Digest of Technical Papers. *1963 Int. Solid State Circuits Conf., Lewis Winner, New York, Feb. 22.*

Hartwig, W. H. (1963b). Novel test techniques measure Q in cryogenic resonant circuits. "Electronics," pp. 43–47. McGraw-Hill, New York.

Hartwig, W. H. (1963c). Application of superconductivity to frequency control. *Proc. Annu. Frequency Contr. Sympo. 17th*, pp. 176–189, USAERDL, Ft. Monmouth, New Jersey.

Hartwig, W. H. (1966a). *Bull. Amer. Phys. Soc.* **II-11**, 52.

Hartwig, W. H. (1966b). *Bull. Amer. Phys. Soc.* **II-11**, 764.

Hartwig, W. H. (1970). Optically-pumped high-Q cavities for photodielectric detectors. *In* "Cryogenics and Infrared Detection" (W. H. Hogan and T. S. Moss, eds.), pp. 205–233. Boston Tech. Publ., Cambridge, Massachusetts.

Hartwig, W. H. (1973). Superconducting Resonators and Devices. *IEEE Proc.* **61**, no. 1, 58–70.

Hartwig, W. H., and Grissom, D. (1965). Dielectric dissipation measurements below 7.2°K. *Low Temp. Phys.* **LT9** (**B**) (*Proc. Int. Conf. Low Temp. Phys., 9th*), pp. 1243–1247. Plenum Press, New York.

Hartwig, W. H., and Hinds, J. J. (1969). Use of superconducting cavities to resolve carrier trapping effects in CdS. *J. Appl. Phys.* **40**, 2020–2027.

Hauser, J. J., and Theurer, H. C. (1965). Superconductivity in Pb-Al superimposed films. *Phys. Lett.* **14**, 270.

Heller, F., Kuntze, M., Mittag, K., and Vetter E. (1970). Experimental investigations on non-resonant beam breakup effects in a superconducting structure. *Proc. Int. Conf. High Energy Accelerators, 1969, Yerevan* p. 682. Armenian Acad. Sci. SSR.

Hietschold, R. (1968). Externer Bericht Nr. 872, Kernforschungszentrum, Karlsruhe.

Hinds, J. J., and Hartwig, W. H. (1971). Material properties analyzers using superconducting resonators. *J. Appl. Phys.* **42**, 170–179.

Johnson, J. B. (1972). M. S. thesis, Univ. of Texas at Austin (unpublished).

Jones, O. M., Judish, J. P., King, R. F., McGovan, F. K., Milner, W. T., and Peebles, P. Z., Jr. (1972). Preliminary measurements on a low phase velocity superconducting resonant cavity. *Particle Accelerators* **3**, 103.

Josephson, B. D. (1962). Possible new effects in superconducting tunneling. *Phys. Lett.* **1**, 251–253.

Kaplan, R., Nethercot, A. H., and Boorse, H. A. (1959). Frequency dependence of the surface resistance of superconducting tin in the millimeter wavelength region. *Phys. Rev.* **116**, 270–279.

Klein, H., and Kuntze, M. (1972). Remarks on superconducting heavy ion linacs. *IEEE Trans. Nucl. Sci.* **NS-19**, No. 2, 304.

Klema, D. (1966). *Proc. AmuAnl Workshop High Voltage Electron Microsc.*, ANL 7275. Argonne Nat. Lab., June 13–July 15.

Kneisel, P., Stolz, O., and Halbritter, J. (1971a). Investigations of the surface resistance of a niobium cavity at S-band. *IEEE Trans.* **NS-18**, 158–159.

Kneisel, P., Stolz, O., Halbritter, J., Diepers, H., Martens, H., Sun, R. K. (1971b). Rf losses of superconducting Nb cavities coated with Nb_5O_5. *Int. Conf. High Energy accelerators, CERN, Geneva.*

Kneisel, P., Stolz, O., and Halbritter, J. (1972). On the variation of RF surface resistance with field strength in anodized niobium cavities. IEEE Conf. Rec. 72 CHO 682-5 TABSC.

Kraft, G. (1972). Critical heat flux of saturated and subcooled helium II in long tubes. IVth *Int. Cryogen. Eng. Conf., 4th Eindhoven, May 24–26.*

Kung, H. H. (1969). M. S. thesis, Univ. of Texas at Austin (unpublished).

Lapostolle, P. M., and Septier, A. L. (1970). "Linear Accelerators." North-Holland Publ., Amsterdam and Wiley, New York.

Leslie, J. D., and Ginzberg, D. M. (1964). Far infrared absorption in superconducting Pb alloys. *Phys. Rev.* **133**, 362.

Lerner, D. S., and Wheeler, H. A. (1960). *I.R.E. Trans. Microwave Theory Tech.* **MTT-8**, 343–345.

London, H. (1934). Production of heat in superconductors by alternating currents. *Nature (London)* **133**, 497.

London H. (1940). High frequency resistance of superconducting tin. *Proc. Roy. Soc. Ser. A* **176**, 522.

London, F. (1961). "Superfluids," Vol. 1. Dover, New York.

Love, R. E., and Shaw, R. V. (1964). Ultrasonic attenuation in superconducting lead. *Rev. Mod. Phys.* **36**, 260.

Macalpine, W. W., and Schildknecht, R. O. (1959). Coaxial resonators with helical inner conductor. *Proc. I.R.E.* **47**, 2099–2105.

Maissel, L. I., and Glang, R. (eds.) (1970). "Handbook of Thin Film Technology." McGraw-Hill, New York.

Markard, E. W., and Perlman, B. S. (1966). Superconducting microwave filters. Radio Corp. of Amer. (journal unknown).

Martens, H., Diepers, H., and Sun, R. K. (1971). Improvements of superconducting Nb cavities by anodic oxide films. *Phys. Lett.* **34A**, 439.

Mason P. V. (1971). Effect of tin additive on indium thin-film superconducting transmission lines. *J. Appl. Phys.* **42**, 97–102.

Mason, P. V., and Gould, R. W. (1969). Slow-wave structures utilizing superconducting thin-film transmission lines. *J. Appl. Phys.* **40**, 2039–2051.

Mattis, D. C., and Bardeen, J. (1958). Theory of anomalous skin effect in normal and superconducting metals. *Phys. Rev.* **111**, 412–417.

Maxwell, E., Marcus, P. M., and Slater, J. C. (1949). Surface Impedance of normal and superconductors at 24,000 megacycles per second. *Phys. Rev.* **76**, 1332.

McAshan, M. S., Schwettmann, H. A., Suelzle, L., and Turneaure, J. P. (1972). Development of the superconducting accelerator. Rep. HEPL 665, Stanford Univ.

McCaa, W. D., and Nahman, N. S. (1969). Frequency and time domain analysis of a superconductive coaxial line using the BCS theory. *J. Appl. Phys.* **40**, 2098–2100.

McMillan, W. L., and Rowell, J. M. (1965). Lead phonon spectrum calculated from the superconducting density of states. *Phys. Rev. Lett.* **14**, 108.

Mesa Instruments (1965). Austin, Texas, private communication.

Meservey, R., and Schwartz, B. B. (1969). Comparison of experimental results with predictions of the BCS theory. "Superconductivity" (R. D. Parks, ed.), p. 141. Dekker, New York.

Miles, J. L., Smith, P. H., and Schonlein, W. (1963). Oscillation and amplification at VHF utilizing electron tunneling between superconducting metals. *Proc. IEEE* **51**, 937.

Miller, P. B. (1960). Surface impedance of superconductors. *Phys. Rev.* **118**, 928–934.

Minet, M. R., and Combet, M. H. A. (date unknown). Filter hyperfrequences intégrés a surtension elevée utilisant des superconductures, Centre Nat. d'Etudes des Télécommun. Centre de Rech. de Lannion, 22 Lannion (France), journal unknown.

Mittag, K., Kuntze, M., Heller, F., and Vetter, J. E. (1969). Experimental results on non-resonant beam break-up effects in a superconducting structure. *Nucl. Instrum. Methods* **76**, 245.

Mittag, K., Hietschold, R., Vetter, J., and Piosczyk, B. (1970). Measurements of loss tangent of dielectric materials at low temperatures. *Proc. 1970 Proton Linear Accelerator Conf.*, p. 257. Nat. Accelerator Lab., Batavia, Illinois.

Montague, B. W. (1963). The application of superconductivity to RF particle separators. CERN Intern. Rep. PSep/63-1.

Mühlschlegel, B. (1959). Die thermodynamischen funktionen des supraleiters. *Z. Phys.* **155**, 313.

Nahman, N. S., and Gooch, G. M. (1960). Nanosecond response and attenuation characteristics of a superconductive coaxial line. *Proc. I.R.E.* **48**, 1852–1856.

Newhouse, V. L. (1964). "Applied Superconductivity," p. 96. Wiley, New York.

Newhouse, V. L., and Edwards, H. H. (1964). An ultrasensitive linear cryotron amplifier. *Proc. IEEE* **52**, 1191–1206.

Passow, C. (1967a). Studie uber einen supraleitenden protonen-linear-beschleuniger im GeV-bereich. Kernforschungszentrum, Karlsruhe.

Passow, C. (1967b). *Proc. Int. Conf. High Energy Accelerators, Cambridge, Massachusetts.*

Passow, C. (1969a). Normale und supraleitende hochfrequenz-beschleuniger als spannungsquelle für electronenmikroskope?, KFK 957, Ges. f. Kernforschung, Karlsruhe, 1969.

Passow, C. (1969b). Instabilitäten und akzeptanzen für teilchen strahlen in linearbeschleunigern mit zwischtankfokussierung. Externer Bericht 3/69-15, Kernforschungszentrum, Karlsruhe.

Passow, C. (1969c). Verhandlungen der Deutsch. *Phys. Ges. Reihe VI* **4**, 505.

Passow, C. (1970). Development of a superconducting proton linear accelerator. *Proc. Int. Conf. High Energy Accelerators, 7th, Yerevan, 1969* p. 605. Acad. of Sci. of Armenia SSR, Yerevan.

Passow, C. (1972a). Explanation of the low-temperature, high-frequency residual surface resistance of superconductors. *Phys. Rev. Lett.* **28**, 427–431.

Passow, C. (1972b). Forschungsvorschlag zum bau eines supraleitenden hochspannungs electronenmikroskops (unpublished).

Passow, C. H., Aupelt, G., Bergötz, R., Heneka, A., Herz, W., Paetzold, H., Schaphals, L., and Spiegel, M. J. (1970). *Proc. ICEC* **3**, Berlin.

Passow, C. H., Aupelt, G., Bergötz, R., Heneka, A., Herz, W., Paetzold, H., Schaphals,

L., and Spiegel, M. J. (1971). Measurement of the heat transport by conductivity of superfluid helium in 4 m long pipes of 1–10 mm diameter. *Cryogenics* **12**, 143.

Passow, C. A., Newhouse, V. L., and Gunshor, R. L. (1972). An investigation of some superconductive microwave devices. IEEE Conf. Rec. 72CHO682-5 TABSC.

Pierce, J. M. (1967). The microwave surface resistance of superconducting lead. PhD Dissertation, Stanford Univ., June.

Pippard, A. B. (1947a). The surface impedance of superconductors and normal metals at high frequencies, Parts I, II, III. *Proc. Roy. Soc.* **A191**, 370.

Pippard, A. B. (1947b). *Proc. Roy. Soc. Ser. A* **191**, 399.

Pippard, A. B. (1950). *Proc. Roy. Soc. Ser. A* **203**, 98.

Rabinowitz, M. (1970a). *Nuovo Cimento Lett.* **4**, 549–553.

Rabinowitz, M. (1970b). *Appl. Phys. Lett.* **16**, 419–421.

Rabinowitz, M. (1971a). Analysis of critical power loss in a superconductor. *J. Appl. Phys.* **42**, 88–96.

Rabinowitz, M. (1971b). Frequency dependence of superconducting cavity Q and magnetic breakdown field. *Appl. Phys. Lett.* **19**, 73–76.

Rabinowitz, M., and Garwin, E. L. (1972). *Appl. Phys. Lett.* **20**, 154–156.

Radhakrishnan, V., and Newhouse, V. L. Noise analysis for amplifiers with superconducting input. *J. Appl. Phys.* **42**, 129–132.

Ramer, O. G., and Walker, G. B. (1969). Superconducting superdirectional antenna arrays. *1969 Southwestern IEEE Conf. Rec.* IEEE Cat. no. 69C16-SWIECO, pp. 15A1–15A7.

Ramo, S., Whinnery, J. R., and Van Duzer, D. (1965). "Fields and Waves in Communication Electronics." Wiley, New York.

Reichert, V. (1971). Dissertation, Karlsruhe Univ.

Reichert, V., and Hasse, J. (1972). Hochfrequenzwiderstand abschreckend kondensierter supraleitender metallschichten. *Z. Phys.* **254**, 10.

Reuter, G., and Sondheimer, E. H. (1948). The theory of the anomalous skin effect in metals. *Proc. Roy. Soc. Ser. A.* **195**, 336–364.

Richards, P. L. (1962). Magnetic field dependence of the surface impedance of superconducting tin at 3KMc/sec. *Phys. Rev.* **126**, 912–918.

Richards, D. C., and Tinkham, M. (1960). Far infrared energy gap measurements in bulk superconducting In, Sn, Ta, V, Pb, and Nb. *Phys. Rev.* **119**, 575.

Rowell, J. M., Anderson, P. W., and Thomas, D. E. Image of the phonon spectrum in the tunnel characteristic between superconductors. *Phys. Rev. Lett.* **10**, 344.

Rymer, T. B. (1966). *Proc. AmuAnl Workshop High Voltage Electron Microsc.*, ANL 7275. Argonne Nat. Lab., June 13–July 15.

Sass, A. R. (1964). Image properties of a superconducting ground plane. *J. Appl. Phys.* **35**, 516–521.

Sass, A. R., and Friedlaender, F. J. (1964). Superconducting-to-normal conducting transitions in strip transmission line structures. *J. Appl. Phys.* **35**, 1494–1500.

Sauter, E. (1971). Stromverteilungen in wendeln endlicher länge. Ges. f. Kernforschung, Karlsruhe.

Schulze, D. (1971a). Pondermotorische stabilitat von hochspannungs-hochfrequenzresonatoren und resonatorregelungssystemen. Dissertation, Karlsruhe Univ. (KFK 1493 Ges. f. Kernforschung, Karlsruhe).

Schulze, D. (1971b). *IEEE Trans.* **NS-18**, 160.

Schwettmann, H. A., Pierce, J. M., Wilson, P. B., and Fairbank, W. M. (1965). The application of superconductivity to electron linear accelerators. *Advan. Cryogen. Eng.* **10**, sect. M-U, 88.

Schwettmann, H. A., Turneaure, J. P., Fairbank, W. M., Smith, T. I., McAshan, M. S., Wilson, P. B., and Chambers, E. E. (1967). Low temperature aspects of a cryogenic accelerator. *IEEE Trans.* **NS-14,** No. 3, 336.

Shapiro, S. (1963). Josephson currents in superconducting tunneling; the effect of microwaves and other observations. *Phys. Rev. Lett.* **11**, 80–82.

Sheahen, T. P. (1966). Rules for energy gap and critical field of superconductors. *Phys. Rev.* **149,** 368.

Sherrill, M. D., and Edwards, H. H. (1961). Superconducting tunneling in bulk Nb. *Phys. Rev. Lett.* **6,** 460.

Shizume, P. K., and Vaher, E. (1962). Superconducting coaxial delay line. *I.R.E. Int. Conf. Rec.*, Part 3—ED MTT, p. 95.

Sierk, A. J., Hamer, C. J., and Tombrello, T. A. (1971). Helical wave guides for heavy particle linacs. *Particle Accelerators* **2,** 149.

Silver, A. H., and Zimmerman, J. E. (1967). Multiple quantum resonance spectroscopy through weakly-connected superconductors. *Appl. Phys. Lett.* **10,** 142–145.

Slay, B. G. (1963). Master's Thesis, Univ. of Texas, Austin (unpublished).

Soderman, D. A., and Rose, K. (1968). Microwave studies of thin superconducting films. *J. Appl. Phys.* **39,** 2610–2617.

Stanford Linear Accelerator Center, (1969). Feasibility Study for a Two-Mile Superconducting Accelerator. Stanford Univ.

Stone, J. L. (1964). Design of a 30 MHz oscillator employing a superconducting resonant circuit. M. S. Thesis, Univ. of Texas (unpublished).

Stone, J. L., and Hartwig, W. H. (1968). Performance of superconducting oscillators and filters. *J. Appl. Phys.* **39,** 2665–2671.

Stone, J. L., and Hartwig, W. H. (1972). "Superconductivity, Book 2," Selected Reprints. Amer. Inst. Phys., New York.

Stone, J. L., Hartwig W. H., and Baker, G. L. (1969a). Automatic tuning of a superconducting cavity using optical feedback. *J. Appl. Phys.* **40,** 2015–2020.

Stone, J. L., Baker, G. L., and Hartwig, W. H. (1969b). An X-band oscillator using a superconducting cavity. *Proc. 1969 Southwestern IEEE Conf.* IEEE Cat. No. 69C16-SWIECO.

Styles, J. B., and Weaver, J. N. (1968). Rep. No. HEPL TN 68-12, Stanford Univ. (unpublished).

Suelzle, L. R. (1968). RF amplitude and phase stabilization for a superconducting linear accelerator by feedback stabilization techniques. HEPL 564, Stanford Univ.

Swihart, J. C. (1961). Field solution for a thin-film superconducting strip transmission line. *J. Appl. Phys.* **32,** 461–469.

Szecsi, L. (1970). Externer Bericht 3/70-8, Kernforschungszentrum, Karlsruhe.

Szecsi, L. (1971). Frequenzabhängigkeit des restwiderstandes von galvanisch aufgebrachten bleischichten im supraleitenden zustand, gemessen bei TEM feldkonfiguration. *Z. Phys.* **241,** 36.

Townsend, P., and Sutton, J. (1962). Investigations by electron tunneling of superconducting energy gaps in Nb, Ta, Sn and Pb. *Phys. Rev.* **128,** 591.

Turneaure, J. P. (1967). Microwave measurements of the surface impedance of superconducting tin and lead. HEPL Rep. No. 507, Hansen Lab., Stanford Univ.

Turneaure, J. P. (1971). Measurements on superconducting Nb prototype structures at 1300 MHz. *IEEE Trans.* **NS-18,** no. 3, 166.

Turneaure, J. (1972). The status of superconductivity for RF applications. IEEE Conf. Rec. 72 CHO 682-5 TABSC.

Turneaure, J. P., and Viet, N. T. (1970). Superconducting Nb TM_{010} mode electron beam welded oavities. *Appl. Phys. Lett.* **16**, 333–335.

Turneaure, J. P., and Weissman, I. (1968). Microwave surface resistance of superconducting niobium. *J. Appl. Phys.* **39**, 4417–4427.

Ulrich, B. T. (1970). *In* "Cryogenics and Infrared Detection" (W. H. Hogan and T. S. Moss, eds.), pp. 120–137. Boston Tech. Publ., Cambridge, Massachusetts.

Ulrich, B. T., and Kluth, E. O. (1972) Josephson junction one millimeter microwave source: coupling outside the dewar. IEEE Conf. Rec. 72CHO682-5 TABSC.

Ulrich, B. T., and Kluth, E. O. (1973). Josephson junction millimeter microwave source and homodyne detector. *Proc. IEEE* **61**, no. 1, 51–54.

Vetter, J. E., Piosczyk, B., Mittag, K., and Hietschold, R. (1970a). *Proc. Proton Linear Accelerator Conf.* Nat. Accelerator Lab., Batavia, Illinois.

Vetter, J. E., Piosczyk, B., Mittag, K., and Hietschold, R. (1970b). Measurements of superconducting helix. Notiz Nr. 118 IEKP, Ges. f. Kernforschung, Karlsruhe (unpublished).

Victor, J. M., and Hartwig, W. H. (1968). Radio-frequency losses in the superconducting penetration depth. *J. Appl. Phys.* **39**, 2539–2546.

Vig, J. R., and Gikow, E. (1973). A superconducting tuner with a broad tuning range. *Proc. IEEE* **61**, no. 1, 122–123.

Walker, G. B., and Haden, C. R. (1969). Superconducting antennas. *J. Appl. Phys.* **40**, 2035–2039.

Wilson, P. B. (1970). *In* "Linear Accelerators" (P. M. Lapostelle and A. L. Septier, eds.), p. 1107. North-Holland Publ., Amsterdam and Wiley, New York.

Wilson, P. B., and Schwettmann, H. A. (1965). Superconducting accelerators. *Particle Accelerator Conf.*, S. 1045, Washington.

Wilson, P. B., Schwettmann, H. A., and Fairbank, W. M. (1964). Status of research at Stanford university on superconducting electron linacs. *Proc. Int. Conf. High Energy Accelerators, Dubna, 1963* p. 535. ATOMIZDAT, Moscow.

Wilson, P. B., Neal, R. B., Loew, G. A., Hogg, H. A., Herrmannsfeldt, W. B., Helm, R. H., and Allen, M. A. (1970). Superconducting accelerator research and development at SLAC. *Particle Accelerators* **1**, 223.

Zimmer, H. (1965). Harmonic generation of microwaves produced by a field-emission cathode in a superconducting cavity. *Appl. Phys. Lett.* **7**, 297, 298.

Special References

Dietrich, I. Supraleitung in der nachrichtenteohnik; also Citron, A., and Passow, C. (1968). Supraleitende hohlräume. *In* "Vorträge über Supraleitung" Phys. Ges. Zürich, Birkhäuser Verlag, Basel and Stuttgart.

Gittleman, J., and Rosenblum, B. (1964). Microwave properties of superconductors. *Proc. IEEE* **52**, no. 10, 1138–1147.

Maxwell, E. (1961). Superconducting resonant cavities. *Advan. Cryogen. Eng.* 125–158.

Newhouse, V. L. (1964). "Applied Superconductivity." Wiley, New York.

Shoenberg, D. (1965). "Superconductivity." Cambridge Univ. Press, London and New York.

Van Duzer, T. (Guest editor) (1973). Special Issue on Applications of Superconductivity *Proc. IEEE* **61**, no. 1, 11 1–144.

Chapter 9

Future Prospects

THEODORE VAN DUZER

Department of Electrical Engineering and Computer Sciences
University of California
Berkeley, California

I. Introduction

It has long been recognized that superconductors have properties that are unique and important for applications. Kammerlingh Onnes' publications on his discovery of the phenomenon of superconductivity already contained discussions of superconducting magnets and suggested the production of intense fields without the use of iron cores. London's remarkable theoretical advances during the 1930s produced enough understanding of the properties of superconductors that the two subsequent decades yielded a fair sprinkling of proposals for applications. Among these was use of the low-resistance property of superconductors for transmission lines and cavities. The understanding of switching between the normal and superconducting states led to attempts to develop superconducting computer memories during the 1950s and 1960s. The theoretical foundations for the behavior of superconductors in strong magnetic fields laid by the Russian physicists Ginzburg, Landau, and Abrikosov and the beautiful microscopic theory of superconductivity by Bardeen, Cooper, and Schrieffer greatly enhanced perspectives that have led to subsequent applications.

The birth of the modern era of superconductor applications occurred in 1961 with the first successful high-field magnet developed by Kunzler *et al.* (1961). Another major spur to applications came shortly thereafter with the theoretical discovery and experimental demonstration of the Josephson effects (Josephson, 1962; Anderson and Rowell, 1963; Shapiro, 1963). The last decade has seen a surge of proposed applications; high-field magnets and ultrasensitive instruments have become available commercially. Superconductors are finding definite applications in maintenance of electric standards and in high-energy physics research. Grand-scale applications of superconductors for train levitation, energy conversion, and power transmission appear possible. Computers using the ultrafast switching of the Josephson junction are showing promise. Detection and mixing, especially in the millimeter-wave range, looks like an attractive application of the ac Josephson effect. It seems clear that superconductors will take an important place in technology.

In judging the viability of an application, the need to refrigerate to temperatures near the boiling temperature of liquid helium must be included today, although considerable work is in progress on higher-temperature superconductors. A large number of alloys and compounds have been studied, and the transition temperatures found suggest that there is an upper limit near 20°K. Recently, a ternary compound niobium–aluminum–germanium was found to have a transition temperature of 20.8°K (Arrhenius *et al.*, 1968). Now success in fabrication Nb_3Ge with a transition temperature of 22.3°K has been achieved (Gavaler, 1973). These higher temperatures raise the possibility of using liquid hydrogen rather than liquid helium as the cryogen and may be the solution to the problem of diminishing helium supplies discussed below. In addition to the work on metallic alloys and compounds, it has been suggested that inorganic compounds offer the possibility of much higher transition temperatures. A recent report reveals the discovery of an organic charge-transfer salt in which an enhancement of conductivity of at least 500 times was seen at 60°K; (Heeger *et al.*, 1973a,b); this has been interpreted as arising from superconducting fluctuations associated with a tendency toward high-temperature superconductivity. The existence of superconductors with appreciably higher temperatures would require a reevaluation of the applications discussed in this chapter. It should be kept in mind, however, that the suitability of a material for applications depends on many factors—not just transition temperature—and practical materials at much higher temperatures are not likely to be in the picture for many years.

There are several manufacturers of helium refrigerators, and they produce a variety of systems ranging from very small closed-cycle units with cooling power on the order of 1 W at temperatures in the range of 10°K for cooling infrared detectors to large units giving hundreds of watts of cooling power at 1.85°K. A system of the latter type is used to cool a superconducting linear accelerator under development at Stanford University. For small commercial refrigerators, such as those used to cool maser amplifiers for the NASA deep-space program, the demonstrated mean time between failures has been in excess of 13,000 hr (see Chapter 5 by Hogan). A review of small commercial refrigerators has been given by Goree (1970). Reliability in larger systems will probably be achieved using turborefrigerators. Studies have been made of the refrigeration requirements of some of the very large systems such as bubble-chamber accelerators and power transmission lines (Jensen, 1972). A tentative design for a superconductive power line using Nb_3Sn conductors would require a cooling power of 1310

W/km. The plan to put a refrigeration station at intervals of 16.1 km would lead to a cooling capacity at 6.2°K of 21 kW each and a required power input to each refrigerator of 2.1 MW. Many thousands of liters of liquid helium would be required per kilometer. A high-energy accelerator proposed for the CERN facility in Geneva will require 97 kW of cooling power at a temperature near 3.5°K to provide helium cooling around a ring of 7-km circumference. The required input power will be 42 MW, and the liquid helium storage capacity is to be 80,000 liters.

It is clear that some applications, such as those mentioned above, will require large quantities of liquid helium. One might reasonably inquire about the available supplies. Helium is found in natural gas and, unless extracted, goes into the atmosphere when the gas is used. The gas fields in southwestern Kansas and the Texas and Oklahoma panhandles contain the largest economically recoverable supply of helium in the United States. In 1962, the Department of the Interior began a conservation program in which helium is purchased from four gas companies and stored in underground chambers. Sources of helium alternative to these helium-rich gases are not very attractive. The cost to remove it from the atmosphere is, at present, extremely prohibitive. The low-power applications of superconductivity will probably be little affected by helium shortages, but the availability of helium for major installations such as power transmission lines must be considered. If the development of higher temperature superconductors continues to bear fruit, the dependence on helium will be reduced.

The major subdivisions of this chapter are assigned to different types of applications roughly in order of decreasing assurance of actual application. For example, the first section is on research applications, where superconductivity is already making an impact, and a later section is on magnetic levitation of trains, which is speculative at present.

II. Research Applications

A. Magnets

The superconducting magnet first became a reality in the work of Kunzler *et al.* (1961) when they succeeded in finding a way to keep the superconductive coil material from being driven into the normal state by the strong magnetic fields which the coil produced. The technique is to have sufficient structural imperfections to prevent the motion of the flux passing through the material. In the several years since this initial achievement, magnet coils have become available commercially with fields up to about 15 T. Coils for fields up to about 8 T are commonly made from the relatively ductile alloys of niobium and titanium in the United States and Western Europe, whereas ternary alloys of Nb–Ti–Zr are favored in the U.S.S.R. and Japan (Stekly and Thome, 1973). For higher fields, the compound Nb_3Sn has been used, but it is hard and brittle and therefore difficult to wind. Considerable work has been expended and will continue in order to create forms (tapes and multifilimentary wires) of Nb_3Sn which are more readily wound into coils. In 1966, a method was developed which permitted the fabrication of V_3Ga in a form suitable for magnetic coils (Tachikawa and Tanaka, 1966). It has about the same critical field (23 T) as Nb_3Sn but has a higher critical current density than Nb_3Sn for fields above about 14–15 T. Since it is more expensive than Nb_3Sn, it will probably not be used in the near future where the field does not exceed 15 T. A magnet to be used for generating fields of 17.5 T will probably be partly wound with V_3Ga and partly with Nb_3Sn, with the former being used for those parts of the coil which are subjected to the highest fields. There are a variety of materials which have the potentiality for use at higher fields and currents which are under investigation; we can doubtless expect a continuing trend toward higher fields. Other current materials research is aimed at magnet wires that can produce time-varying fields without excessive losses; these are multifilament superconductors imbedded in a copper matrix. The resulting wire is twisted about its axis with the rate of twist dependent on the expected rate of change of magnetic field through the wire.

We can comment now on some of the research applications of superconducting magnets which can be expected to be of importance in the coming years. Superconducting magnets are presently in use and show definite promise of widespread future application in the study of biological molecules using nuclear magnetic resonance (NMR) spectroscopy. The high sensitivity and resolution of this instrumentation yield data not available by other means. NMR measurements can be done under conditions similar to the natural physiological environment of the molecules. The measurements help clarify the reaction mechanisms involved in the biological roles of molecules (Wuthrich and Shulman, 1970). Magnets for 15 T will become routinely available; high fields and temporal stability are the important features for this work.

Extreme temporal stability is important for the deflection and focusing coils in very high-resolution electron microscopes. However, it is doubtful that there is sufficient advantage over well-stabilized normal-metal coils to justify the expense of refrigeration for most microscope applications. Probably the largest-ranging set of applications of superconducting magnets currently being developed relates to high-energy physics. There are very persuasive arguments for using superconducting magnets, and this has led to their funding. In high-energy physics, superconducting magnets are used exclusively when fields greater than 2 T are required or when moderate fields (0.1 T) over large volumes are needed. These are applications when conventional systems do not satisfy the requirements. Whenever either conventional *or* superconducting magnets would work, the tendency has been to use a conventional magnet for better reliability. Magnets for bubble chambers used for identification and analysis of particle interactions have been and are being constructed. The largest is that at the Argonne National Laboratory; it has an inside diameter of 4.8 m and will produce a field of 1.8 T. The one at CERN in Geneva is slightly smaller but has a much higher field (3.5 T). The large investment required for the construction of these magnets will be offset by power savings on the order of $1,000,000 annually. Scores of other smaller dipole and quadrupole magnets to select, guide, and focus beams may be employed in the continuing improvement of the performance of accelerators. The stronger fields obtainable with superconducting magnets give shorter focal lengths, which are advantageous in studying short-lived particles. This is an application which already exists in some places and shows almost certain prospects for further development. Considerable effort is presently being directed toward development of the main synchrotron magnets which are used to keep the particle beam in a circular orbit during acceleration. The magnets require a high degree of dimensional stability and must operate in a pulsed mode with the attendant losses in a radiation environment. The development of these magnets, now in the prototype phase, is the key to several proposed new and updated systems (Stekly and Thome, 1973).

B. Microwave Cavities

Another research application is the use of superconducting microwave cavities for linear accelerators. In a major project at Stanford University, a 150-m set of Nb cavities is to be used to accelerate electrons for particle physics studies. The system is still in the development stage, and difficulty is being experienced as a result of magnetic breakdown and field emission in the cavities at fields lower than the design values. The original design called for an accelerating gradient of 13.1 MeV/m; it now appears that progress in achieving gradients greater than about 3.3 MeV/m will be slow. Consideration is being given to recirculating the electrons. Development efforts are in progress in other high-energy physics laboratories on heavy-particle accelerators which use cavities containing helices. For example, at the Argonne National Laboratory the aim is to use 115 helical units to construct an accelerator 40 m long (Industrial Research, 1972). See Chapter 8 by Hartwig and Passow for more details. Cavity systems for separating different kinds of particles are also being developed.

C. Detectors and Mixers

Largely in connection with radio astronomy applications, there has been an effort, involving scores of researchers, to make use of Josephson devices for mixing and detection in the millimeter and submillimeter range (Richards *et al.*, 1973). Most of this work has concentrated on the dependence of the heights of the steps induced in the I–V characteristic on the amplitude of the rf signal applied. Figure 1 shows the I–V characteristic of a typical Josephson point contact with various applied rf signal levels. Constant-voltage steps are induced at voltages having the values $nhf/2e$, where f is the frequency of the applied rf signal and n is an integer. If the amplitude of the rf signal is varied, the step heights, including that at $V = 0$, vary correspondingly, as seen in Fig. 1. A broad-band rf signal, such as

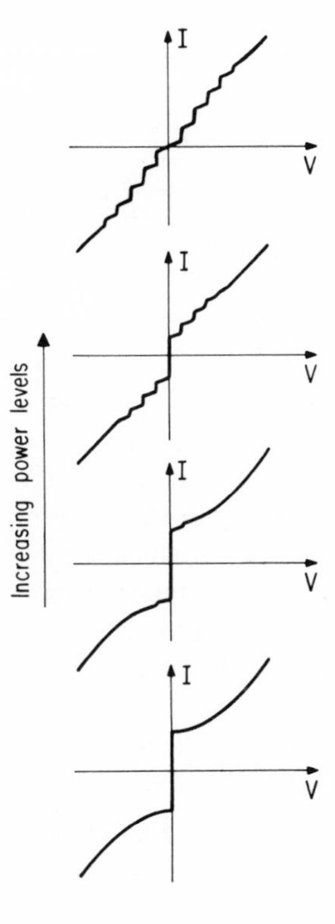

FIG. 1. I–V characteristics of a point-contact Josephson junction with increasing levels of applied rf current (after Richards *et al.*, 1973).

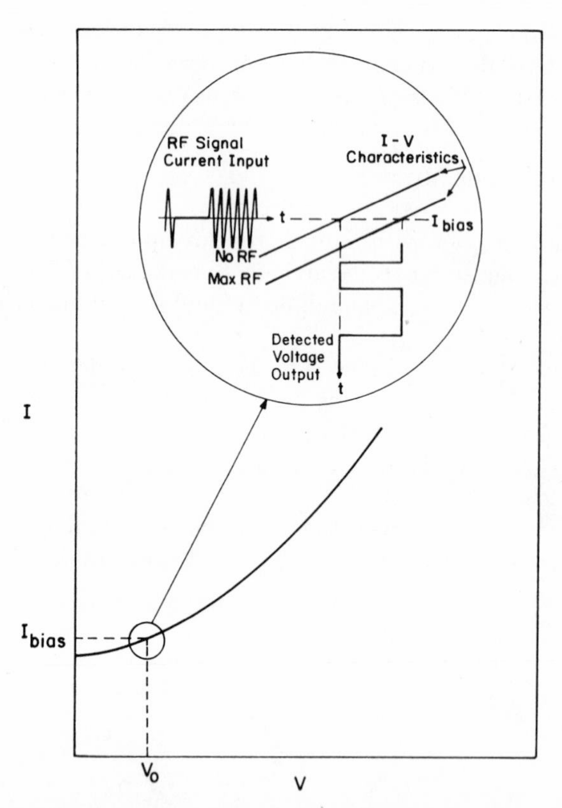

FIG. 2. I–V characteristics of a point-contact Josephson junction operating in a detection mode.

would be typical in radio astronomy, leads to complicated step structures, but all frequencies affect the step at $V = 0$. In a typical mode of operation for radio astronomy, the device would be biased at a current above the dc value of I_c as shown in Fig. 2. The incoming signal would be chopped between the antenna and the detector. This would lead to a modulation of the step height at $V = 0$ which would be proportional to the signal power. As can be seen from the insert, there would be a resulting modulation of the device voltage. The theoretical prediction (Auracher and Van Duzer, 1972) of the possibility of conversion gain in a mixing mode with an external oscillator has recently been confirmed by experiment (Taur *et al.*, 1973). The most successful device configuration to date has been the point contact; an application at 1 mm wavelength has given performance comparable with the best competing devices (Ulrich, 1970). The point contact has the disadvantage of requiring adjustment, and so there have been attempts to use constricted thin films and tunnel junctions. An important problem in the application of these devices is matching their low impedances to the wave-guiding system. The tunnel junction suffers from excessive shunt capacitance, and the constricted thin film from excessive shunt conductance. The use of photo- or electron-beam lithography may make possible fabrication of tunnel junctions of small enough cross section and, hence, capacitance. Another configuration which promises to be of importance is a planar structure in which a small gap (1 μ or less) between two regions of thin-film superconductor is overlaid with bismuth film (Ohta *et al.*, 1973). The impedance of this device apparently could be high enough for efficient coupling. Whether the Josephson devices will prove to be more advantageous for millimeter-wave and far-infrared mixing and detection than the competing devices such as Schottky barriers and the semiconductor bolometer remains to be seen in the next few years. Another type of detector that may prove to be important is the superconducting bolometer (see Chapter 4 by Rose *et al.*; also Clarke *et al.*, 1973). It is of importance to note that the competing devices must also be cooled to liquid helium temperatures so that refrigeration is not a negative factor for the Josephson devices or thin-film bolometers.

D. Sensitive Instrumentation

The invention and development of the superconducting quantum interference device (SQUID) has made possible magnetic field measurements of sensitivity on the order of 10^{-15} T/Hz$^{1/2}$. The SQUID is a superconducting loop containing either one or two weak links (Josephson devices) depending on whether the bias on the loop is rf or dc, respectively. In practice, the major emphasis has been on the rf system, and so we shall restrict explanation of SQUID operation to that type. The loop, containing a single junction, is magnetically coupled to the resonant coil of an rf oscillator (typically, 30 MHz). The relation of peak rf current induced in the loop to the critical current of the junction determines the amount of loading the loop places on the coil to which it is coupled. The lower the critical current, the greater is the loading.

The critical current is a periodic function of the sampled field intensity, having a period of Φ_0/A, where Φ_0 is the flux quantum and A is the loop area. Therefore, the loading is a periodic function of the sampled field intensity.

The variation of loading on the oscillator is detected and processed to provide a measure of the sampled field. Usually, the SQUID is used in a feedback mode in which the change of critical current caused by a small change in the ambient field is detected and a field is fed back to drive the total flux through the loop to zero. The current producing the feedback field is monitored in a way which permits counting individual flux quanta or gives an analog output.

The flux linking the sensing loop is often multiplied by use of a flux transformer which gives the positive effect of a larger loop without the disadvantage of a correspondingly large inductance in the SQUID loop.

The sensitivity of the SQUID magnetometer is so great that the natural fluctuations in the earth's magnetic field are orders of magnitude above the minimum detectable fields. To avoid this difficulty, the measurements can be made in a heavily shielded enclosure. Alternatively, a pair of counter-wound coils can be arranged to measure the gradient of the ambient field. In this gradiometer scheme, use is made of the fact that the earth's field and other spurious fields are produced by sources far from the measurement point and are, therefore, spatially uniform and induce no net effect in the pair of coils. Fields produced by sources close to the measurement coils have appreciable spatial variation and can be measured.

Essentially all uses of SQUIDs to date are in research, although some are destined to develop into tools for more applied work. Measurement of weak magnetic properties of materials has been, and will continue to be, a valuable research application. We have already seen uses which include the measurement of susceptibilities of moon rocks (Goodman *et al.*, 1973) and magnetic biochemical compounds (Hoenig *et al.*, 1973). Magnetocardiography, which measures the weak magnetic fields produced by currents in the heart, is in a research phase at present, as little experience has been accumulated in interpreting the cardiographs. It is necessary to place the patient in a heavily shielded room to eliminate the noise of the earth's field and other man-made fields or to use the gradiometer method mentioned above. So far, it has been necessary to operate gradiometers in environments without nearby electric machinery. This work is in such an early stage that several years of research probably will be required before the SQUID magnetocardiograph can move into clinical use. Important advantages of a magnetocardiogram over the electrocardiogram include the ability to obtain absolute readings of the body's electrical activity, independence of changes in body weight, and the fact that no contacts need be made to the body. The last is probably the factor which will put pressure behind the development of magnetocardiography as a clinical tool.

SQUIDs can also be employed as a part of very sensitive voltmeters. The voltage drives a current through a resistor of known value, and the current produces a magnetic field which is measured by the SQUID magnetometer. These voltmeters have been applied to measurements on thermoelectric emf's, Hall voltages, electrical noise, flux creep in superconductors, proximity effects in superconducting-normal-superconducting (SNS) sandwiches, the Josephson voltage–frequency relation, and chemical potentials in nonequilibrium superconductors (see Chapter 1 by Silver and Zimmerman; also Clarke, 1973). All of these measurements have been made inside strong magnetic shielding. It is nearly essential that such high-resolution measurements be restricted to voltages originating in a cryogenic environment.

SQUID magnetometers are now available from at least three companies. They are sometimes accompanied by superconductive magnetic shields which provide the necessary magnetically quiet environment. Aside from magnets, these instruments and shields represent the only currently significant commercial products that depend on superconductivity.

There are also potential applications to measurement of small currents, resistance, and inductances. Other research applications will develop as more workers in various fields become aware of the capabilities of these instruments to measure weak magnetic properties and minute electrical quantities

E. Gravimeters

Prothero and Goodkind (1968) have developed an instrument which employs a 25.4-mm superconducting sphere which is levitated and has suspension that does not change over the long periods needed to measure gravitational effects. The sphere is levitated by an inhomogeneous magnetic field maintained by persistent currents in a set of superconducting coils. A capacitance detector and a feedback system maintains the sphere in a constant position. Comparing the signals in the capacitance detection system with measurements of the magnetic field variations allows separation of the effects of gravitational variations. Fractional changes of gravitational acceleration of 10^{-10} can be observed. The system has been in use for several years in measuring earth tides. Apparatus similar to this but on a much larger scale is being developed to attempt to measure gravity waves.

III. Electrical Standards

The importance of superconductivity in basic standards is attested to by the extensive activity in the world's standards laboratories. The Josephson effect has already taken an official position in maintainance of the United States legal volt. The United States National Bureau of Standards is at work on a number of possible new standards employing the Josephson effect and flux quantization.

A. Voltage

On July 1, 1972, the United States National Bureau of Standards (1972) adopted a new procedure for maintaining the legal volt that employs the Josephson relations. Recent work has shown that $2e/h$ can be determined in terms of a particular "as-maintained" unit of voltage to one part in 10^6—or better. The new basic premise is that the physical factor $2e/h$ has a definite value which is taken as that determined by frequency measurement and the Josephson relation, with the reference group of standard cells as of that date:

$$2e/h = 483.593420 \text{ THz}/V_{\text{NBS}}$$

As discussed in Section II.C, an rf signal applied to a Josephson junction induces steps in the I–V characteristic at voltages $V_n = (nh/2e)f$, where f is the frequency of the rf signal and n takes on integer values. By current biasing to a current within one step, a voltage precisely related to the frequency of the applied signal is obtained; this is then compared with the standard cell in a divider-bridge arrangement of $1/10^8$ accuracy. The frequency of the applied rf signal gives the voltage of the cell. Standards laboratories of other countries have similar systems under development. It seems likely that a single international value will be adopted for $2e/h$ for use in maintaining the volt.

Finnegan and Denenstein (1971) have suggested a secondary standard for voltage which would use the energy gap of a tunnel junction. Actually, many junctions would be connected in series to give a more convenient voltage. The current would be biased to the steepest part of the junction I–V characteristic where the voltage would have the minimum sensitivity to the current. The current could be controlled by a current-ratio instrument to be discussed below, and the stability of the voltage would exceed that of the current.

B. Current

A scheme has been proposed by Meservey (1968) to create an absolute ampere standard. The current to be measured produces a magnetic field by passing through a carefully constructed solenoid, and the field is measured by a SQUID loop placed inside the solenoid. The current is measured in terms of the flux passing through the SQUID ring, the area of the ring, and the geometrical factors for the solenoid. The determination of the geometrical factors and the area to a sufficient accuracy would be a formidable undertaking at present.

Although the Meservey proposal for an absolute current standard may not be possible because of the need for precise dimensional measurements, it does offer a means for making very precise measurements of current and voltage ratios. The response can be expected to be highly linear over the full current range (from 10^{-10} A to a few amperes). The current ratios could be converted to voltage ratios using a stable cryogenic resistor; these voltage ratios should be commensurate with the current-ratio measurements (Sullivan, 1972; Kamper and Sullivan, 1972). Ratio measurements such as these will probably play an important role in the future of precision metrology.

C. RF Power and Attenuation

The SQUID magnetometer described in Section II.D has been adapted to measure rf magnetic fields having frequencies up to 1 GHz (Kamper *et al.*, 1972, 1973). The 30-MHz oscillator in the SQUID system used to measure dc and low-frequency fields is replaced here by a 9-GHz oscillator. The basic nature of the SQUID operation is maintained as long as the oscillator frequency is appreciably higher than the frequency of the signal to be measured. The signal to be measured is inductively coupled to the SQUID. Measurements published to date have given only ratios because of an unknown insertion loss in an attenuator which provided the 30-MHz test signal. The attenuator was calibrated by the National Bureau of Standards Calibration Service. Excellent agreement was found between the attenuation determined from the SQUID measurements and that of the calibrated attenuator. The SQUID has the important advantage of being broad band, whereas the attenuators are designed for a narrow frequency range. This type of system is not limited to attenuation measurements, but can also be applied to absolute measurements of rf current, phase difference, and power. Kamper and Sullivan (1972) suggest that SQUID systems may ultimately replace several of the national rf electrical standards.

Work is in progress to extend the frequency range of measurable signals to 18 GHz. The performance goals also include measurement of power over the range of -120 to $+30$ dBm with an accuracy of ± 0.1 dB and the measurement of attenuation over a range of 160 dB with an accuracy of ±0.001 dB. It may be necessary to use a series of three complementary SQUIDs installed in series along the coaxial rf input line in order to cover the desired ranges.

D. Temperature

Superconductive devices are playing an important role in thermometry at low cryogenic temperatures. For example, the United States National Bureau of Standards Office of Standard Reference Materials has offered for sale a set of superconductor rods of various high-purity materials with a mutual inductance coil set. The transition temperatures of these rods provide fixed points from 0.5 to 7.2°K, reproducible to within 1 m°K, which can be used to calibrate other thermometers.

Another very promising kind of thermometer measures the thermal noise in a small ($\approx 10^{-5}\ \Omega$) resistor by applying it to a Josephson junction in a low-inductance loop. The average voltage across the resistor, resulting from an applied direct current, biases the junction so that it oscillates at about 30 MHz. The noise voltage in the resistor causes a corresponding fluctuation of the Josephson frequency. A counter may be used to determine the variance of the frequency fluctuations, which can be shown to be directly proportional to temperature. All of the factors in the proportionality constant are either known or can be determined readily so that the Josephson noise therometer can be used to measure absolute temperature.

A current review of low-temperature thermometry using SQUIDs has been given by Giffard *et al.* (1972), and comparisons are made with the above-described thermometer. It can be expected that these developments will lead to new temperature standards.

E. Frequency

Recent work on using high Q ($\approx 10^9$) superconducting cavities to stabilize microwave oscillators has given some exciting results. Stein and Turneaure (1972) stabilized an 8.6-GHz Gunn-effect oscillator with fractional frequency fluctuations of 1.5×10^{-14} for 10-sec averaging times and a frequency drift of 1.6×10^{-14}/hr for a 32-hr interval. This oscillator is the most stable in existence for averaging times less than 50 sec. Turneaure (1973) predicts another order-of-magnitude improvement of stability. Possible applications cited by Stein and Turneaure are (1) attempts to measure the fine-structure constant, (2) a compact clock for experiments to test general relativity, (3) measurement of gravitational waves, (4) improvement of the cesium atomic clock and the proposed hydrogen storage beam atomic clock, and (5) provision of an ultrastable X-band source of radiation for measuring the frequencies of infrared lasers. It seems clear that superconductive cavities will be used to enhance the stability of frequency standards in the future.

IV. RF Signal Processing and Transmission

A. Coaxial Transmission Lines

Superconducting coaxial transmission lines have been constructed which give losses orders of magnitude below conventional lines over wide bandwidths; they are of particular interest for high-fidelity, ultrashort (100-psec) pulse propagation. Early work by Nahman and his associates demonstrated a line which preserved a pulse rise time of less than 500 psec for a pulse traveling several hundred meters. Hoshiko and Chiba (Kamper and Sullivan, 1972) have proposed an information system employing a cable of 200 individual superconducting coaxial lines, each of about 1-mm diam. The system would have a bandwidth of 3 GHz (or 300 Gbits/sec) and would require repeaters at intervals of 25 to 50 km. Recent work by Chiba *et al.* (1973) has demonstrated considerable progress with the problem of spurious reflections, and they have produced a line with less than 1-dB attenuation per kilometer at 1 GHz. The attenuation did not depend on the choice of superconductor and increased with frequency according to the law for the dielectric, thus indicating its domination of the losses. The proposed superconductive transmission system must compete with the oversized cylindrical guides developed by Bell Telephone Laboratories and the emerging optical guiding systems employing low-loss dielectric fibers. Since the superconducting lines require cooling, it is possible that their only applications will be found in systems where refrigeration is already available, as in connection with superconducting power-transmission lines.

B. Resonators

There have been numerous efforts over the last decade to apply superconductive resonators to communication systems. Much of this work has been in the HF and VHF ranges, but also in the microwave range. Although the advantage of higher Q's achieves the desired aim of narrow bandwidth, not much use has yet been made in practical communication systems except in conjunction with masers and parametric amplifiers. Development of the uses of the Josephson junction for detection and mixing will lead to further use of microwave resonators and related system components. It has been pointed out that a superconducting superdirective antenna array would offer advantages over the normal equivalent by virtue of reduced losses and narrowed bandwidth (see Chapter 8).

One can generally expect that the future will see more use of superconducting resonators in communication systems which already require helium refrigeration for other reasons. One exception to this may be the very stable oscillators presently under development and discussed in Section III.E. The high stability may be sufficient justification for refrigeration in some communications applications.

C. Oscillators

Chapter 8 by Hartwig and Passow reviews various reports of oscillators employing superconductive circuits; the oscillator short-term stabilities have improved from values of 10^{-7} to 10^{-14}. The latter work is that of Turneaure reported in Section III.E.

Another source of oscillations which has not yet been developed much is the ac Josephson effect. If a dc voltage V_0 is applied across a Josephson device, there will be an ac component of the current through the device at a frequency $f = 2eV_0/h$. For a single junction, the maximum power that can be expected from this oscillation is only about 10^{-8} W. For example, Elsley and Sievers (1972) have measured 10^{-9} W from a point contact in the frequency range 60–390 GHz.

There is evidence that arrays of junctions have the superradiant property; that is, the available power from the array increases as the square of the number of elements. This has been shown theoretically by Tilley (1970) and experimentally for an array of two junctions by Finnegan and Wahlsten (1972). Clark (1968, 1973) has constructed two-dimensional arrays of many junctions by pressing together up to 100 spheres of 1-mm diam. Array work is in a very preliminary stage, and it is to be expected that the technology of large-scale integration of electronic circuits will be brought to bear to make arrays of thousands of junctions which will be able to provide a substantial power output at tunable millimeter-wave frequencies.

D. Miscellaneous Circuit Elements

A number of passive elements such as tunable inductors (which use the kinetic inductance of the paired electrons), strip lines, and linear and nonlinear delay lines have been studied. They have a variety of possible applications, including pulse compression, filtering, and delay.

Generally speaking, however, these signal-processing devices are not advantageous enough in themselves to warrant refrigeration, but they may be employed in a system that requires cryogenic temperatures for more essential reasons.

E. Amplifiers

A variety of principles have been used to make superconductive amplifiers (see Chapter 2 by Goree and Hesterman). The most important classes are the thin-film cryotron amplifier, amplifiers working on related principles but employing Josephson devices, and parametric amplifiers. The thin-film cryotron amplifier is limited to frequencies below about 50 MHz by heating effects. The other two types are operable into the microwave range. The first two types have found important uses in research measurements but have not been used in communications. The superconductive parametric amplifier has not shown itself to be sufficiently advantageous to find its way into communications systems, but the development effort was somewhat limited compared with other devices which have been brought into application. There may be potential for completely superconductive low-noise receivers using the mixers to be discussed in the next section with superconductive IF amplifiers.

F. Millimeter-Wave Mixing

There is little likelihood that there will be much application of superconductive mixers in communication systems for frequencies below about 50 GHz, since other types of mixers have sufficiently low noise temperatures. On the other hand, low-noise detectors in the millimeter-wave range are at a premium, and the inconvenience and cost of refrigeration may not prevent the use of superconductive devices. At present, most of the emphasis is on Josephson devices for use in receivers for radio astronomy, and this has been discussed in Section II.C. As pointed out there, a variety of types of Josephson devices are under investigation, and the work is still in a fairly early stage.

Another development employing a superconductor with potentially important impingement on mixing in communications systems is the superconductor–semiconductor contact diode (McColl *et al.*, 1973). The nonlinearity of the Schottky barrier is enhanced by the superconductor's energy gap. The initial tests have been made in a video detection mode at

10 GHz and have given an NEP (noise equivalent power) of 2×10^{-15} $W/Hz^{1/2}$, which seems to be the lowest value reported in the literature for video detection. Mixing measurements are in progress.

G. Extremely-Low-Frequency Detection

At very low frequencies, antennas must be inconveniently large to have reasonable efficiencies, and, for frequencies below 100 Hz, it is generally conceded that a sensitive magnetometer would be a better receiver than a conventional loop antenna and radio receiver. The SQUID gradiometer may find use in this application (Kamper and Sullivan, 1972).

V. Computer Components and Systems

This is an application with a very large commercial potential and a comparably large required investment, and one with strong competition from well-developed technologies. Factors of speed and cost are closely compared, and investments are made with extreme caution in the computer field.

A. Nontunneling Elements

The first cryotron was developed by Buck (1956) and contained switching elements made of a high T_c (niobium) wire wrapped around a lower T_c (tantalum) wire so that a current through the former would cause the latter to go into the normal state. Connection of these elements in a loop permitted switching the current from one leg to the other to represent 1's and 0's; the switching time was of the order of 100 μsec. It was soon appreciated that the switching time and cost of fabrication could be decreased by making thin-film analogs of Buck's cryotron. There followed a decade during which a variety of thin-film memory elements were studied, and considerable effort was expended on the development of memory systems. One particularly simple system, employing circulating currents, excited in a continuous film as the memory was abandoned because of inability to produce a sufficiently uniform film. Some of the other memory cells were developed to a rather high state, but switching times failed to reach values which would make the superconductive memory, with its need for refrigeration, economically advantageous. All work on nontunneling cryotron systems has been terminated. However, the extensive level of understanding of cryotron circuits which was developed has not been wasted; we shall see in the next section that a very promising memory is being developed which is similar to earlier cryotrons, but with important advantages.

B. Tunneling Cryotron

The tunneling cryotron invented by J. Matisoo (1967) utilizes one junction as a switch in each leg of a loop (Fig. 3). The control of that switch is exercised by the current in the overlying control line. If the current passing through a junction is made to exceed a certain critical value I_c, the junction must switch into the finite-voltage quasiparticle state as indicated by the line A–B in Fig. 4. In practice, I_c is reduced below the gate current to achieve the switching by applying a pulse to the control line. When the switched junction goes into the finite-voltage state, that voltage drives the current into the other (zero-resistance) side of the loop. The loop is bistable because a voltage is required to provide the $L(dI/dt)$ necessary to steer the current to the other leg. The configuration shown in Fig. 3 has current gain, which means that the steered gate current exceeds the value of the control current. Gain is required to allowed the gate current to be used to control other cells.

Zappe and Grebe (1973) have made the most recently reported current-transfer measurements on a model of the tunneling cryotron and have studied the switching times of isolated junctions. In both cases, the junction size was about $10^4\ \mu^2$. They found junction switching times as low as 60 psec and current-transfer times as low as 300 psec even with large loop

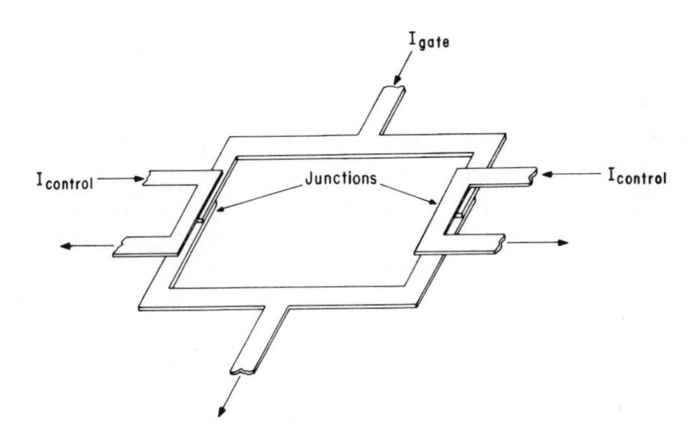

FIG. 3. Basic tunneling-cryotron flip-flop circuit (after Van Duzer and C. W. Turner, 1972).

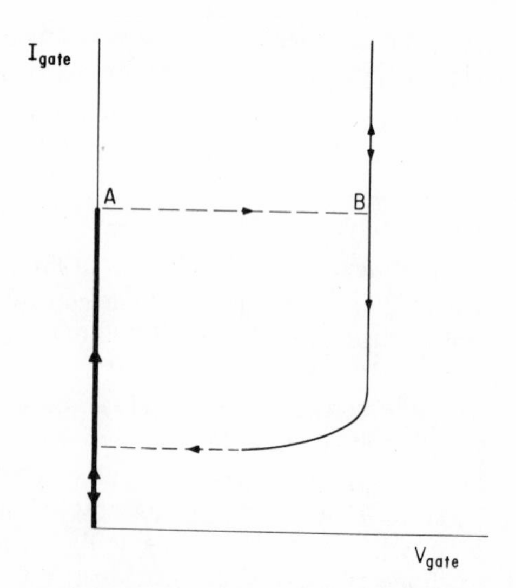

FIG. 4. *I–V* characteristic of a Josephson tunnel junction showing the zero-voltage to finite-voltage transition (A–B).

inductance ($>10^{-11}$ H). These numbers become even smaller upon reducing the size of the components. Jutzi *et al.* (1973) have measured single-junction switching times as low as 34 psec for junctions with dimensions of $8.5 \times 20\ \mu$ and 32 psec for junctions with dimensions of $1.25 \times 3.1\ \mu$. The loop can be scaled down, also, yielding a lower inductance and hence a lower current-transfer time.

These remarkably fast switching times are spurring the development of a prototype memory at IBM's Thomas J. Watson Research Center (Anacher, 1969). This project has been in progress for several years, with the first period being spent on developing the technology for fabricating reproducible junctions having the desired electrical properties and being capable of withstanding thermal cycling. The next few years should be important in establishing the viability of this technology. If successful, it may constitute one of the largest applications of superconductivity, in economic terms. It would also stimulate a great deal of other computer-oriented applications of superconductivity, since justifying the refrigerator is a major hurdle.

C. Shift Register

Fulton *et al.* (1973) have described a shift resister which employs single flux quanta as bits. They have analyzed and constructed versions in both continuous-junction and discrete-junction forms. Flux quanta are injected by application of current pulses at one end of the register. They are shifted by application of pulses at points along the resister. The time required to shift the position of a flux quantum can be as little as 10 psec. No systems work using this device has been reported; its utilization probably depends on the acceptance of the tunneling cryotron, as argued above.

VI. Energy Conversion and Power Transmission

Steady progress has been made during the past decade in using superconductive machines and cables for electric energy conversion and power transmission. This was made possible by the development of intrinsically stable superconductors in the early 1960s. Superconducting magnets will be required in essentially all magnetohydrodynamics (MHD) applications and in some of the controlled-thermonuclear-fusion systems (Stekly and Thome, 1973). Motors and generators employing superconducting coils to minimize their size and weight are of interest for ship and aircraft use and for large fixed installations such as central power stations. Definite advantages are offered in the form of reduced power loss by superconducting transmission cables where very heavy loads are to be carried. It is not expected that any of these applications, with the possible exception of small- to moderate-size motors or generators, will find application in practice within the next decade. However, they are all potentially major applications, and work on them can be expected to continue for quite some time.

A. MHD Power Generation

The aim of this work is to generate directly electric power by passing an energetic conducting gas or liquid metal through a magnetic field. Whereas the liquid metal machine does not require a superconducting magnet and experimental gas systems usually use normal magnets, it has been shown that superconducting coils will be required in essentially all applications (Stekly and Thome, 1973). The first superconducting MHD magnet was completed in the United States in 1966, and others have been made since then in Japan and Germany. The current state of the art of high-current-density MHD magnets is represented by a saddle magnet under development which will produce a central field of 5 T uniform over length of 0.9 m in a warm-bore Dewar with an inner diameter of 0.18 m.

Interest in MHD has cooled somewhat in the past few years as a result of some doubt about its competitiveness with other energy sources such as gas turbines (Bogner, 1972).

B. Controlled Thermonuclear Fusion

Optimism about thermonuclear fusion is running high, and many of the systems require the use of superconducting coils. Magnetic fields are required in most schemes to confine the reacting plasma, and that task necessitates complex configurations for the coils and intense fields. In some cases, the coil must be levitated. All of the coils which might be used for power reactors are enormous, as are the problems of structural support, cooling, etc. The construction of these magnets will continue as long as the optimism about the possibility of fusion holds up, since it is unlikely that there will be any abatement of our need for electric power. For example, it is expected that several large toroidal-coil machines will be built in the next 10 years having bores of several meters and the highest possible magnetic fields (Taylor, 1972). Taylor indicates that 25 years is a reasonable estimate of the time required to demonstrate practicality, assuming that scientific feasibility is shown within 10 years.

There is a continuing need for improved magnetic materials which are rugged enough to stand the strains and cheap enough to be economically possible. It can be expected that large needs such as fusion magnets will put on continuing pressure to support materials research.

C. Energy Storage

High-field superconducting coils can store large quantities of energy in the magnetic field. There have been proposals for, and experiments on, the application of this energy-storage capacity to two different situations. In one, the coils will supply pulsed currents primarily for fusion (Laquer *et al.*, 1972), but also wherever capacitors are presently used (Lucas *et al.*, 1972). It is expected that the cost per joule will be an order of magnitude lower than when using capacitors. In the higher levels of energy storage (megajoules to tens of megajoules), there is a potential for achieving energy densities of thousands of joules per pound, which is far higher than available from capacitors (Stekly and Thome, 1973). The other use is for peak power shaving in power distribution systems (Hassenzahl *et al.*, 1973). This application seems somewhat more speculative, but cost estimates suggest that it may be practical.

D. Motors and Generators

Development work has been in progress for several years on large dc homopolar motors in England, and the 1960s saw initiation of work on ac generators in several countries. Potential savings in cost, weight, size, and efficiency have provoked considerable current interest in the use of superconducting machines for ship propulsion and for central-power-station installations. Both ac and dc machines may be used for ship propulsion, and central power stations usually require ac generators. Generators for aircraft also have been considered.

The use of superconducting generators and motors to effect a power transmission between the gas turbines and a ship's propellers would provide a degree of flexibility of operation not available from mechanical transmissions and would permit the use of fixed-pitch propellers. Furthermore, the precise alignment needed for mechanical drives currently requires placement of turbines deep in hulls with consequent large duct volumes. The use of an electrical transmission system with light, small, superconducting motors would add design flexibility and ease alignment requirements. A small ship propulsion system, involving a gas turbine of about 1000 hp and a dc superconducting generator and motor, with appropriate interconnections, is under development (Levendahl, 1972). One of the major problems with dc superconducting machines is the commutation of current from the rotor; the best solution appears to be liquid metal contacts, although a new type of brush constructed of metal-plated carbon fibers shows promise (Appleton, 1973). Fox and Hatch (1972) have discussed the possible application of ac synchronous machines to ship propulsion.

Of the various possible ways of employing superconducting coils in ac synchronous generators, the generally agreed-upon configuration is a rotating superconducting field winding inside and a normal-metal stationary armature outside (Mole *et al.*, 1973). The choice of synchronous operation with a rotating superconducting field coil minimizes the ac fields on the superconductor which would lead to power losses. The choice of the interior rather than exterior superconducting field is based on reduction of ampere turns and magnetic shielding and improvement of armature design. The armature must be of normal metal, as it is subjected to strong ac fields, and the ac losses in a superconducting coil would cause too much helium vaporization. The chosen configuration requires the rotor to be refrigerated and to have rotary seals. This and the fact that the forces must be borne by the coils themselves (rather than an iron structure, as in conventional machines) are principal problems in the development of ac generators.

Only for relatively large ac generators is the capital cost of a superconducting machine expected to be lower than that of the conventional equivalent. Appleton and Anderson (1972) have indicated that 500 MW is the cost breakpoint, and a report by the British Science Research Council (1972) suggests 1000 MW. Successful ac generators as large as 5 MW (Mole *et al.*, 1973) have been operated under experimental conditions. The British Science Research Council (1972) made a thorough survey of work in progress in 1972 and suggested a likely timetable of development of superconducting ac generators to include the construction and operation of a 60-MW prototype in the 1976–1982 period, followed by a 600-MW machine in the mid-1980s and a 1000-MW model in the late 1980s. Production of large units for central power stations would not take place before the 1990s. Even with careful analysis, extrapolation of technology that far into the future is a brave undertaking, as the timetable can be affected in many ways.

E. Power Transmission

One of the most obvious possible applications for superconductivity is in the transmission of electric power. The need to refrigerate the line imposes the first restriction; it must be an underground line, and undergrounding is expensive. Furthermore, the requirement of a refrigeration system over many miles of line imposes an additional technological and economic burden. Nevertheless, the power requirements of the near future are so enormous that the superconducting line must be considered as one of the alternatives.

There are several experimental projects underway at present; these are restricted to very short line lengths because of the great cost of constructing longer lines. Most of the work is aimed at ac transmission, which is preferable for short transmission lengths to avoid the power-conditioning equipment required at each end of a dc line to match it to the rest of an ac system. Superconductors carrying ac current are not lossless, but the losses are low enough to make them attractive. The principal choices of material to date are niobium and niobium–tin. The former has lower ac losses, but the latter can be operated at a higher temperature, and the refrigeration requirements are reduced. The analysis by Forsythe *et al.* (1972) indicates that the system economics work out best if one uses the material with the highest possible T_c. Meyerhoff (1972) has reported a 12-year program (currently funded for the first $3\frac{1}{2}$ years) aimed at developing a three-phase ac line. At present, they consider niobium coated on a normal-metal backing to be the best material but will examine alternatives. Some of the factors that must be considered in the development of a competitive superconducting line are means for joining the conductors and allowing for differential contraction, development of ways to match the line to the system, handling of faults, and cryogenic systems development.

The power requirements in the United States are presently estimated to continue to double every 10 to 12 years, leading to the load in the year 2000 being several times its value in 1974. Furthermore, the growing public concern for environmental esthetics results in mounting objections to overhead lines. Pollution regulations may lead to the location of future generating stations 20–50 miles from major load centers. We can expect more undergrounding of cables at higher capacity and over longer distances and efforts to narrow the competitive disadvantage of underground transmission (Corry, 1972). Although there are many difficult problems in the development of a superconducting power line, the pressing needs require that effort be made to determine its competitive position. Present indications are that the superconducting line will be the most economical for loads over 2000–3000 MVA. We can expect work to continue on this problen for at least several years.

VII. Transportation

In the 1960s, it became clear that very difficult problems would arise by the 1980s in connection with air transportation in concentrated population areas such as the Northeast corridor in the United States, and high-speed trains came under consideration. Since it is very difficult to achieve the tolerances necessary for wheel-on-rail systems for speeds in excess of about 240 km/hr (150 mph), other support systems were examined. The first to be studied and the most highly developed today is the air-suspension vehicle. Later, the idea of using magnetic suspension was considered, and there are now two principal competing schemes. In one, normal electromagnets on board the vehicle are attracted upward to a ferromagnetic rail. This system is not stable, and so magnet currents are controlled by servo systems to maintain the desired clearance of a few centimeters. In an alternative scheme, first proposed by Powell and Danby (1966), repulsive magnetic forces from superconducting magnets provide the support. There is an interaction of the magnetic field with the induced eddy currents in the roadbed. A principal difference between the two magnetic approaches lies in the increased clearance in the repulsive system. The attractive system will have a clearance of a few centimeters, whereas the repulsive arrangement can have separations in the range of 10–30 cm at high speeds.

Several different configurations have been proposed for the repulsive system's roadbed, which is a major cost factor. Most of these have been either some arrangement of coils or continuous conducting sheets. Cost considerations favor the latter. Maintenance of lateral position is achieved by arrangements similar to those used for lift. The strongest contender for the drive system is a linear induction motor, but other approaches are being considered (Thornton, 1973). See Chapter 7 by Buchhold for more details on lift and lateral control.

This is a period where many new ideas for repulsive magnetic levitation are appearing, and some experiments are in progress in the United States, Europe, and Japan. The Japanese National Railway (JNR) has already demonstrated the levitation of a four-man vehicle using a roadbed consisting of normal-metal coils (Thornton, 1973), and they are now studying the use of an aluminum ladder-shaped structure. The JNR goal is to develop a train with a maximum speed of 340 mph (550 km/hr) by 1975 (Stekly and Thome, 1973). Until one of the other competing above-mentioned technologies is shown to be clearly more favorable, one can expect to see continued development of train levitation by superconducting magnets.

VIII. Miscellaneous

A. Medical Instrumentation

The major impact of superconductivity on medicine has been in the areas mentioned above in research. These include biochemical nuclear magnetic resonance (NMR) studies using superconducting magnets. SQUID instrumentation is being developed which may contribute to improving NMR capability. The SQUID gradient magnetometer, although still in the research phase, may become a clinical tool as more understanding of magnetocardiograms is accumulated. Since no leads need be connected to the patient, the making of the cardiogram would be simpler than for EKG and would facilitate mass screening of patients, as has been done with mobile x-ray machines (Goodman *et al.*, 1973).

A system has been proposed in which catheterization of channels in the human body would be achieved by controlling the movement of a permanent-magnet-tipped catheter by strong fields from a set of superconducting magnets. The aim is to reach certain parts of the body which would otherwise be available only by major surgery. An example of the possible use of the proposed system is the controlled catherization of the intracranial internal carotid artery and its divisions, which have not been previously reached because of the complex configuration of the vessels.

B. Water-Pollution Control

Experiments have been conducted at the Massachusetts Institute of Technology National Magnet Laboratory in which polluted water was seeded with a magnetic material and then passed through a superconducting magnetic high-field separator. The coliform count was reduced from 10,000/ml to drinking-water standards in one pass through the separator. It was estimated that a magnet the size of the Argonne National Laboratory bubble chamber could purify the entire water supply of a city the size of Boston at a fraction of the capital cost of the present system (see Chapter 6 by Iwasa and Montgomery).

C. Prospecting

Means exist for detecting ore bodies buried to a depth of 100–150 ft below the earth's surface; a new device which uses superconducting coils promises to double the depth of detectability (Morrison and Dolan, 1973). A large-diameter coil is housed in a toroidal Dewar suspended from a helicopter with its axis horizontal at a height of about 100 ft above the ground. The superconducting coil is connected in series with a resonating capacitor in one arm of a bridge circuit and is energized by an ac signal. When passing over a conducting body, eddy currents and, hence, losses are produced in the body, and these result in a change of the resistance in that arm of the bridge circuit. Superconducting coils are necessary because the change of resistance would be too small to be detected if the coils were of normal wire. There will actually be three coils in separate Dewars on the same axis and spaced by 4 ft. The diameter of one coil will be 3 m, and the other diameters will be 1 m. They will be connected in separate bridge circuits and driven with frequencies which will be approximately 40, 200, and 2000 Hz. The use of a set of frequencies allows depth differentiation and, therefore, distinction between ore bodies and other lossy objects such as surface water. Initial testing on a stationary model has been successful.

IX. Summary

We have discussed what amounts to an explosion of applications of superconductivity occurring mostly in the late 1960s as the potentials of superconducting magnets and Josephson devices began to be appreciated. In some cases, the use of superconductors is securely entrenched, such as magnet coils in research applications and the Josephson effects in precision metrology. The low-loss property also has proved its importance in the stabilization of oscillators. In many other situations, such as superconducting windings in motors and generators and Josephson devices for millimeter-wave detection and computer components, the prospects look very favorable, but the practical application is not yet certain. Some applications, such as magnetic levitation of trains and power transmission by superconducting cables, presently must be considered speculative, though of considerable importance.

Many of the applications would be put in a much better competitive position if appreciably higher temperature operation were possible. For example, the refrigeration costs would decrease to about 10% of those for 4.2°K if operation at 40°K were possible. Although this is mostly true for the high-power applications, higher-temperature materials also would facilitate the use of Josephson devices by permitting employment of less expensive, small, closed-cycle refrigerators. Even apart from raising the operating temperatures, materials properties play other important determining roles, as exemplified by the importance of losses in accelerator cavities, pulsed coils, and transmission lines.

Entire new classes of applications of Josephson devices may emerge if attention is given to them by circuit theorists. New levels of reliability and circuit integration could result if the methods currently used in semiconductor technology were brought to bear in a larger way than at present. The technology of fabricating these devices must be considered to be at an early stage today.

Finally, an apology should be made here both to the readers and to the discoverers for those applications which were not included. It is hoped that the most important ones are represented so that a reasonably accurate view of the future prospects is given.

Acknowledgments

I would like to express my thanks to several colleagues (especially Dr. M. Nisenoff) who gave me helpful information used in preparing this chapter and to Miss Cheryl Brown and Miss Lynda Janks for their work in preparing the manuscript. The loving support of my wife, Janice, and of Jeff, Eric, and Leslie, which came in spite of missed opportunities to share precious days and evenings of California summer sunshine, is greatly appreciated.

References

Anacher, W. (1969). *IEEE Trans. on Magn.* **MAG-5,** 968.

Anderson, P. W., and Rowell, J. M. (1963). *Phys. Rev. Lett.* **10,** 230.

Appleton, A. D. (1973). *Proc. IEEE* **61,** 106.

Appleton, A. D., and Anderson, A. F. (1972). *Proc. 1972 Appl. Superconduc. Conf., Annapolis, Maryland, 1–3 May,* p. 136. IEEE Pub. No. 72 CH0682-5-TABSC.

Arrhenius, G., Corenzwit, E., Fitzgerald, R., Hull, G. W., Jr., Luo, H. L., Matthias, B. T., and Zachariasen, W. H. (1968). *Proc. Nat. Acad. Sci. U. S.* **61,** 621.

Bogner, G. (1972). *Proc. 1972 Appl. Superconduct. Conf., Annapolis, Maryland, 1–3 May* p. 241. IEEE Pub. No. 72CH0682-5-TABSC.

British Science Research Council (1972). Report on Superconducting ac Generators. A spec. rep. of the Elec. and Syst. Eng. Committee, Sci. Res. Council, State House, High Holborn, London, WC1 R 4TA.

Buck, D. (1956). *Proc. IRE* **44,** 482.

Chiba, N., Kashiwayanagi, Y., and Mikoshiba, K. (1973). *Proc. IEEE* **61,** 124.

Clark, T. D. (1968). *Phys. Lett. A* **27,** 585.

Clark, T. D. (1973). *Phys. Rev. B* **8,** 137.

Clarke, J. (1973). *Proc. IEEE* **61,** 8.

Clarke, J., Hoffer, G. I., and Richards, P. L. (1973). Paper presented at the *Int. Conf. Detect. Emission Electromagn. Radiat. Josephson Junctions, Perros-Guirec, France, 3–5 September* (1974). *Rev. Phys. Appl.* **9,** 69.

Corry, A. F. (1972). *Proc. 1972 Appl. Superconduct. Conf., Annapolis, Maryland, 1–3 May* p. 160. IEEE Pub. No. 72CH0682-5-TABSC.

Elsley, R. K., and Sievers, A. J. (1972). *Proc. 1972 Appl. Superconduct. Conf., Annapolis, Maryland, 1–3 May* p. 716. IEEE Pub. No. 72CH0682-5-TABSC.

Finnegan, T. F., and Denenstein, A. (1971). *Metrologia* **7,** 167.

Finnegan, T. F., and Wahlsten, S. (1972). *Appl. Phys. Lett.* **21,** 541.

Forsyth, E. B., Garber, M., Jensen, J. E., Morgan, G. H., Britton, R. B., Powell, J. R., Blewett, J. P., Gurinsky, D. H., and Hendrie, J. M. (1972). *Proc. 1972 Appl. Superconduct. Conf., Annapolis, Maryland, 1–3 May* p. 202. IEEE Pub. No. 72CH0682-5-TABSC.

Fox, G. R., and Hatch, B. D. (1972). *Proc. 1972 Appl. Superconduct. Conf., Annapolis, Maryland 1–3 May* p. 33. IEEE Pub. No. 72CH0682-5-TABSC.

Fulton, T. A., Dynes, R. C., and Anderson, P. W. (1973). *Proc. IEEE* **61,** 28.

Gavaler, J. R., (1973). *Appl. Phys. Lett.* **23,** 480.

Giffard, R. P., Webb, R. A., and Wheatley, J. C. (1972). *J. Low Temp. Phys.* **6,** 533.

Goodman, W. L., Hesterman, V. W., Rorden, L. H., and Goree, W. S. (1973). *Proc. IEEE* **61,** 20.

Goree, W. S. (1970). *Rev. Phys. Appl.* **5,** 3.

Hassenzahl, W. V., Rogers, J. D., and McDonald, T. E. (1973). Presented at IEEE Int. Convention—INTERCON, March 26–30, 1973, New York.

Heeger, A. J., Garito, A. F., Coleman, L. B., Cohen, M. J., Sandman, D. J., and Yamagishi, F. G. (1973a). Paper presented at Amer. Phys. Soc. Meeting, San Diego, California, March 23, 1973.

Heeger, A. J., Garito, A. F., Coleman L. B., Cohen, M. J., Sandman, D. J., and Yamagishi, F. G. (1973b). *Phys. Today* **26,** 17.

Hoenig, H. E., Wang, R. H., Rossman, G. R., and Mercereau, J. E. (1972). *Proc. 1972 Appl. Superconduct. Conf., Annapolis, Maryland, 1–3 May* p. 570. IEEE Pub. No. 72CH0682-5-TABSC.

Industrial Research (May 1972). V1.

Jensen, J. E. (1972). Paper presented at the Cryogenic Soc. of Amer., Helium Div. Session, Chicago, Illinois, 3–5 October.

Josephson, B. D. (1962). *Phys. Lett.* **1**, 251.

Kamper, R. A., and Sullivan, D. B. (1972). U. S. Nat. Bur. of Std. Tech. Note 630.

Kamper, R. A., Simmonds, M. B., Adair, R. T., and Hoer, C. A. (1972). *Proc. 1972 Appl. Superconduct. Conf., Annapolis, Maryland, 1–3 May* p. 696. IEEE Pub. No. 72CH0682-5-TABSC.

Kamper, R. A., Simmonds, M. B., Adair, R. T., and Hoer, C. A. (1973). *Proc. IEEE* **61**, 121.

Kunzler, J. E., Buehler, E., Hsu, F. S. L., and Wernick, J. H. (1961). *Phys. Rev. Lett.* **6**, 89.

Laquer, H. L., Lindsay, J. D. G., Little, E. M., and Weldon, D. M. (1972). *Proc. 1972 Appl. Superconduct. Conf., Annapolis, Maryland, 1–3 May* p. 98. IEEE Pub. No. 72CH0682-5-TABSC.

Levedahl, W. J. (1972). *Proc. 1972 Appl. Superconduct. Conf., Annapolis, Maryland, 1–3 May* p. 26. IEEE Pub. No. 72CH0682-5-TABSC.

Lucas, E. J., Punchard, W. F. B., Thome, R. J., Verga, R. L., and Turner, J. M. (1972). *Proc. 1972 Appl. Superconduct. Conf., Annapolis, Maryland, 1–3 May* p. 102. IEEE Pub. No. 72CH0682-5-TABSC.

Matisoo, J. (1967). *Proc. IEEE* **55**, 172.

McColl, M., Millea, M. F., and Silver, A. H. (1973). *Appl. Phys. Lett.* **23**, 263.

Meservey, R. (1968). *J. Appl. Phys.* **39**, 2598.

Meyerhoff, R. W. (1972). *Proc. 1972 Appl. Superconduct. Conf., Annapolis, Maryland, 1–3 May* p. 194. IEEE Pub. No. 72CH0682-5-TABSC.

Mole, C. J., Brenner, W. C., and Haller, H. E. III. (1973). *Proc. IEEE* **61**, 95.

Morrison, H. F., and Dolan, W. M. (1973). Presented at the 43rd Annu. Int. Soc. of Exploration Geophys. Meeting, Mexico City, 21–25 October.

Ohta, H., Feldman, M. J., Parrish, P. T., Chiao, R. Y. (1973). Paper presented at the *Int. Conf. Detect. Emission Electromagn. Radiat. with Josephson Junctions, Perros-Guirec, France, 3–5 September* 1974. *Rev. Phys. Appl.* **9**, 61.

Powell, J. R., and Danby, G. T. (1966). Presented at the ASME Winter Annu. Meeting, New York, Railroad Div., Tech. Rep. 66-WA/RR-5.

Prothero, W. A., Jr., and Goodkind, J. M. (1968). *Rev. Sci. Instrum.* **39**, 1257.

Richards, P. L., Auracher, F., and Van Duzer, T. (1973). *Proc. IEEE* **61**, 36.

Shapiro, S. (1963). *Phys. Rev. Lett.* **11**, 80.

Stein, S. R., and Turneaure, J. P. (1972). *Proc. Int. Conf. Low Temp. Phys., 13th, Boulder, Colorado, 21–25 August.* Vol. 4, p. 535. Plenum Press, N.Y. (1974).

Stekly, Z. J. J., and Thome, R. J. (1973). *Proc. IEEE* **61**, 85.

Sullivan, D. B. (1972). *Proc. 1972 Appl. Superconduct. Conf., Annapolis, Maryland, 1–3 May. p.* 631. IEEE Pub. No. 72CH0682-5-TABSC.

Tachikawa, K., and Tanaka, Y. (1966). *Jap. J. Appl. Phys.* **5**, 834.

Taur, Y., Claassen, J. H., and Richards, P. L. (1973). Paper at the *Int. Conf. Detect. Emission Electromagn. Radiat.* with *Josephson Junctions, Perros-Guirec, France, 3–5 September* 1974. *Rev. Phys. Appl.* **9**, 263.

Taylor, C. E. (1972). *Proc. 1972 Appl. Superconduct. Conf., Annapolis, Maryland, 1–3 May.* p. 239. IEEE Pub. No. 72CH0682-5-TABSC.

Thornton, R. D. (April 1973). *IEEE Spectrum* **10,** 47.

Tilley D. R. (1970). *Phys. Lett. A* **33,** 205.

Turneaure, J. P. (1973). Private communication.

Ulrich, B. T. (1970). *Proc. Int. Conf. Low Temp. Phys., 12th, Kyoto, Japan, 4–10 September.* p. 867.

Van Duzer, T., and Turner C. W. (December 1972). *IEEE Spectrum* **9,** 53.

U. S. National Bureau of Standards (1972). *NBS Tech. News Bull.* **56,** 159.

Wüthrich, K., and Shulman, R. G. (April 1970). *Phys. Today* **23,** 43.

Zappe, H. H., and Grebe, K. R. (1973). *J. Appl. Phys.* **44,** 865.

Index for Volume II

D

Q

R

W

Y

Z

Index for Volume I

D

N

S

U

V

Z

A
B 5
C 6
D 7
E 8
F 9
G 0
H 1
I 2
J 3